Ecologies of Writing Programs

Writing Program Administration
Series Editors: Susan H. McLeod and Margot Soven

The Writing Program Administration series provides a venue for scholarly monographs and projects that are research- or theory-based and that provide insights into important issues in the field. We encourage submissions that examine the work of writing program administration, broadly defined (e.g., not just administration of first-year composition programs). Possible topics include but are not limited to 1) historical studies of writing program administration or administrators (archival work is particularly encouraged); 2) studies evaluating the relevance of theories developed in other fields (e.g., management, sustainability, organizational theory); 3) studies of particular personnel issues (e.g., unionization, use of adjunct faculty); 4) research on developing and articulating curricula; 5) studies of assessment and accountability issues for WPAs; and 6) examinations of the politics of writing program administration work at the community college.

Books in the Series

Ecologies of Writing Programs: Program Profiles in Context edited by Mary Jo Reiff, Anis Bawarshi, Michelle Ballif, & Christian Weisser (2015)

A Rhetoric for Writing Program Administrators edited by Rita Malenczyk (2013)

Writing Program Administration and the Community College by Heather Ostman (2013)

The WPA Outcomes Statement—A Decade Later, edited by Nicholas N. Behm, Gregory R. Glau, Deborah H. Holdstein, Duane Roen, & Edward M. White (2012)

Writing Program Administration at Small Liberal Arts Colleges by Jill M. Gladstein and Dara Rossman Regaignon (2012)

GenAdmin: Theorizing WPA Identities in the 21st Century by Colin Charlton, Jonikka Charlton, Tarez Samra Graban, Kathleen J. Ryan, and Amy Ferdinandt Stolley (2012). *Winner of the CWPA Best Book Award*

ECOLOGIES OF WRITING PROGRAMS

PROGRAM PROFILES IN CONTEXT

Edited by Mary Jo Reiff, Anis Bawarshi, Michelle Ballif, and Christian Weisser

Parlor Press
Anderson, South Carolina
www.parlorpress.com

Parlor Press LLC, Anderson, South Carolina, USA

© 2015 by Parlor Press
All rights reserved.
Printed in the United States of America

S A N: 2 5 4 - 8 8 7 9

Library of Congress Cataloging-in-Publication Data

Ecologies of writing programs : program profiles in context / Edited by Mary Jo Reiff, Anis Bawarshi, Michelle Ballif, and Christian Weisser.
   pages cm -- (Writing Programs Administration.)
   Includes bibliographical references and index.
   ISBN 978-1-60235-511-8 (pbk. : alk. paper) -- ISBN 978-1-60235-512-5 (hardcover : alk. paper)
   1. English language--Rhetoric--Study and teaching (Higher) 2. Environmental literature--Authorship--Study and teaching (Higher) 3. Writing centers--Administration. 4. Natural history--Authorship--Study and teaching (Higher) 5. Ecology--Authorship--Study and teaching (Higher) 6. Interdisciplinary approach in education. 7. Academic writing--Study and teaching (Higher) 8. Nature study. I. Reiff, Mary Jo., editor. II. Bawarshi, Anis S., editor. III. Ballif, Michelle, 1964- editor. IV. Weisser, Christian R., 1970- editor.
   PE1479.N28E36 2015
   808'.0420711--dc23
                              2015007500

1 2 3 4 5

Writing Program Administration
Series Editors: Susan H. McLeod and Margot Soven

Cover image:
Cover design by   .
Printed on acid-free paper.

Parlor Press, LLC is an independent publisher of scholarly and trade titles in print and multimedia formats. This book is available in paper, cloth and eBook formats from Parlor Press on the World Wide Web at http://www.parlorpress.com or through online and brick-and-mortar bookstores. For submission information or to find out about Parlor Press publications, write to Parlor Press, 3015 Brackenberry Drive, Anderson, South Carolina, 29621, or email editor@parlorpress.com.

Contents

Acknowledgments *vii*

Writing Program Ecologies: An Introduction   3
    Mary Jo Reiff, Anis Bawarshi, Michelle Ballif, and Christian Weisser

Part I. The Contested Ecologies of FYC Programs: Negotiating between Stability and Change   19

1 The *Kairotic* Moment: Pragmatic Revision of Basic Writing Instruction at Indiana University-Purdue University Fort Wayne   22
    Sara Webb-Sunderhaus and Stevens Amidon

2 Standardizing English 101 at Southern Illinois University Carbondale: Reflections on the Promise of Improved GTA Preparation and More Effective Writing Instruction   41
    Ronda Leathers Dively

3 Taking the High Road: Teaching for Transfer in an FYC Program   68
    Jenn Fishman and Mary Jo Reiff

4 Intractable Writing Program Problems, Kairos, and Writing-about-Writing: A Profile of the University of Central Florida's First-Year Composition Program   91
    Elizabeth Wardle

Part II. Remapping Interdisciplinary Ecologies: WAC and WID Programs   121

5 The Writing Intensive Program at the University of Georgia   125
    Michelle Ballif

6 Back to the Future: First-Year Writing in the Binghamton University Writing Initiative, State University of New York   142
    Kelly Kinney and Kristi Murray Costello

7 Imagining a Writing and Rhetoric Program Based on Principles of Knowledge "Transfer": Dartmouth's Institute for Writing and Rhetoric   163
    Stephanie Boone, Sara Biggs Chaney, Josh Compton, Christiane Donahue, and Karen Gocsik

Part III. Claiming Disciplinary Locations: The Undergraduate Major in Rhetoric and Composition *195*

8 Diverse Lessons: Developing an Undergraduate Program in Rhetoric, Writing, and Culture at Texas A&M *198*
   Stephanie L. Kerschbaum and M. Jimmie Killingsworth

9 Reflections on the Major in Writing and Rhetoric at Oakland University *211*
   Lori Ostergaard, Greg A. Giberson, and Jim Nugent

10 The Case for a Major in Writing Studies: The University of Minnesota Duluth *228*
   David Beard

Part IV. Interconnected Sites of Agency: Situating Assessment within Institutional Ecologies *249*

11 Self-Assessment as Programmatic Center: The First Year Writing Program and Its Assessment at California State University, Fresno *252*
   Asao B. Inoue

12 Utilizing Strategic Assessment to Support FYC Curricular Revision at Murray State University *282*
   Paul Walker and Elizabeth Myers

Part V. Third Spaces: Creating Liminal Ecologies *305*

13 A Collaborative Approach to Information Literacy: First-year Composition, Writing Center, and Library Partnerships at West Virginia University *308*
   Laura Brady, Nathalie Singh-Corcoran, Jo Ann Dadisman, and Kelly Diamond

14 The Peer-Interactive Writing Center at the University of New Mexico *334*
   Daniel Sanford

15 Writing the Transition to College: A Summer College Writing Experience at Elon University *363*
   Jessie L. Moore, Kimberly B. Pyne, and Paula Patch

Index *397*
About the Editors *399*

Acknowledgments

Anyone who has administered a writing program knows about the complex, multidimensional, interconnected challenges involved in coordinating such programs. The challenges are intensified when working to cultivate and sustain a culture of innovation, often leaving very little space and time to reflect on, write about, and make that work available for the intellectual archive of composition studies. So most of all we want to thank the contributors whose profiled programs have made this book possible: Stevens Amidon, Michelle Ballif, David Beard, Sara Biggs Chaney, Stephanie Boone, Laura Brady, Josh Compton, Kristi Murray Costello, Jo Ann Dadisman, Kelly Diamond, Ronda Leathers Dively, Christiane Donahue, Jenn Fishman, Greg A. Giberson, Karen Gocsik, Asao B. Inoue, Stephanie L. Kerschbaum, M. Jimmie Killingsworth, Kelly Kinney, Jessie L. Moore, Elizabeth Myers, Jim Nugent, Lori Ostergaard, Paula Patch, Kimberly B. Pyne, Mary Jo Reiff, Daniel Sanford, Nathalie Singh-Corcoran, Paul Walker, Elizabeth Wardle, and Sara Webb-Sunderhaus.

A special thank you to David Blakesley for his encouragement and support of this project. From the book's earliest conceptualization to publication, David's big picture ability to synthesize reviewer feedback in light of the book's goals while also attending to editorial details has guided this project. As well, we would like to thank Susan McLeod and Margot Soven, the Writing Program Administration series editors at Parlor Press, for encouraging us to sustain connections across chapters that helped to strengthen the book's coherence. For her outstanding work in preparing the concordance for the book's index, no easy task, we thank Jennifer Lin LeMesurier.

Completing a project of this scale does not happen without sacrifice of time and energy. For their understanding and support, we thank our families and friends.

A portion of the proceeds of this book will go to Surfrider Foundation USA, a grassroots nonprofit environmental organization that works to protect the world's oceans, beaches, rivers, and waterways. Check them out at www.surfrider.org.

ECOLOGIES OF WRITING PROGRAMS

Writing Program Ecologies: An Introduction

Mary Jo Reiff, Anis Bawarshi, Michelle Ballif, and Christian Weisser

More than three decades ago, Marilyn Cooper proposed an ecological model of writing, "whose fundamental tenet is that writing is an activity through which a person is continuously engaged with a variety of socially constituted systems" (367). That is, Cooper suggested that writers are not solitary agents but are instead enmeshed in complex, circulative relationships with other writers, texts, contexts, ideas, and exigencies. Cooper's "The Ecology of Writing" speculated that "all the characteristics of any individual writer or piece of writing both determine and are determined by the characteristics of all other writers and writings in the systems" (368). This perspective, which has contributed to a fundamental shift in our conceptions of the discursive situation, envisions writing not as an individual act, nor even as a social activity, but as a network, a system, a web—an ecology.

To be clear, Cooper's work is neither the first nor the most comprehensive exploration of ecology in writing studies, yet it does signal recognition of complex relationships and dynamic connections that has marked the work that would follow. Many scholars have since drawn upon and extended Cooper's notion of an ecology of writing, and our conceptions have developed and evolved in significant ways. Most notably, scholars have addressed growing awareness of the system itself as the locus of meaning rather than the individual actors or units within it. Consequently, much of the current theoretical work in writing studies works from an inherently ecological perspective, envisioning writing as bound up in, influenced by, and relational to spaces, places, locations, environments, and the interconnections among the entities they contain.

It is safe to say that the ecological model has become central to our conceptions of what writing is, how it functions, and how it emerges in and through systems.

However, few scholarly works have connected this notion of the ecology of writing with the often more pragmatic work of writing programs and writing program administration. Writing programs are still imagined by some as the utilitarian end point of writing studies, somehow "separate" from the theoretical study of writing itself. We talk about the complexity of writing in our scholarly journals, we postulate theories of writing as ecological, complex, dynamic, and interrelational, and yet when it comes to the programs we help to create and maintain in our universities and other sites of practice, we have difficulty seeing them in the same ecological light. We do not intend to suggest that contemporary writing programs are isolationist, since most WPAs would be quick to note the many influences, benefactors, and contributors who help in the continued development of their programs, both on their respective campuses and in the field. However, acknowledging those influences is not quite the same thing as recognizing the ways in which writing programs themselves are complex ecologies. Nor do we see a dearth of meaningful scholarship about writing programs and their administration, since the wealth of scholarly works about such programs that precede this book situate it in a rich ecology of meaning; however, again, this is not quite the same thing as seeing the ecologies of the programs themselves. Instead, we are suggesting that writing programs are quintessentially discursive and material ecologies because they emerge through complex networks of interrelations, depend upon adaptation, fluidity, and the constant motion of discursive systems, are generative and constitutive of diverse rhetorics and discourses, and exhibit a range of other ecological characteristics.

The purpose of this collection is to highlight the ways in which writing programs—like all discursive systems—are ecologies. They are not just *like* ecologies, nor are they simply useful metaphoric examples of ecologies. They are ecologies themselves, in every sense of the word. From first-year composition programs, to undergraduate writing majors, to writing across the curriculum programs, to undergraduate and graduate programs in rhetoric and composition—writing programs are complex ecological networks. In many ways, the scientific framework for understanding ecology—that of biological relationships between living organisms and their environments—is insufficient to explain the complexity of interaction exhibited in writing programs and other discursive, material units. Because discursive systems themselves are different phenomena from so-called "natural" environmental structures, they require different methods of understanding. As Sidney Dobrin suggests, "writing studies requires a more complex kind of complexity than has yet been proposed simply because writing systems are different, more complex kinds of systems than complexity theories, systems theories, and ecological theories have worked to engage thus far" (*Ecology* 9). Consequently, it is necessary to create new tools, new frameworks, and new methodologies to understand discursive-material ecologies. Such an ecological perspective enables us to account for, among other things, the networked agency at play in WPA work, as recent ecological approaches "compel us to think of causation in far more complex terms; to recognize that phenomena are caught in a multitude of interlocking systems and forces and to consider anew the location and nature of capacities for agency" (Coole and Frost 9). This collection, with its profiles of writing programs within their fluid, dynamic, and relational contexts, contributes to our understanding of writing programs as complex ecologies.

With this understanding, we identify four ecological characteristics as particularly relevant in conceptualizing the ecologies of writing programs: *interconnectedness, fluctuation, complexity,* and *emergence.* This is far from an exhaustive list of ecological attributes; they are simply the most pertinent in our present conceptions of the ecology of writing programs. These four characteristics are drawn from more comprehensive examinations of the ecology of writing by scholars such as Margaret Syverson, Byron Hawk, Sidney Dobrin, Collin Brooke, Jenny Edbauer, and others—all excellent sources for further study. Our purpose is not to forward a more holistic method of conceptualizing the ecology of writing; it is merely to highlight a few of the particular ways in which writing programs are ecological. We encourage readers to consider the programs profiled in this collection through the lens of these four ecological attributes—indeed, we guide readers to such considerations in our introduction to each section of the book.

Writing Programs Are Interconnected

Ecologies are characterized by interconnections, networks, and relationships. Ecologies consist of multiple intertwining parts that act on one another in various ways. An ecological perspective shifts the emphasis away from the individual unit, node, or entity, focusing instead on the network itself as the locus of meaning. All of the acts, actors, and objects in an ecology are connected, both in space and time, and the interactions among them reverberate throughout and beyond the system itself. This concept is so central to the notion of ecology that prominent ecologist Barry Commoner designated it as the first of the Four Laws of Ecology: "Everything Is Connected to Everything Else" (*The Closing Circle* 16).

When applied to writing more generally, and to writing programs more specifically, it is easy to see the ways in which discourses, rhetors, texts, utterances, and material (and immaterial) objects form such networks of dynamic interaction. Syverson, for example, proposes that an ecology is "a kind of meta-complex system composed of interrelated and interdependent complex systems and their environmental structures and processes" that identifies the interdependencies and dynamism in writing as a system (5). All of the writing programs profiled in this collection exhibit interconnectedness in some way. They are intertwined with departments, divisions, and colleges; majors, minors, and concentrations; colleagues, organizations, and scholarship; administrators, faculty, and students; proposals, websites, and reports; offices, buildings, and campuses; legislative decisions, budgets, state mandates, and accreditation requirements—each in varying levels of influence at diverse points in time yet influenced and interconnected always. The authors of these profiles are quick to note such connections, pointing out the ways in which the network of affiliations has shaped and continues to reshape their programs in constructive and detrimental ways.

Writing Programs Fluctuate

Another central attribute of ecologies, corollary to their interconnected nature, is that ecologies are in constant flux and continual transformation. Ecologies evolve over time, are influenced and transformed through actions and activity, and fluctuate, grow, and wither as a result of internal and external forces. Cooper's metaphor of writing as a web is an applicable starting point here, "in which anything that affects one strand of the web vibrates throughout the whole" (370), yet the web metaphor fails to capture the complexity of motion and malleability inherent in such systems. In other words, the web metaphor infers a static system of pathways, whereas fluidity, change, and, at times, volatility are central to ecologies. The characteristically evolving structure of these systems is partly a survival mechanism; the individual parts of ecologies—and consequently the ecologies themselves—change and reorganize over time to adapt to problems and improve their ability to interact with other parts and other ecologies.

Byron Hawk, drawing upon Mark C. Taylor, refers to "complex adaptive systems" in which change arises through multi-layered interaction, negotiation, and transformation. Hawk writes that such systems are "never static, they produce larger scale behavior, texts, and structures from the movement and interactions of smaller parts" (835). Writing, then, can be seen as a fluid system in which adaptation and evolution are the norm. Writers are both transformative and transformed agents in ecologies, part of the complex, unstable network of discourse. Discourses adapt and are adaptive, changing as a result of the diverse relationships between individuals, texts, and environments. In this sense, the post-process movement in composition studies is ecological (and has been identified as such), since it postulates not a fixed endpoint in the production of discourse, but a fluid and dynamic state of motion.

Our awareness of the inherent change and fluidity in writing programs prompted us to invite authors to include "Where We Are Now" codas at the end of their program profiles. As the codas demonstrate, writing programs are in a constant state of flux. The profiles represent a snapshot of each respective program, a gaze backward, while the programs themselves continue to change and develop. Many of the authors are quick to note the inherent change and fluidity of their programs. In fact, a mark of success in a writing program is its ability to transform, adapt, and evolve as a result of interactions and influences, as WPAs enact their agency in relation to the complex set of agencies at work on and around writing programs. Repeatedly, the collected program profiles document the ways that writing programs adapt to constraints and external impositions while strategically re-appropriating them to their advantage.

WRITING PROGRAMS ARE COMPLEX

Complexity refers to the intricate interweaving of discrete aspects in apparently chaotic systems. Complexity arises when an increasing number of variables interact to form an ecology. In many ways, complexity and systems theories are intermingled with recent ecological inquiries. In the sciences, complexity is a move away from the traditional reductionist paradigm, in which scientists tried to understand and describe the dynamics of systems by studying and describing their component parts. Complexity theory postulates that those components are strongly interrelated, self-organizing, and dynamic. Rain forests, immune systems, and the World Wide Web are often cited as examples of complex systems, and complexity is central to contemporary scientific theory. In fact, in an interview in 2000, Stephen Hawking predicted that "the next century [21st] will be the century of complexity."

Writing theorists have begun to unpack the ways in which discourse is a complex system. A complex view of writing moves away from the simple rhetor/audience dichotomy, incorporating a range of other variables and contexts that account for the temporal, spatial, material, and ambient dimensions of writing performances. In addition, recent scholarship addressing complexity and writing has helped to highlight the ways in which earlier theories were overly static or stable, recognizing that the complexities of writing "are so diverse and divergent that we may never be able to fully account for all of the facets and functions of writing, particularly as writing endlessly fluctuates as a system" (Dobrin, *Postcomposition* 143). Writing programs embody and enact this complexity, as they are systems of interaction in which order, patterns, and structure (be it learning outcomes, assessment mechanisms, training and professional development programs, curricular infrastructure) arise and create meaning through ongoing, interactive, contingent performances. One goal of the program profiles in this collection is to shed light on the patterns of meaning within each program, with an eye toward unpacking the complexity of writing program development, maintenance, assessment, and transformation that others may face.

WRITING PROGRAMS ARE EMERGENT

Emergence refers to the ways in which unique and coherent structures, patterns, and properties evolve during the process of self-organization in complex systems. Ecologies are seen as emergent when the actors or objects within them form more complex behaviors as a collective. In other words, emergence is the tendency toward greater development and evolution, in which the whole creates something that could not be generated by an individual part, nor even through the combined attributes of all parts. Emergence has been tied to both scientific and philosophical evolutionary theory. John Stuart Mill used the simple example of water to illustrate emergence: "The chemical combination of two substances produces, as is well known, a third substance with properties different from those of either of the two substances separately, or of both of them taken together" (371). Emergence has been used more recently to explain stock market behavior, the migration of birds, and even what happens at a rock concert. Emergence is what self-organizing, complex ecologies produce.

Writing is both an emergent attribute of a complex human society as well as an ecology in which emergence occurs. From a seemingly simple alphabetic system, ideas, concepts, literary movements, genres, and styles emerge. As Syverson suggests, emergence is the "self-organization that arises globally in networks of simple components connected to each other and acting locally—readers, writers, and texts, for instance" (183). Writing programs are emergent in that they create something new from the shared perspectives and interactions among faculty, administrators, and students. Writing programs evolve, as evidenced in these program profiles, and new ideas and ways of thinking surface as a result. The very concept of a "writing program" is an emergent attribute of the contemporary university system, and such programs will give way to new structures, patterns, and developments in the future.

These ecological attributes (interconnectedness, fluctuation, complexity, and emergence) are evident to various degrees within each of the programs profiled in this book. In what follows, we first situate the program profiles within their publication history in the journal *Composition Forum*. Then we locate this book in relation to recent scholarship in writing program administration, before providing an overview of the sections and program profiles that follow.

Situating the Program Profiles

Each of the profiles featured in this collection was first published in the journal *Composition Forum* as part of its Program Profiles section. Shortly after Christian Weisser's appointment as editor in 2005, *Composition Forum*'s Program Profiles section was made a regular feature of each published volume. Although programs and courses had previously been described in the older, print version of the journal, the feature was standardized to focus specifically on programs rather than a variety of curricular issues that did not fit the confines of a traditional essay. With the appointment of Michelle Ballif as the Program Profiles Editor and with the publication of her inaugural, standardized program profile (reprinted in this volume) in 2006, the Program Profiles section became a venue for showcasing exemplary programs and highlighting the *scholarly* contribution to our field that such program development and administration demonstrates. When Ballif moved to assume the role of Managing Editor of the journal in 2009, Anis Bawarshi and Mary Jo Reiff assumed the editorship of this important feature. Under their editorship, the profiles have grown in complexity and have developed various emergent features, including rich appendices, detailed course offerings, syllabi samples, and other documents substantiating and illuminating the varied, discursively and materially networked ways in which program developers and administrators respond to complex institutional, public, and personal demands. With this ecological emphasis, the profiles highlight the key attributes of fluctuation and complexity—the "complex adaptation and the constant motion of discursive systems" (Weisser 68) that include not only individual writers but also the "technologies" of the classroom, the electronic media informing such classrooms, the administrative and institutional contexts, the material conditions, and the public and political constituencies that rally both criticism and support for writing programs.

As agents and players in this complex system—and as part of an additional "discursive system" of tenure and promotion demands—the authors of these program profiles articulate the emergence, interconnectedness, fluctuation, and complexity of their writing programs, including not only the institutional constraints but also—as our call for submissions makes clear—the history, context, and goals of the program; the theory informing the program; the structural interrelationships; and reflection on lessons learned and advice to be offered to other program developers and administrators. Purposively, the Program Profiles feature has sought to advance a notion of writing programs in its most inclusive definition, including first-year composition programs, professional writing programs, writing across the curriculum and/or writing in the disciplines programs, writing centers, graduate programs in rhetoric and composition, and undergraduate major or certificate programs in writing.

In tandem with exploring and acknowledging the various "locations of composition" or the "ecologies of interaction" (Weisser 69), the Program Profiles feature has also sought to acknowledge the *scholarship* of programmatic development and administration. As is well known, such administrative work is often devalued (or not recognized) in tenure and promotion decisions. As Jeanne Gunner argues, "WPAs have often been blindsided in the actual review process when they discover, too late, that much of their scholarly administrative work has been discounted. The risks are especially high in traditional literature departments, where not only scholarly administrative work but any work in rhetoric-composition may be seen as second-class, as not true scholarship" (321). The Program Profiles feature allows writing program administrators to publish scholarship regarding the development, administration, and assessment of programs—academic work that might otherwise have gone unnoticed, perhaps locally, at an individual's home institution, where the context, as Gunner suggests, may not be supportive of the scholarly value of administration or program development. Additionally, by acknowledging writing program administration as scholarly work, published program profiles situate the local work of program administration within a broader disciplinary ecology, where such work can participate in larger questions and conversations about the study and teaching of writing. In this way, program profiles contribute to our field's collective knowledge while acknowledging the situated enactment of that knowledge within local ecologies.

DISCIPLINARY LOCATION OF THE BOOK

With its ecological framework and focus on locating writing programs within their fluid, dynamic, and relational institutional contexts, this book takes its place among a rich body of existing scholarship on writing program administration. Within these dynamic disciplinary ecologies, writing program administration has been (re)positioned as intellectual work, beginning fifteen years ago with the Council of Writing Program Administrators Executive Committee position statement on "Evaluating the Intellectual Work of Writing Administration" (1998), a statement that recognized the disciplinary and scholarly expertise that informs program creation, curricular design, faculty development, program assessment, and program-related textual production. Since then, the field has seen the emergence of a rich and growing body of scholarship that grows out of and supports this intellectual work, from historical studies of writing program administration (L'Eplattenier and Mastrangelo; McLeod), to essential resources and references on writing program theories and practices (Brown and Enos; Ward and Carpenter), to critical studies of working conditions and ethical practices of WPA work (Enos and Borrowman; McGhee and Handa; Strickland and Gunner), to the activist work of WPAs (Adler-Kassner, Rose and Weiser), to writing program administration at small colleges (Gladstein and Regaignon), to international perspectives on writing program administration (Thaiss, Bräuer, Carlino, Ganobcsik-Williams, and Sinha). Our book seeks to contribute to and enrich these and other existing broad perspectives on the history, theory, and practices of writing programs by offering close examinations of writing programs as they are located within specific institutional contexts.

Because program profiles describe writing program administration as situated within a particular program's history, context, and goals, they provide a thick description of individual programs that elucidates existing scholarship. While current scholarship provides a more comprehensive overview of the issues, responsibilities, and opportunities of writing program administration, program profiles focus specifically on how WPAs, located within particular institutional contexts, manage and enact these issues and responsibilities—negotiating budgets, legislative mandates, personnel, curriculum development and revision, assessment, new technologies, changing student demographics, teacher training and supervision, etc.—while being responsive to research and theory within the field of rhetoric and composition. This examination of particular programs in context complements broader perspectives on the history, theory, and practices of writing program administration, shifting the focus to how theories get enacted in particular programs and how histories and practices are enabled and constrained by particular institutional locations and contexts. With the focus of program profiles on constraints or challenges to developing a program, this book also extends important critical discussions of the working conditions of WPAs, highlighting material and managerial matters, along with the conflicting cultural and institutional issues that shape WPA work.

OVERVIEW OF THE BOOK

The following chapters examine how and where writing programs are located (from FYC sites to the disciplinary sites of undergraduate majors in rhetoric and composition studies), how the activities of WPAs can carve out new spaces for collaborative relationships and interactions (with writing centers, libraries, etc.), and how WPAs reposition programs (and are themselves repositioned) as they undertake curricular revision and explore new sites for writing program administration. In Part I, "The Contested Ecologies of FYC Programs: Negotiating between Stability and Change," we establish a framework for understanding the ecological attributes of interconnectedness, fluctuation, and emergence, with a focus on how first-year writing programs participate in larger institutional systems and respond to institutional changes. The program profile by Sara Webb-Sunderhaus and Stevens Amidon at Indiana University-Purdue University Fort Wayne describes a state-mandated elimination of the remedial or basic writing course and the emergence of a new curriculum and new methods of self-placement to adapt and respond to these changes. In her profile of the program at Southern Illinois University Carbondale, Ronda Leathers Dively also highlights the negotiation between stability and change and the effect on the interactions among and positioning of WPAs, GTAs, and FYC students. This negotiation of the interconnected network of affiliations within writing program ecologies is further highlighted in the program revisions described by Jenn Fishman and Mary Jo Reiff, who interacted with multiple stakeholders to transform their second-semester composition course at the University of Knoxville-Tennessee from a literature-based to an inquiry-based curriculum. The program profile by Elizabeth Wardle further focuses on the emergence and development of an approach to teaching "writing about writing" at the University of Central Florida, capitalizing on *kairotic* "opportunity spaces" to locate FYC within the knowledge of the field.

In response to research in composition studies that describes writing development as on-going and connected to disciplinary identities and ways of knowing, Part II, "Remapping Interdisciplinary Ecologies: WAC and WID Programs," features program profiles that describe attempts to distribute writing instruction across disciplinary ecologies, reflecting the key ecological principle of interconnectedness and emphasizing the interactions, relationships, and interdependencies of sites within writing program ecologies. In her profile of the Writing Intensive Program at the University of Georgia, Michelle Ballif explores the way in which writing-intensive courses in the disciplines build on the first-year composition experience, and she emphasizes the interdependency of these experiences. Alternatively, driven by a top-down mandate, the program profile of SUNY Binghamton, by Kelly Kinney and Kristi Costello, focuses on the development of an "autonomous" writing program that moved first-year writing outside of English, relocating it as a "university writing initiative." The ecological characteristic of emergence is further reflected in the self-organized and uniquely situated Institute for Writing and Rhetoric at Dartmouth University, which is building a fluid, interconnected (writing and speech) program structured to foster transfer, as explored by the institute faculty (Stephanie Boone, Sara Biggs Chaney, Josh Compton, Christiane Donahue, and Karen Gocsik).

Moving from writing across the disciplines to writing within the discipline of rhetoric and composition studies, Part III, "Claiming Disciplinary Locations: The Undergraduate Major in Rhetoric and Composition," highlights the ecological hallmarks of emergence, interconnectedness, and fluctuation and includes program profiles that describe the development of undergraduate writing majors: the conditions that gave rise to these programs, the resources needed to launch and sustain them, and the larger effects these programs have on institutional ecologies. In their program profile published in 2007, Stephanie L. Kerschbaum and M. Jimmie Killingsworth note the dearth of undergraduate programs in the field and describe the emergence and development of the undergraduate rhetoric concentration within the English department at Texas A&M University. With this chapter as a starting point, more recent profiles of undergraduate programs follow and update this perspective, including a profile by Lori Ostergaard, Greg A. Giberson, and Jim Nugent that describes the development and implementation of a new writing and rhetoric major at

Oakland University and focuses on the complex interrelationships and interconnected issues involved in creating an undergraduate program. Moving from a focus on interconnectedness to fluctuation, David Beard, in his profile of the University of Minnesota-Duluth program, describes the shifting landscapes and ecological debates surrounding the new writing studies major.

Shifting from disciplinary to institutional locations and the high stakes work of program assessment, Part IV, "Situating Assessment within Institutional Ecologies," presents two programs that have developed assessment practices that are responsive to institutional exigencies, student needs, and curricular integrity. As WPAs work to locate their programs and curricular goals, program-driven assessment occupies a central role within complex ecologies of curricular development and revision, outcomes, teacher training, etc. In his profile of California State University, Fresno, Asao B. Inoue demonstrates how assessment measures become part of the fluid framework of program development, curricular revision, and systems of institutional accountability, or what he describes as a "culture of assessment." Within this culture of assessment, Paul Walker and Elizabeth Myers, in their profile of Murray State University, describe how they negotiated complex variables—such as labor conditions and university-wide general requirements—and adapted to the constraints of accountability while strategically using assessment to their advantage to transform their writing curriculum and to facilitate the emergence of new curricular structures and developments.

Finally, Part V, "Third Spaces: Creating Liminal Ecologies," features profiles that explore interconnected institutional spaces that operate outside and alongside writing programs. These "third spaces" are hybrid, overlapping spaces that support multiple literacies and engage student writers in exploring multiple discourses and navigating different disciplinary and writing contexts. Often seen as institutional "support structures" for writing programs, places like writing centers and libraries reveal the complexity and interconnectedness of writing programs and can play a more dynamic role in writing program ecologies and institutional systems of writing. A program profile of FYC, library, and writing center partnerships at West Virginia University (Laura Brady, Nathalie Singh-Corcoran, Jo Ann Dadisman, and Kelly Diamond) examines how professional and disciplinary boundaries can be redrawn and how new relationships emerge within institutional ecologies of writing. In addition, a profile of an innovative, peer-interactive model of tutoring at University of New Mexico's writing center by Daniel Sanford demonstrates how the goals of writing tutoring can be realigned to more effectively embody collaborative, process-oriented views of writing and to reposition writing tutors and the writing center as a crucial part of the writing environment and culture of writing. The Elon College profile—by Jessie L. Moore, Kimberly B. Pyne, and Paula Patch—describes a transitional space within a larger ecology of writing—a summer transition program that creates a network of institutional affiliations that support and prepares students for FYC and access to "college capital."

Collectively, the program profiles that follow reveal the dynamic inter-relationships as well as the complex rhetorical and material conditions that writing programs inhabit—conditions and relationships that are constantly in flux as WPAs negotiate constraint and innovation. The organization and grouping of each section highlights these inter-relationships, and the section introductions further contextualize the ways in which each program profile relates to and shapes other work in the field. Similarly, the new addendum to each program profile—a retrospective "Where We Are Now" coda written by the author(s) of the piece—adds valuable reflection and perspective on each profile and on the program it describes. Overall, as published within *Composition Forum,* each program profile tells its own local story, but collected, framed, and organized in the following collection, they reveal the larger ecologies that influence and are influenced by writing programs.

WORKS CITED

Adler-Kassner, Linda. *The Activist WPA: Changing Stories about Writing and Writers.* Logan: Utah State UP. 2008. Print.

Brooke, Collin Gifford. *Lingua Fracta: Towards a Rhetoric of New Media.* Cresskill, NJ: Hampton P, 2009. Print.

Brown, Stuart C., and Theresa Enos. *The Writing Program Administrator's Resource: A Guide to Reflective Institutional Practice.* Mahwah: Erlbaum, 2002. Print.

Commoner, Barry. *The Closing Circle: Nature, Man, and Technology.* New York: Knopf, 1971. Print.

Coole, Diana, and Samantha Frost. *New Materialism: Ontology, Agency, and Politics.* Durham: Duke UP, 2010. Print.

Cooper, Marilyn. "The Ecology of Writing." *College English* 48.4 (1986): 364–75. Print.
Council of Writing Program Administrators. "Evaluating the Intellectual Work of Writing Administration." 1998. Web. 15 May 2012.
Dobrin, Sidney I. *Ecology, Writing Theory, and New Media: Writing Ecology.* New York: Routledge, 2012. Print.
—. *Postcomposition.* Carbondale: Southern Illinois UP, 2011. Print.
Edbauer, Jenny. "Unframing Models of Public Distribution: From Rhetorical Situation to Rhetorical Ecology." *Rhetoric Society Quarterly* 35.4 (2005): 5–25. Print.
Enos, Theresa, and Shane Borrowman. *The Promise and Perils of Writing Program Administration.* West Lafayette: Parlor P, 2008. Print.
Gladstein, Jill M., and Dara Rossman Regaignon, eds. *Writing Program Administration at Small Liberal Arts Colleges.* Anderson, SC: Parlor P, 2012. Print.
Gunner, Jeanne. "Professional Advancement of the WPA: Rhetoric and Politics in Tenure and Promotion." *The Allyn & Bacon Sourcebook for Writing Program Administrators.* Ed. Irene Ward and William J. Carpenter. New York: Longman, 2002. 315–30. Print.
Hawk, Byron. "Toward a Rhetoric of Network (Media) Culture: Notes on Polarities and Potentiality." *JAC* 24.4 (2004): 831–50. Print.
Hawking, Stephen. "'Unified Theory' Is Getting Closer, Hawking Predicts." Interview in *San Jose Mercury News* (23 Jan 2000), 29A.
L'Eplattenier, Barbara, and Lisa Mastrangelo, eds. *Historical Studies of Writing Program Administration: Individuals, Communities, and the Formation of a Discipline.* West Lafayette: Parlor P, 2004. Print.
McGee, Sharon James, and Carolyn Handa, eds. *Discord and Direction: The Postmodern Writing Program Administrator.* Logan: Utah State UP, 2005. Print.
McLeod, Susan H. *Writing Program Administration.* West Lafayette: Parlor P, 2007. Print.
Mill, John Stuart. *A System of Logic Ratiocinative and Inductive.* London: John W. Parker and Son, 1872. Print.
Rose, Shirley K, and Irwin Weiser, eds. *The Writing Program Administrator as Theorist: Making Knowledge Work.* Portsmouth: Boynton/Cook/Heinemann, 2002. Print.
Strickland, Donna, and Jeanne Gunner, eds. *The Writing Program Interrupted: Making Space for Critical Discourse.* Portsmouth: Boynton Cook, 2009. Print.
Syverson, Margaret. *The Wealth of Reality: An Ecology of Composition.* Carbondale: SIU Press, 1999. Print.
Thaiss, Chris, Gerd Bräuer, Paula Carlino, Lisa Ganobcsik-Williams, and Aparna Sinha, eds. *Writing Programs Worldwide: Profiles of Academic*

Writing in Many Places. Fort Collins, CO, and Anderson, SC: The WAC Clearinghouse and Parlor P, 2012. Print.

Ward, Irene, and William Carpenter, *The Allyn and Bacon Sourcebook for Writing Program Administrators.* NY: Longman, 2002. Print.

Weisser, Christian. "Ecology." *Keywords in Writing Studies.* Ed. Peter Vandenberg and Paul Heilker. Logan: Utah State UP, 2015. 61–71. Print.

Part I. The Contested Ecologies of FYC Programs: Negotiating between Stability and Change

Given the institutional positioning of first-year composition (FYC) as a university-wide course requirement, FYC programs participate in dynamic interrelationships with multiple stakeholders—from state legislatures, to faculty colleagues, to writing program instructors (adjunct faculty and GTAs), to FYC students—demonstrating well the essential ecological characteristics of interconnectedness and fluctuation. The program profiles featured in this section describe how first-year writing programs participate in larger institutional systems and how WPAs, located within particular institutional contexts, respond to and enact institutional changes—how they negotiate legislative mandates, curricular development and revision, teacher training and supervision—while being responsive to evolving research and theory within the field of rhetoric and composition. As WPAs navigate material conditions that are constantly in flux, and as they work to reposition writing programs and relocate writing instruction within a network of institutional affiliations, they must constantly negotiate between constraint and innovation, stability and change.

In "The *Kairotic* Moment: Pragmatic Revision of Basic Writing Instruction at Indiana University-Purdue University Fort Wayne" (published in 2011), Sara Webb-Sunderhaus and Stevens Amidon describe how they responded to a state-mandated elimination of the basic writing course at IPFW by developing a new curriculum and new methods of self-placement. Within the complex ecologies of their institution, the authors had to negotiate challenges of retention and student success as they revised the program to create a credit-bearing basic writing course, to shape a curriculum based on the Council of Writing Program Administrators Outcomes Statement, and to institute guided self-placement.

Illustrating the fluctuation and emergence within these institutional ecologies, in their "Where We Are Now" coda, the authors later reflect on continuing state mandates that present challenges but also opportunities for redefining "remedial" course models. As they negotiate legislative mandates, assessment measures, and new statewide general education requirements, the authors continue to consider effects on the curriculum and on placement models, noting that within the complex system of writing program ecologies, "the one thing we can count on . . . is uncertainty and instability."

In the face of uncertainty and in the midst of shifting locations and responsibilities, Ronda Leathers Dively explains how WPAs might adapt to fluctuation and change within writing program ecologies by identifying emergent structures, patterns, or "common places" that contribute to a collective vision, such as defining common writing outcomes, creating a common experience for students, and establishing shared knowledge among teachers. In "Standardizing English 101 at Southern Illinois University Carbondale: Reflections on the Promise of Improved GTA Preparation and More Effective Writing Instruction" (published in 2010), Dively describes the move to a standardized English 101 curriculum at SIUC, the theoretical and practical implications for the move, and the political and logistical challenges encountered by the writing studies staff. As she notes in her "Where We Are Now" coda, Dively has since stepped away from the writing studies directorship, and this remote perspective has enabled her "to view standardization not as an all-or-nothing prospect but, rather, as a continuum for teacher support"—a constant negotiation between the WPA's concern for consistent and coherent writing program objectives and the valuing of teacher autonomy and innovation.

This ecological attribute of fluctuation or negotiation between stability and change, constraint and innovation, is also highlighted in the program revisions described in "Taking the High Road: Teaching for Transfer in an FYC Program," a profile published in 2008 by Jenn Fishman and Mary Jo Reiff. Fishman and Reiff transformed their second-semester composition course from a literature-based to an inquiry-based course and designed a curriculum geared toward helping students acquire the rhetorical knowledge and skills vital to communicating effectively in multiple contexts. Illustrating the ecological trait of emergence, they created something new from the shared perspectives and interactions among faculty, administrators, and students.

They detail their negotiation of new general education requirements for writing, feedback from instructors and students, and research on writing knowledge transfer, as they worked to build a program based on the fluid transfer of rhetorical knowledge and the development of multiple methods of inquiry. In their "Where We Are Now" coda—written from their positions in new institutional contexts—they affirm this approach to "teaching for transfer," which supports a potentially sustainable curriculum even as programmatic and institutional ecologies shift and change.

In the midst of these complex and fluctuating institutional contexts, WPAs may find themselves in the position of seeking out emergent or opportune moments for relocating and repositioning writing instruction. In her 2013 profile, "Intractable Writing Program Problems, Kairos, and Writing-about-Writing: A Narrative about the University of Central Florida's Composition Program," Elizabeth Wardle explains how, in the midst of negotiating the constraints of contingent faculty labor, curricular objectives, and institutional assessment, she capitalized on *kairotic* "opportunity spaces" to locate FYC within the disciplinary knowledge of the field through a writing-about-writing curriculum. In her "Where We Are Now" coda, she later reflects on the ways in which complex, external forces continue to exert influence on the program, with institutional hierarchies that reinforce a two-tiered system of tenure-track and non-tenure track faculty and with legislative changes to general education that threaten writing requirements. Illustrating the ecological characteristic of complexity and the dynamic development and transformation of writing programs, the following profiles also demonstrate how—within the context of the shifting landscapes and contested ecologies of FYC programs—WPAs are increasingly called upon to enact their agency in relation to the complex set of agencies at work within writing program ecologies.

1 The *Kairotic* Moment: Pragmatic Revision of Basic Writing Instruction at Indiana University-Purdue University Fort Wayne

Sara Webb-Sunderhaus and Stevens Amidon

In the fall of 2008, a series of events emerged which created the conditions for a major overhaul of the writing program here at Indiana University-Purdue University Fort Wayne (IPFW), a regional, comprehensive university with a population of more than 14,000 students.[1] We—the director of writing (Steve) and the basic writing course coordinator (Sara)—were faced with a state mandate that public, four-year institutions take immediate steps towards the elimination of so-called "remedial" (i.e., non-credit bearing) courses in the curriculum. The Indiana Commission on Higher Education (ICHE) recommended that such non-credit, developmental courses be eliminated from the curriculum by 2011; at that time such courses would only be offered through the state's growing community college system (Indiana Commission on Higher Education, 2008). Basic writing courses have long been viewed—and dismissed—by some as a place of remediation (Bernstein; Bloom; Grego and Thompson; Hull et al.; Rose, "Language"; Soliday; Troyka). Because so much has already been written about these efforts to eliminate basic writing instruction from four-year universities (see among many others Goen-Salter, Smoke, Stevens, Stygall, and Wiener), we will simply note here that Indiana's move towards removing basic writing courses from its four-year universities is in keeping with national trends of "outsourcing" basic writing to two-year colleges.

As we explain below, this mandate did not change the fact that our campus continues to serve many students who are less likely to be retained than their peers and who are underprepared by traditional standards to achieve our general education writing outcomes in a single, fifteen-week semester course. IPFW only requires one semester of composition for all students, and this semester of instruction is simply not enough for a significant portion of our first-year students, many of whom are inexperienced writers and some of whom are returning to school after many years in the workforce or at home. Furthermore, their lack of preparedness is not uniform enough that a single, fifteen-week course could address the needs of these students to gain familiarity with a variety of genres. We have found that too many of our students are inexperienced in making the transition from familiar to unfamiliar writing genres; any experience and familiarity they may have with writing does not transfer when they are asked to write outside of the genres they find most comfortable. While the issue of genre knowledge and transfer is complex and the subject of much current debate in composition studies, we are persuaded by Amy Devitt's argument that "writers use the genres they know when faced with a genre they do not know. These genres are not in fact transferable [. . . but] they help writers move into a new genre; they help writers adjust their old situations to new locations" (222). Many of our students, however, have done such little writing that they have limited awareness and experience of multiple genres.

Since there is such a diversity of writing experience and preparation among our students, we cannot assume that all of our students are familiar with any particular "old location." Some of our students come to us with a great deal of experience writing five-paragraph themes or simple argumentative essays, but they have never written a narrative; others are quite comfortable writing narratives but are unfamiliar with thesis-driven arguments. Devitt writes, "Each genre a writer acquires well increases the likelihood of having a genre to use with a situation more similar to the new genre" (222), yet some of our students have very few genres to call upon as they face the various rhetorical situations of their first year of college. Thus, one semester of instruction is simply not enough time to allow for the growth and development of writers limited by this lack of experience.

After the state issued its mandate in 2008, we quickly realized that we had to create a new, credit-bearing, basic writing course that would meet the needs of our most at-risk students, a course that would not be identified as "remedial" and would thus circumvent the vagaries of the state legislature. While the state's actions were extremely troubling, we viewed the directives with which we were faced not as a threat to our "turf," but as an opportunity to re-make basic writing instruction in ways that would be more effective for both students and faculty at our institution. What emerged was a new basic writing course, theoretically sound and pedagogically appropriate, that provides some students with an additional semester of writing instruction and that meets the needs of our institution, our students, and our faculty to a far greater degree.

WHO WE ARE AND WHO WE SERVE

IPFW is a joint, regional, four-year campus of two large, Research 1 institutions (Indiana and Purdue Universities), and our writing program is part of the Department of English and Linguistics, the largest department in our college and one of the largest at our university. A major problem facing our university is retention. *The Journal Gazette,* a local newspaper, reports that students entering IPFW in 1999 had a four-year graduation rate of 4 percent and a six-year graduation rate of 18 percent. These rates are the second-lowest of any four-year public institution in the state (Soderlund). First-year retention rates are also lower than peer institutions' benchmarks. According to William Baden, senior analyst in the university's Office of Institutional Research and Analysis, of the Fall 2006 entering class, only 60 percent were enrolled by Fall 2007. First-year retention rates for basic writers are lower still. For students enrolled in basic writing courses who began their college education in the fall semesters of 2003–2007, the retention rate was 56.7 percent; for first-year writing students, the retention rate was 64.6 percent (Baden). While these low retention and graduation rates are affected by students who transfer to other institutions (especially Purdue-West Lafayette and Indiana-Bloomington) and by students who are enrolled part-time, those two factors alone cannot completely explain our university's poor student success rates.

Our programmatic data also showed a disturbing pattern of high DWF (drop, withdraw, fail) rates. In 2006 our basic writing (BW) program enrolled 2300 students. Eighty percent of those students passed one of the basic classes. Of the students who completed the two-course sequence (one of the BW courses and the FYC course), 70 percent passed FYC. However, when we included students who dropped the basic writing or first-year writing course at some point in the sequence, that FYC pass rate dropped to around 50 percent. We were losing half of our first-year writing students to failure or attrition—a highly disturbing phenomenon.

The diversity of the seventeen-county region from which we draw students is a factor in complicating retention and student preparedness for our curriculum. A little less than half of our students come from Fort Wayne; the rest come from rural Allen County and the sixteen-county region surrounding the city (Office of Institutional Research). The level of underpreparedness we see in our first-year students doesn't appear to be simply a matter of students from suburban schools succeeding while students from urban schools fail. For example, some students from our local, urban, public school system—Fort Wayne Community Schools (FWCS), the largest feeder system for our program and the second largest school district in the state—compete on at least equal footing with students from the most well-funded suburban and private schools in our area, while other FWCS students do not; many times, issues of race and class emerge here, as students from the poorest and majority-minority high schools in the district tend not to succeed at the same rate as students from FWCS high schools with more racial and socio-economic integration. However, virtually all FWCS students outperform their peers from some of the rural, outlying school districts, as our faculty's teaching experiences have shown that these rural students are far more likely to come to our program with virtually no experience composing drafts of even moderate length (i.e., 4–5 pages) and can be completely overwhelmed by the rhetorical tasks asked of them in college. IPFW students who responded to the 2009 National Survey of Student Engagement (NSSE) showed that 20 percent of first-year students completed a paper of 20+ pages in their first year, and 9 percent completed more than ten papers between 5 and 19 pages in their first year. This volume of writing is a challenging task to underprepared students.

PAST AS PROLOGUE: THE HISTORY OF OUR PROGRAM

Due to the diverse needs of our student population, our department has always offered more than one entry point into the writing curriculum. Until the Fall 2008 semester, we used the College Board's Accuplacer test to involuntary place students. This multiple-choice test used a series of questions focusing on grammar and sentence structure to place students. Students who performed well placed into our FYC course, which is required by the university's general education curriculum. Students who performed poorly were placed into a non-credit, BW course. Students on the borderline took the FYC course along with a two-credit, studio-style course that was designed to give these students extra instruction and support as they grappled with the demands of the FYC course. A few students bypassed FYC entirely, earning credit through school-based programs or AP exams. Many of those students took our second-level composition course in the first year.[2] This "bypass" option is still available to students, and most of them continue to take second-level writing during their first year.

Involuntary placement into so-called "remedial," non-credit courses frustrated many of our students (as well as their parents, who sometimes pay the tuition bill), resulting in attitudes that mitigated success in those courses. Another challenge we faced was that the various sections of the basic writing course did not always offer consistency in instruction, leading to frustration among students and faculty. When Sara arrived at IPFW in the fall of 2006, there was no defined curriculum or philosophy of instruction that was consistent across the many sections of the basic writing course and few, if any, program-wide outcomes. The writing program did not articulate its outcomes until 2007; as a result, for several years the course did not have a clear identity or sense of purpose, other than the rather nebulous notion of "preparing" students for our first-year writing course.

This lack of clarity resulted in instructors—almost all TAs and adjuncts—using a variety of texts and assignments, some of which overlapped with those texts and assignments used in FYC; yet other assignments were not aligned with the theoretical understandings and best practices of our field and instead resembled skill-and-drill workbook-style approaches that emphasized grammatical correctness and focused on sentences and paragraphs. Due to the overlap in some sections, a number of students who moved from BW to FYC understandably felt as if they were taking the same course twice, a fact they greatly resented. Other students complained that their section was much harder or easier than those of their peers, claims that were not without merit. We, too, had noticed that some sections of the course stood out as extremely undemanding or challenging. Grading practices were also an issue; there were no programmatic grading rubrics, and understandings of what passing work looked like varied from instructor to instructor. In short, the course was a hodgepodge of writing instruction that did not serve the needs of our students or our writing program.

Further compounding the problem of consistency was instructor turnover. Sara is the only tenure-line faculty member in IPFW's department of English and linguistics who regularly teaches basic writing. The vast majority of our basic, first-year, and second-level writing courses are taught by part-time instructors and TAs, who often cycle in and out of our writing program—about 25 percent leave the program each year. Furthermore, almost all of our part-time instructors have terminal MAs, and some of them have not kept abreast of current theoretical and pedagogical developments in composition studies. Our TAs are MA students who, by the very nature of their position, are just learning about composition theory and pedagogy and are inexperienced instructors. Generally, adjuncts who expressed interest in basic writing were assigned to teach the BW courses, but in many instances, especially in the case of the studio course (explained in more detail below), instructors asked for these courses simply to maximize their teaching load at IPFW. Steve (as director of writing) usually has about sixty adjuncts and TAs to oversee in a semester, and while we have many fine instructors working for us, it's all too easy for a person or problem to slip through the cracks. We do provide seminars devoted to writing pedagogy at the beginning of each fall semester and periodically during the school year, but other than the mandatory session at the beginning of the academic year, attendance at many of these seminars is disappointing. We know that some of our adjuncts teach courses at other institutions in our area, and their ability to volunteer their time for pedagogical improvement is limited.

Kairos Enacted: The Program Revision

All of these issues—state mandates, retention, diversity of student and faculty preparedness, and student and faculty dissatisfaction—created the kairotic conditions for a major overhaul of the writing program. The problems we faced were too complex to simply fix in piecemeal fashion by changing assignment sequences or textbooks. The approach we took was threefold; we will outline it here, with further explanation to come in forthcoming subsections:

1. We needed to replace the non-credit course which the state had identified as remedial as well as the unsuccessful studio course

that aimed to "mainstream" some basic writers into FYC but failed to do so successfully, due to a variety of factors.
2. Rather than involuntarily placing students in courses that, rightly or wrongly, the students perceived as remedial and for which they received no credit, we wanted to empower students to accurately self-place themselves into courses that best fit their needs.
3. We wanted a curriculum that offered consistency of outcomes but that wasn't narrowly defined by a standardized, one-size-fits-all syllabus. We had contingent faculty and teaching assistants, many of whom were excellent instructors, who would have had difficulty making such a transition.

The Creation and Elimination of Basic Writing Courses

Because our department is relatively large and has diverse offerings, the director of writing—who is appointed by the chair of the English and linguistics department—manages the administration of our basic, first-year, and second-level writing courses (i.e., the writing program). The composition committee designs and implements programmatic policies and advises the director of writing. A pivotal event that shaped the new direction of our writing program's curriculum was the decision of the director and committee to eliminate the two-credit, studio course that had been designed to serve students who didn't require as much intervention in order to succeed in FYC. There were several reasons this course was failing. First of all, registration management and scheduling issues placed students from multiple FYC sections into a single studio section. Since those students were on multiple course calendars, it was very difficult for instructors to prepare instructional materials that helped students. Too many of the studio course instructors were relying on error-based approaches that placed too much emphasis on sentence-level grammar and punctuation exercises. Although there had been for some time a consensus inside and outside the writing program that this studio course was not successful in improving outcomes, removing the course proved difficult, due to administrative turnover and objections from some faculty, who enjoyed teaching the course in spite of its problems. The state mandate provided the necessary impetus to remove the course from the curriculum.

In replacing both the studio course and the old, non-credit BW course with our new, credit-bearing BW course, we chose to standardize our approach to basic writing in order to prevent the wide variation in pedagogical approaches we saw. Some basic writing teachers took a process-based approach in this course similar to that of our FYC course, but once again too many instructors were still using archaic and discredited pedagogical approaches with a heavy emphasis on grammar worksheets, vocabulary drills, and punctuation exercises. The new BW course features a course cap of eighteen students[3] and has the same course outcomes as FYC; the achievement of these outcomes is stretched over two semesters, however.[4] While the BW curriculum will later be discussed in further detail, our course design was built on the philosophy of stretch programs articulated by Greg Glau, which includes:

- a view of basic writing students as capable and intelligent but lacking experience in the kinds of writing expected at the university level . . . ;
- a belief that in order to learn to write, any writer must *write*, receive feedback on that writing, and then revise her work, over and over . . . ;
- a belief that students should receive course credit for their college work;
- and the notion that beginning writers, since they lack experience in writing, need more *time* to learn to work with and to develop appropriate writing strategies. (80)

In spite of this shared philosophy, we do not identify our basic writing program as a stretch program, because our new BW course is a separate course with books and assignments distinct from those of the FYC course. In other words, although our course outcomes are stretched over two semesters, our methods for assisting students in meeting those outcomes are not. Given the transience of our faculty and student populations, the traditional stretch course described by Glau did not seem viable for our campus and its particular set of issues and needs. Furthermore, as discussed in the introduction, our basic writing students need experience with a variety of genres as they grapple with unfamiliar rhetorical situations. Stretching the four major assignments of one course over two semesters would not accomplish our goal of assisting students in developing their generic knowledge as broadly and deeply as possible.

The Move to Guided Self-Placement

In 2007, Stuart Blythe, who was at that time the director of writing at IPFW, successfully proposed replacing the university's placement vehicle for writing courses, the *Accuplacer* test, with a guided system of self-placement. One of the earliest examples of self-placement we examined was at Grand Valley State University, which used a brochure with a series of student prompts to guide the student to their placement decision. The prompts asked students to evaluate their own readiness and confidence in their ability to succeed in first-year writing (Royer and Gilles). We liked the use of prompts, but we believed that our advisors and administrators would be more likely to have confidence in the placement recommendations if they were tied to more empirical measures. We realized that the Daly-Miller test for writing apprehension asked the same kinds of questions as GVSU's prompts—questions that evaluated students' experience as writers, confidence in their writing ability, and their own judgment of preparedness for college writing. It also had a reliability rate of about 90 percent during a thirty-year history of use in research studies.[5]

A more recent example of guided self-placement, at Southern Illinois University-Carbondale, used prompts for initial placement but also used a diagnostic essay in the first week of class "as a check on the process" (Blakesley 18). We rejected the use of the essay because one, we wanted to demonstrate trust in the students' self-efficacy to make their placement decision, and two, research suggests that timed essays are not reliable predictors of success in FYC. We also weren't convinced our large group of instructors would be able to consistently make placement decisions, and we were concerned that some of our instructors would agonize over the decision, which we felt was best left to the students themselves (Haswell, Huot).[6]

After much consideration and study, we designed a system of guided self-placement that offers students two options into which they place themselves after reviewing information to help them with their placement decision: (1) students who feel they are ready for the general education writing course place themselves in FYC; or (2) students who feel they need more time to complete the FYC outcomes place themselves in a two-course sequence that begins with our new credit-bearing, basic writing course, after which they take FYC. Our students complete an online placement instrument,[7] which combines data that includes (1) students' high school class standing; (2) students' SAT scores (we actually use the math score since it correlates more reliably with FYC success at our institution than the verbal score), and students' level of writing apprehension as determined by the Daly-Miller test. A combined score is generated, and students are informed that the placement vehicle recommends that they take either the BW or FYC course. Students then make the final decision.

The design of the self-placement vehicle was based upon a statistical analysis of scores and high school grades commonly used in placement systems. The students' class standing and SAT math scores were chosen because they were readily available and were the only data which an analysis by our Office of Institutional Research found to meet commonly accepted standards of statistical significance in correlation studies. We added the Daly-Miller test because we also wanted to factor in student confidence, which as we noted earlier, can play a major role in some students' success. At our university, while we have found that a high degree of writing apprehension correlates with student failure in our courses, so, too, does a low level of writing apprehension combined with a poor academic record. The use of statistically valid measures in our self-placement design played a major role in ensuring support from upper-level administration for our proposed changes.

The Development of the New Basic Writing Curriculum

As the new coordinator of the basic writing program and an instructor of the revised course, Sara developed a revised curriculum, drawing on her work with basic writers at the University of Cincinnati and The Ohio State University. Utilizing the theoretical model of Bartholomae and Petrosky's *Facts, Artifacts, and Counterfacts: Theory and Method for a Reading and Writing Course*, the new basic writing course stresses the connection between reading and writing, as both are foundational skills for success in college. All instructors of the course (with the exception of some who are teaching in learning communities or teaching non-native speakers of English) now share a common theme, texts, and assignments. We want to honor our instructors' knowledge of students and pedagogical expertise, as well as insure that they have a voice in, and feel ownership for, what they teach. Therefore, the theme is chosen by the instructors and changes every year. Recent themes have included education, American politics, and the millennial generation. All instructors use our writing program's handbook and the same rhetoric (Barbara Fine Clouse's *A Troubleshooting Guide for Writers*). In addition, after choosing a theme for the course, the instructors select one of two book-length expository texts that address the theme in some way; for example, instructors used Jean Twenge's *Generation Me* or Neil Howe and William Strauss's *Millennials Rising* as the expository text for the millennial theme. The expository text selection changes each year, in conjunction with the theme. Homework assignments, including short summaries of the assigned readings, also reinforce the reading-writing connection and seek to enhance reading comprehension skills while further developing students' writing abilities.[8]

Our general course outcomes state that students who complete the BW course should be able to demonstrate their competence in the four following areas:

- rhetorical knowledge, including the ability to focus on a purpose and audience; to respond appropriately to different kinds of rhetorical situations; to adopt appropriate voice, tone, and level of formality; and to write in several genres.

- critical thinking, reading, and writing, including the ability to use writing and reading for inquiry, learning, thinking, and communicating; to manage a writing assignment as a series of tasks, including finding, evaluating, analyzing, and synthesizing appropriate primary and secondary sources; and to integrate one's own ideas with those of others.

- writing processes, including the use of multiple drafts to create and complete a successful text; the development of flexible strategies for generating, revising, editing, and proof-reading; and participation in collaborative and social processes that require the ability to critique one's own and others' works.

- knowledge of conventions, including the ability to follow common formats for different kinds of genres; to practice appropriate means of documenting one's work; to control such surface features as syntax, grammar, punctuation, and spelling.

Thus, in order to develop the students' rhetorical knowledge and knowledge of conventions, the major writing assignments of the course ask students to engage in four kinds of rhetorical tasks: a personal narrative, a textual analysis, a self-designed research project, and a self-reflection that assesses their performance in the course and their development as writers.[9] All of these assignments require the students to engage in critical thinking, reading, and writing. Students' writing processes are developed through the use of revision, a foundational part of the course, as all of these assignments are revised multiple times. Peer review (whether face-to-face or electronic) and instructor commentary (in the form of written feedback, conferences, or both) are offered in various stages of the composing process.

The sequence of assignments is designed to gradually move students' writing from personal to public purposes and audiences, as well. The course begins with a narrative that asks students to write about a personal experience related to the theme, but by the third assignment—the self-designed research project—the students are required to write for a public audience of their choice, using the genre and format that is most appropriate for the rhetorical situation they have devised. While instructors are expected to utilize the four rhetorical tasks when designing the major assignments and to integrate the course theme and texts, our curriculum offers instructors the flexibility to design their particular sections (including the assignments) in ways that build on their pedagogical strengths while also meeting the course outcomes. We believe this flexibility encourages our instructors to buy into the course and invest in the programmatic goals. The assignment design also encourages students to take more ownership of their writing.

The first two assignments (the narrative and textual analysis) impose audience, genre, and length requirements; the later assignments push students towards taking responsibility by choosing their own audiences and genres, once again building the students' repertoire of generic tools they can bring to varying rhetorical situations. Finally, the first three assignments address the common theme; these three writing tasks are unique to the new BW course and are not repeated in FYC, a situation that sometimes occurred in the old basic writing course. However, these assignments do help prepare students for the assignments and outcomes they will be expected to meet in FYC.

In an attempt to better coordinate our efforts across sections, Sara set up a listserv that is exclusive to our basic writing instructors, where we can (and do) share our assignments, our experiences teaching the course, and challenges we are facing. She also established teaching circles, small groups of instructors who choose to meet together to collaborate. As part of the mandatory fall workshop discussed previously, Sara leads a workshop on teaching basic writing at IPFW, and throughout the academic year she also hosts workshops on topics ranging from using electronic response and conferences to recognizing and adjusting to the differences between BW and FYC students. Because of the previously mentioned difficulties many of our instructors have with attending these workshops, Sara confers with individual or small groups of instructors via email and meetings as well, and many instructor resources are available online. Updating this website[10] is one of our projects for Summer 2011. Finally, during the Spring 2011 semester, Sara is teaching a graduate seminar on teaching marginalized populations of students, particularly basic writers; this seminar has high enrollment and will offer many of our TAs the opportunity to further explore, develop, and theorize their basic writing pedagogy. All of the measures we've described here have enabled the course and its instructors to develop a shared sense of identity and purpose that we are told did not really exist before these changes were made.

Where We're Going, Where We've Been: Reflections on the Changing Nature of Basic Writing

The factors we have considered in analyzing the new curriculum are student satisfaction, success, and retention rates, since they were major contributors to our decision to give students more decision-making power in their curricular placement. Many of our advisors and administrators, and even some of our faculty, were skeptical about this change, despite the fact that guided self-placement had been successfully implemented at other universities, including ones with a similar mission to IPFW. They doubted that students would elect to voluntarily take the new basic writing course when they could bypass it.

This concern has proven to be unfounded. In the first two years since implementation, we have seen about the same 30/70 ratio of students placing themselves into the basic writing course that we saw while using Accuplacer. What we have also seen is a great deal less student dissatisfaction with the new basic writing course, as compared to the old, non-credit course and the studio course. Evidence for this can found in improved student evaluations of basic writing faculty and the curriculum as compared to the same evaluations for the courses that were eliminated. Looking at 2002 student evaluations as a baseline and 2009 evaluations as an endpoint, we have seen overall student satisfaction in composition courses increase from 3.86 to 4.14 on a five-point Likert scale. Another indicator of student satisfaction with the course is that grade appeals for the new course have dropped by about 75 percent as compared to the previous courses.

We have also seen major improvement in programmatic retention and success rates. Whereas basic writing students previously withdrew from the university at a rate that meant less than 60 percent of students completed FYC successfully during their first two semesters, today that percentage has risen to around 70 percent. We have also seen a decrease in students failing basic writing. During the last five years that the old basic writing and studio courses were offered, the combined DWF rate of these courses ranged from a low of 40.46 percent to a high of 55.93 percent. In comparison, during the first two years of our new basic writing course, the DWF rate has averaged 31.05 percent (Baden). While we only have two years of data so far, these preliminary findings are very promising.

Furthermore, self-placement has not led to lower grades in FYC. The percentage of students dropping, failing, or withdrawing from FYC has remained steady at around 25 percent. Quite simply, self-placement has not led to an increase in FYC failure rates that some at our university feared, while BW failure rates have improved markedly. These data indicate to us that self-placement is at least as effective as Accuplacer, if not more so.

The trends that led in part to our own programmatic changes—the outsourcing of basic writing instruction to community colleges, flatlining budgets, and the linking of university funding to retention and graduation rates—are only likely to accelerate in the current educational climate. Our concern as compositionists is that these pressures have led, and will continue to lead, universities to abandon basic writers, as some universities have already raised admission standards in an attempt to shut out underprepared students who are more likely to adversely impact these rates. Our success in developing a course for basic writers at our university gives us hope that all is not lost for basic writers at institutions like ours.

Of course, convincing curriculum committees and administrators that a basic writing course that requires college-level work can be developed and successfully implemented is challenging. We believe we succeeded because we kept the needs of our stakeholders in mind while designing a course that, in keeping with the best practices of our field, transformed basic writing instruction from an error-based approach to a course that emphasizes rhetorical genres and the writing process. We also believe that, in order to be successful, significant attention has to be paid to instructor preparation, continuing support, and investment;

without such moves, instructors are more than likely to return to approaches that are mistakenly thought to be "tried and true." We have implemented these ideals in our program.

While the statistical data encourage us that we are on the right track, we are far more impressed with the reports of satisfaction we are receiving from students, faculty, and administrators. Given the changing demographics of our university, rapid growth in enrollments, and the increasing diversity of our student population, statistical data by itself cannot adequately measure the effectiveness of the program and these changes. The fact that our students, instructors, advisors, and administrators are satisfied with the changes is the measure that matters most.

Some challenges, however, remain. The trend to outsourcing writing instruction is not limited to remedial education—some officials in our state would like to see most, if not all, first year writing instruction delivered through the community college system, a new online governor's university, or through high-school based, dual-credit instruction. In the future, we expect to see even more diversity in the level of preparation for college writing among our entry-level students. This may lead us to consider an even more nuanced placement system, and multiple entry points into our second year course, which, although not a general education course, is required by the course of study of 80 percent of our graduates. As we address these challenges—and others we will probably face, but can't yet predict—the lessons we learned in revising our placement system and the BW curriculum will help us face that uncertain future.

Coda: Where We Are Now

Our writing program here at IPFW reflects its complex ecology and continues to fluctuate as we face new challenges, such as institutional needs and even more state mandates, in large part due to the program's interconnectedness. Our most recent task from our dean was to develop a reading course that fully integrates reading and writing in ways appropriate to a college-level course.

Several years ago the previous dean transferred responsibility for reading instruction to our study skills center, the Center for Academic Support and Advancement (CASA). The course created by CASA was not successful; it had one of the highest DWF (drop, withdraw, fail) rates of any course in the university. This is not surprising given the history described in our profile. The initial issue that drove us to develop

W129—the state mandate to eliminate remediation—ultimately led to the success of that course, because this mandate gave us the institutional leverage to craft a college level, credit-bearing course. In contrast, CASA's course kept the remedial model. When a course's assessment is a standardized measurement of high school reading levels and the course syllabus is constructed around an online reading program advertised as offering "skill remediation," then clearly the remedial model is still at work.

The reading course that has emerged benefitted from the lessons we learned while developing W129. However, similar to the legislative pressures Elizabeth Wardle describes in this volume, a new statewide general education structure threatens to undo our work. To protect both W129 and the new reading course from further state interference, we may eventually have to combine the two courses, making W129 a five-hour, general education course that integrates writing and reading. We fear that if W129 and the reading course remain outside the general education framework, they will not survive the state's purge. Another possible approach is to create a two-semester sequence based on the stretch model that would become our standard FYC experience. Our current FYC course would then become an accelerated option. Thus, in the very near future, the writing program may fluctuate yet again and migrate to another model of basic writing instruction.

This possibility raises additional issues, such as placement, that further illustrate the interconnected nature of our work. Our successful self-placement model in writing and the placement test currently used for the reading course may be superseded by the new PARCC (Partnership for Assessment of Readiness for College and Careers) exam. This exam, which Indiana has selected as its assessment instrument for the national secondary Common Core Curriculum, designates students as "College Ready" or "Not College Ready," and the rumblings we hear from the state indicate that it may be designated as the required measure for writing placement at all state universities. The state had originally committed to begin implementation of this standardized instrument in 2014. A recent rebellion against the Common Core Curriculum in our state, which has featured an unusual alliance between pro-education forces on the left and states' rights advocates on the right, has resulted in a one-year moratorium on the implementation of this plan. However, the powerful testing lobby is stepping up its efforts to undermine opposition to one-size-fits-all, standardized testing.

The one thing we can count on in Indiana is further fluctuations, uncertainty, and instability, thanks to the continued tinkering with higher education by our state leaders. And while it is easy to identify the negatives and unintended consequences of ill-conceived legislative mandates, it is also easy to stereotype educators as out-of-touch intellectuals who demand state monies but resist state oversight. We will continue to work through these issues and strive to put the needs of our students first as we inhabit complex, competing ecologies.

Notes

1. Sara is grateful to IPFW's Office of Research and External Support for awarding her a 2010 Summer Research Grant, which helped ease the writing of this profile.

2. Most, but not all, IPFW students are required to take a writing course beyond first-year composition; this requirement varies among the programs and colleges. Furthermore, not all students who are required by their college to take such a course enroll in the English department's course, as some departments offer their own "writing for majors" course that fulfills this requirement. Even English majors do not take the generalized second-level writing course so many IPFW students take; instead, they take a writing course designed specifically for English majors.

3. The former, non-credit bearing course also had a cap of 18; the studio-style course was capped at 22. Due to record enrollment during the Fall 2009 and 2010 semesters, the dean of our college raised the cap of the new basic course to 20. The cap reverted to 18 for the Spring 2010 and 2011 semesters, however.

4. For a detailed iteration of our writing program's outcomes, see http://compositionforum.com/issue/23/ipfw-appendices.php#appx1.

5. Readers interested in exploring the Daly-Miller test can find the test here: http://www.csus.edu/indiv/s/stonerm/The%20Daly-Miller%20Test.htm. A scoring guide can be found at http://www.csus.edu/indiv/s/stonerm/daly_miller_scoring.htm. See Blythe et al. for a deeper discussion of our institution's efforts to institute self-placement.

6. For a discussion of the history and philosophies of these basic writing programs, see Nicole Pepinster Greene and Patricia McAlexander's *Basic Writing in America: The History of Nine College Programs*.

7. See https://webapp1.ipfw.edu/pls/appdb1/f?p=194:1:1925717166566138 for the online placement instrument.

8. See http://compositionforum.com/issue/23/ipfw-appendices.php#appx2 for Sara's most recent syllabus and daily schedule.

9. See sample assignments at http://compositionforum.com/issue/23/ipfw-appendices.php#appx3.

10. See http://new.ipfw.edu/departments/coas/depts/english/resources/ for the website.

Works Cited

Baden, William. "Re: English W129 Data." Message to the author. 9 July 2010. E-mail.

Bernstein, Susan Naomi. "Social Justice Initiative for Basic Writers." *BWe: Basic Writing e-Journal* 7.1 (2008): n pag. Web. 7 June 2010.

Blakesley, David. "Directed Self-Placement in the University." *WPA: Writing Program Administration* 25.2 (2002): 9–39. Print.

Bloom, Lynn Z. "A Name with a View." *Journal of Basic Writing* 14.1 (1995): 7–14. Print.

Blythe, Stuart, et al. "Exploring Options for Students at the Boundaries of the 'At-Risk' Designation." *WPA: Writing Program Administration* 33.1–2 (2009): 9–28. Print.

Daly, John and Michael Miller. "The Empirical Development of an Instrument to Measure Writing Apprehension." *Research in the Teaching of English* 12 (1975): 242–249. Print.

Devitt, Amy. "Transferability and Genres." *The Locations of Composition*. Eds. Christopher J. Keller and Christian R. Weisser. Albany: State University of New York Press, 2007. 215–228. Print.

Glau, Gregory R. "The 'Stretch Program': Arizona State University's New Model of University-Level Basic Writing Instruction." *WPA: Writing Program Administration* 20.1–2 (1996): 79–91. Print.

Goen-Salter, Sugie. "Critiquing the Need to Eliminate Remediation: Lessons from San Francisco State." *Journal of Basic Writing* 27.2 (2008): 81–105. Print.

Greene, Nicole Pepinster, and Patricia J. McAlexander, eds. *Basic Writing in America: The History of Nine College Programs*. Creskill, NJ: Hampton Press, 2008. Print.

Grego, Rhonda, and Nancy Thompson. "Repositioning Remediation: Renegotiating Composition's Work in the Academy." *College Composition and Communication* 47.1 (1996): 62–84. Print.

Haswell, Richard H. *Writing Placement in College: A Research Synopsis*. N.p. Nov. 2004. Web. 30 June 2010.

Hull, Glynda, et al. "Remediation as Social Construct: Perspectives from an Analysis of Classroom Discourse." *College Composition and Communication* 42.3 (1991): 299–329. Print.

Huot, Brian. *(Re)Articulating Writing Assessment for Teaching and Learning*. Logan, UT: Utah State University, 2002. Print.

Indiana Commission on Higher Education. "Reaching Higher with Ivy Tech Community College of Indiana: Focusing on the Role of Community Colleges, June 13, 2008." Web, 28 November 2010.

"Indiana University-Purdue University Fort Wayne College Portrait." *College Portraits*. American Association of State Colleges and Universities and the Association of Public and Land-grant Universities, (2010). Web. 9 July 2010.

Lotkowski, Veronica A., Stephen B. Robbins, and Richard J. Noeth. *The Role of Academic and Non-Academic Factors in Improving College Retention*. Iowa City: ACT, 2004. Print.

National Survey of Student Engagement. *National Survey of Student Engagement (NSSE) 2009 Institutional Report, IPFW*. Indiana University-Purdue University Fort Wayne. n.d. Web. 1 July 2010.

Office of Institutional Research, Indiana University-Purdue University Fort Wayne. *2009–2010 Statistical Profile*. Indiana University-Purdue University Fort Wayne. Nov. 2009. Web. 23 May 2010.

Penrose, Ann M. "Academic Literacy Perception and Performance: Comparing First-Generation and Continuing-Generation College Students." *Research in the Teaching of English* 36.4 (2002): 437–461. Print.

Rodriguez, Richard. *Hunger of Memory: The Education of Richard Rodriguez*. Boston: Godine, 1982. Print.

Rose, Mike. *Lives on the Boundary: The Struggles and Achievements of America's Underprepared*. New York: Free Press, 1989. Print.

—. "The Language of Exclusion: Writing Instruction at the University." *College English* 47.4 (1985): 341–59. Print.

Royer, Daniel J. and Roger Gilles. "Directed Self-Placement: An Attitude of Orientation." *College Composition and Communication* 50.1 (1998): 54–70. Print.

Soderlund, Kelly. "IPFW Grad Rates 2nd Lowest in State." *The Journal Gazette*. 11 Aug. 2008: A1+.

Smoke, Trudy. "What Is the Future of Basic Writing?" *Journal of Basic Writing* 20.2 (2001): 88–96. Print.

Soliday, Mary. *The Politics of Remediation: Institutional and Student Needs in Higher Education*. Pittsburgh: University of Pittsburgh Press, 2002. Print.

Stevens, Scott. "Nowhere to Go: Basic Writing and the Scapegoating of Civic Failure." *Journal of Basic Writing* 21.1 (2002): 3–15. Print.

Stygall, Gail. "Unraveling at Both Ends: Anti-Undergraduate Education, Anti-Affirmative Action, and Basic Writing at Research Schools." *Journal of Basic Writing* 18.2 (1999): 4–22. Print.

Troyka, Lynn Quitman. "Defining Basic Writing in Context." *A Sourcebook for Basic Writing Teachers*. Ed. Theresa Enos. New York: Random, 1987. 2–15. Print.

Wiener, Harvey S. "The Attack on Basic Writing—and After." *Journal of Basic Writing* 17.1 (1998): 96–103. Print.

2 Standardizing English 101 at Southern Illinois University Carbondale: Reflections on the Promise of Improved GTA Preparation and More Effective Writing Instruction

Ronda Leathers Dively

Catalysts and Contexts for Change

For some time, the rhetoric and composition faculty at Southern Illinois University Carbondale (SIUC) debated the question of whether or not to standardize their version of English 101—the initial phase of a two-course composition requirement—which serves approximately 2,500 undergraduates and employs approximately sixty graduate teaching assistants (GTAs) per year.[1] Finally, in the Fall of 2006, the initial version of a common syllabus began driving English 101 in support of its longstanding, overarching purpose: to familiarize students with various genres of public discourse, engaging them in an array of writing scenarios that, through attention to process, collectively seek to sharpen their rhetorical sensibilities and expand their repertoire of composing strategies. Indeed, the general purpose of English 101 had never been in question, for it seemed to function effectively as a broad introduction to concepts, skills and practices that would be revisited in variously focused, subsequent composition and writing intensive courses across the curriculum. What *was* in question was the extent to which that purpose was being realized in all the numerous sections of this foundational course—a concern that sparked our thinking about standardization.

Those rhetoric and composition faculty members who initially questioned the move to a standardized syllabus in favor of the status quo (which recommended but did not mandate a given approach to English 101) felt strongly about the need for our incoming GTAs to think their way through the challenges of designing their own courses for the positive impact that doing so would have on their level of engagement and their development as writing instructors. In addition, for proponents of the status quo, the potential ills of indoctrination loomed large (see Welch; Martin and Paine 222), rendering a minimally restrictive model of preparation and oversight especially appealing.

In contrast, faculty members who favored standardization privileged a high level of continuity in instruction for purposes of more assuredly and regularly foregrounding "best practices," and they fretted over expending English 101 instructional time waiting for new GTAs to find their way naturally toward these goals. The movement for change did acknowledge the crucial roles that teacher autonomy and the opportunity for pedagogical discovery play in a teacher's development. Nonetheless, it chose to limit them initially, believing, as Kelly Kinney and Kristi Murray Costello do, "that [a shared syllabus] is a responsible choice for a writing program staffed primarily by graduate students with limited exposure to rhetoric and writing studies and little experience in teaching process-based, genre focused composition." Besides, though the GTAs would be required to follow a 101 curriculum created by instructors with a more specialized knowledge base and years of classroom experience, it was understood that subsequent course assignments would allow greater opportunity for autonomy and discovery. In light of the new GTAs' gradual increase in autonomy as well as a growing sense of the profound influence they would have over the undergraduates they helped usher into the university (Nyquist and Wulff 34), the movement for change in the direction of a standardized curriculum banked on the assumption that, in the end, they might actually welcome a significant degree of direction.

Of course, the status quo never did assume that our incoming GTAs could design and execute a composition course without any formal preparation. Indeed, for many years the SIUC Writing Program has flourished (and still does) on the foundation of an eight-day orientation seminar, or "Pre-Semester Workshop" (PSW), that introduces GTAs to the objectives of English 101, familiarizes them with various strategies for teaching and assessing writing, involves them in grade-norming sessions, alerts them to on-campus programs and services that might impact their instruction, etc. In addition, GTAs in English at SIUC have, for years, been required to complete a graduate seminar in teaching composition. This seminar, English 502, builds upon the information and experiences constituting PSW as it introduces them to "best practices" and current theory in composition instruction, invites them to interrogate their instructional dispositions, and expands their repertoire of pedagogical methods. The vast majority meet this requirement during their first semester at SIUC while they are teaching English 101; GTAs who enter our program having completed such a graduate seminar elsewhere are exempted from this requirement.

Although the status quo—through the PSW and English 502 and based on a set of department-sanctioned objectives for English 101 (see Appendix)—advocated for certain types of assignments, models of assignment sequencing, methods of response and assessment, and strategies for teacher-student/student-student interactions, it did not impose genres, topics, or sequences of assignments, nor did it prescribe classroom activities at any level. Therefore, despite the fact that there was some effective teaching going on in English 101, there existed considerable disparity between the course's numerous sections, as was substantiated through classroom observations, collection of example syllabi, and core curriculum assessment.[2] This disparity between sections ignited concern because, to some degree, it appeared to limit the potential for knowledge transfer between English 101 and other general composition courses and, as follows, between the general composition sequence and writing intensive courses across the curriculum (Nelms and Dively 223–224).[3] This disparity also weakened claims to a valid and reliable course assessment.

Another potentially undesirable situation existing under the status quo was that, during any given fall semester, many of our thirty-five or so new GTAs were initially very dependent in designing their courses on advice, assignments, and strategies provided by our sixty or so experienced GTAs. This state of affairs, in many cases, served the new GTAs well since our program was populated by many talented instructors and there were many successful model assignments floating around the GTAs' offices. In other cases, however, the new GTAs were misled by individuals who, instead of consistently enacting approaches advocated by the program, were motivated by interests that did not coincide with program objectives. Most of these cases were not spitefully motivated; for example, many of our graduate students in creative writing and literary studies wanted to teach composition by means of their primary passion—literature. Although our program objectives did not prohibit the occasional use of a literary work (e.g., a poem or short story to help introduce strategies for analysis), some of these GTAs were overzealous in their use of such vehicles at the exclusion of public or academic expository discourse.

Even when the new GTAs were not pressing the boundaries of genre expectations under the status quo, it seemed they were all too often settling for syllabi that did not fully address course objectives and/or did not effectively cohere. That is, for lack of time and knowledge about the kinds of assignments most conducive to achieving course objectives or about strategies for productively sequencing composition courses, some of them resorted to cobbling together a string of loosely related assignments simply because they were available from a peer at a moment of desperation and/or sounded "fun." Of course, this situation did not result from a lack of effort in preparing these inexperienced GTAs, as is evidenced by the requirements of PSW and English 502, nor did it result from laziness on the part of the GTAs (after all, having never taught before, most were teaching two sections of English 101 and taking two graduate seminars). Rather, it resulted, at least in part, from a dearth of resources and opportunity to educate these new teachers *before* they entered the classroom[4]—coupled with an approach to teacher training that believed strongly in the power of an organically wrought sense of course ownership to compensate in significant ways for an initial lack of specialized knowledge in composition pedagogy.

As the previous paragraphs indicate, several perceived vulnerabilities in the status quo prompted our move to a standardized curriculum—a move we presumed in summary, then, would heighten potential for a valid and reliable course assessment; increase opportunities for knowledge transfer as subsequent courses could explicitly reference lessons and assignments that *all* 101 students had encountered; reduce the workload of inexperienced GTAs as they were acclimating to the many demands of graduate school; and channel 101 instruction more effectively in the direction of "best practices." Even after the decision for change was made, however, the route traveled in implementing a common English 101 syllabus was rather convoluted as a result of staffing problems, political conflicts within the department and college, and the shuffling of administrative positions at both levels. As a result, the first phase of development occurred under the auspices of a literature professor who accepted a brief, yearlong stint as the director of writing studies (DWS [SIUC'S equivalent to a WPA]) amidst shifting departmental and college dynamics. He was advised by a writing studies committee of English department faculty from various sub-disciplines, as well as by a newly appointed assistant DWS, who had recently graduated with a PhD from SIUC in rhetoric and composition.[5] Together, these individuals implemented the initial version of SIUC's standardized English 101 curriculum.

During that transitional year (while I happened to be serving as the director of undergraduate studies), I was teaching English 502 with a class of nearly forty new GTAs (the unusually high number resulting from the staffing problems referenced earlier). From this position situated between the writing program administration and the GTAs, I became aware of the angst experienced by both parties as they rode the transition to a standardized curriculum. A year later, as the newly appointed DWS, my perspective was enhanced as I worked (along with the assistant DWS who was then beginning his second year in that position) to substantially revise the initial version of the common syllabus, as well as the PSW and English 502, both of which had not yet been retooled to support the new curriculum. Now, having completed almost four years as DWS, I reflect on our program's transition to a common English 101 syllabus with an elaborated sense of the challenges we faced, the advantages we immediately enjoyed, and the benefits we anticipate are yet to come. My hope is that these reflections will provide program leaders who are considering standardization with ideas for building a rationale and strategies for easing the transition.

The Continuum of Freedom and Control: Theorizing Our Way to Balancing the Interests of Undergraduates, GTAs, and the Writing Studies Office

Wanda Martin and Charles Paine have aptly characterized the struggle that WPAs face as they work to manage the needs of all who are impacted by their approach to preparing and supervising GTAs:

> On the one hand, we want to give these teachers—experienced as well as new ones—as much free rein as possible to discover and practice what works best for them and their students. . . . On the other hand, we have our own beliefs about what constitutes good writing and good writing instruction, we want to articulate and practice a relatively coherent and stable philosophy of writing, and we are obliged to ensure a degree of consistency across all sections of first-year English. (222)

Elaborating on this struggle, they point to a tension that seems particularly relevant to the process of generating a standardized syllabus: The tension "between the institutional imperative to control the program—be in charge, know what's going on in the classrooms, make multiple sections consistent in content and grading—and the human necessity of letting go to promote individual responsibility and prevent WPA heart attacks" (222). In the midst of our own grapplings with institutional imperatives and the desire to promote some level of individual responsibility, a question that my assistant and I wrestled with for some time was one of just how strictly we wanted to control not only the parameters of the major essay assignments but also the day-to-day classroom activities. Similar to the process of assignment construction—which can be conceived on a continuum from "well-defined" or directed (i.e., all variables of the rhetorical scenario are specified) to "ill-defined" or open (i.e., few if any variables are specified) (see Carey and Flower qtd. in Dively 60–61)—the process of designing a standardized syllabus as we viewed it (standardization precluding utter freedom) evoked a continuum between "scripted," where every assignment, activity, reading and word spoken by the instructor would be pre-determined, and "guided," which would require instructors to make selections from menus of assignments, activities, readings, etc.

The analogy to assignment construction is instructive in that the advantages and disadvantages that characterize extremes of that continuum are applicable to those involved in syllabus construction. For example, directed assignments may prevent writers from capitalizing on their strengths and from harnessing the potential gains of sudden insight and serendipity. Thus, writers might not be able to realize their best efforts because they are compelled to write in an unfamiliar genre. Or, they might miss an opportunity for generating something fresh or special because avenues of thinking and research could be restricted from informing each other in ways that lead to discovery or illumination

(Csikszentmihalyi and Sawyer 341). Applying this realization to teaching, successful instructors know that tight strictures on pedagogy might prevent them from employing activities or pedagogical strategies that mesh with their personalities and sources of confidence. Further, these instructors know that such strictures may prevent unanticipated convergences of phenomena that present opportunities for the kinds of energizing learning activities that move beyond the usual or mundane. We've all had the experience of a discussion assuming a life of its own, begging us to abandon our agenda for the moment to explore the possibilities of the unexpected. This is the hallmark of creativity, the willingness to break free from ruts in our thinking and practice so as to investigate new and promising pathways (see Ward, Finke, and Smith 111, 165–166). When such opportunities are stymied, the capacity for effective instruction is diminished.

As the previous paragraph suggests, certain manifestations of "scripted" syllabi could discourage if not prevent shining moments associated with effective teaching, just as strings of directed assignments could tie composition students to what might feel like uninspired, unfulfilling bouts of drudgery. But, of course, directed assignments do have their purposes. Without some direction, students would likely choose the path of least resistance, sticking with what they know or what seems comfortable, not only because doing so is less taxing, but also because the risks for receiving unfavorable evaluations are lower. As a result, their development as writers could be arrested, their repertoire of strategies and skills remaining limited.

Just as direction has its place in the education of writers, so does it have its place in the education of teachers. Indeed, direction in the form of a standardized syllabus not only helps to ensure that instruction, on large scale, remains consistent with program objectives, but it also drives inexperienced instructors beyond their own comfort zones, *compelling* (rather than merely *encouraging*) them to experiment with models and strategies for effective composition instruction that are informed by scholarship in the discipline. In other words, they are obliged (not merely invited) to develop their "composition literacy" or knowledge of the academic community that informs their teaching (Griffith 4) so that their capacity to reflect critically on their pedagogical practices, to enact appropriate practices in future contexts, and to articulate the rationale behind those practices will grow.

Our own quest to negotiate the advantages and disadvantages of freedom and control left us with a syllabus that rested somewhere between the extremes of the "scripted"-"guided" continuum. Working with a core curriculum mandate that English 101 students compose at least six essays and opting for a portfolio assessment system, we settled for a list of five multiply drafted essays (four of them to be submitted in the portfolio for evaluation at semester's end) and one timed (though not impromptu) essay in response to an argumentative text completed during the final exam period.[6] Within these parameters, SIUC's standardized syllabus prescribed the major essay assignments in a way that allowed the GTAs little room to alter them. While the assignments and the approach to assessment were stipulated so as to reflect core program values, the schedules of daily activities for each major unit enabled considerable flexibility (except for the first unit, which was highly regimented by class meeting for the purpose of providing a firm foundation as the new GTAs found their footing). For units two through six, however, instead of *daily* outlines listing assigned readings and approved activities (the latter in the order in which they should occur), the syllabus presented *unit* outlines that included weekly menus of suggested readings and activities specifically relevant to the assignment at hand.

The decision to prescribe the nature and sequence of major essay assignments was based on the premise that they are the cornerstone of any composition course and that at least this level of standardization would be sufficient for meeting many of the objectives we sought to achieve. Though we realized that prescription at the assignment and sequence levels might feel somewhat confining to students and GTAs, we worked diligently to counter this effect by foregrounding opportunities for choice throughout the course. More specifically, though the assignments were directive with regard to genre and general topic, students would be able to determine their own foci within these parameters (e.g., the particular event portrayed in the literacy narrative, the ad to be analyzed relevant to the ad analysis). As the course progressed, topics would become more broadly cast so as to allow for increasing freedom with regard to a particular focus and, thus, more intensive engagement with invention strategies. Moreover, the students would enjoy some degree of choice with regard to the essays they submitted in their final portfolios.

These elements of choice to be enjoyed by the students would potentially carry benefits for the instructors as well. Specifically, such freedoms would invite a reasonable variety of themes in the student writing that the GTAs received—a motivating factor when one is facing large stacks of essays to assess over a weekend. In addition, because there would exist a level of freedom relevant to the specific focus for a given assignment, the GTAs would experience the sense of engagement that comes with helping students locate controlling ideas that excite and challenge them. This sense of engagement would potentially intensify as the course progressed and the opportunities for student choice increased. But for the GTAs, freedom and choice would most readily be realized in the crafting of the units once they moved beyond the highly regimented first unit. Once beyond that unit, not only would the GTAs be able to select readings from the menus of suggested textbook chapters and model essays listed on the assignment handouts for each unit, but they would be welcome to introduce supplementary materials they presumed would be helpful in highlighting unit objectives. Moreover, they would be able to select the exercises and activities—whether drawn from the list of suggestions provided on the unit schedules or from their growing repertoire of ideas from English 502 and discussions with peers—that they felt would be most edifying for their students at given points in the unit. Theoretically, then, while the extent to which the syllabus was directed would undoubtedly feel somewhat limiting to students and teachers, the fact that the course was guided, as opposed to scripted, promised to help students and teachers realize at least some sense of ownership and the kind of growth that accompanies choice and experimentation.

Crucial to maintaining a productive balance of freedom and control—as the latter's appearance is inevitably dominant in any standardized context—would be the solicitation of feedback from those who were actually delivering the course and who could relay the reactions of those who were receiving it. Although my staff and I were committed to the idea of a standardized syllabus and the pedagogical principles on which the syllabus was based, and although we had devoted countless hours to designing the course and composing the assignments and the unit plans, we undertook this venture fully expecting that the syllabus would forever remain a work in progress. That attitude led me to craft a final assignment for English 502 that asked GTAs to critique the standardized syllabus by proposing an alternative assignment and/or arguing for an alternative assignment sequence. In addition to that gesture, at the end of each semester, my staff and I held informal focus groups for GTAs who were teaching the standardized curriculum. In the context of these meetings, those who responded to an open call would meet with me or a staff member to answer specific questions and voice their concerns about the common syllabus—feedback that would help us conceive of possible revisions. After all, the GTAs would have first-hand knowledge of how students were reacting to the assignments; how difficult the assignments were to articulate; how effective the sequence seemed to be; how certain suggested readings, activities, and informal exercises served to engage or befuddle, etc. Not only did we stand to gain valuable information from these conversations, but also we believed that inviting this kind of feedback from the GTAs would help offset the negative effects of feeling "controlled." If we actively sought their now experienced voices on pedagogical matters (and, indirectly, the voices of their students) and were open to implementing their suggestions, we figured that any resistance they harbored would become less intense, that their collective sense of investment in the course would deepen, and that their motivation levels would increase. And, of course, we assumed that the anticipated effects of such outreach—an even better course and more invested, motivated teachers—would be felt by the English 101 students.

Yet another consideration relevant to the continuum of freedom and control in the move toward a standardized syllabus centered on the nature and content of the PSW. Prior to the writing program's adoption of the standardized syllabus, this eight-day orientation seminar offered an impressive variety of sessions, many of which were proposed and executed by experienced GTAs still in the program. The PSW, then, provided a flexible venue in which developing composition instructors could share what they were learning about and through teaching, with the opportunity to pursue areas of particular interest. Upon moving to the standardized curriculum, my staff and I wanted to preserve this aspect of the PSW to the degree that we could, but we also realized that standardization in the curriculum begged standardization in the PSW. In other words, with the continuity provided by the prescribed unit assignments, the prescribed sequence of units across sections of 101, and the menu of suggested readings and activities, would come the capacity—and, essentially, a pedagogical imperative—for gearing the PSW toward detailed preparation in teaching the components of the standardized syllabus. Without this change in emphasis, another potential benefit of standardization, the specificity and depth of instructor preparation that can occur when all are teaching the same material, could not be fully realized.

In an attempt to capitalize on the advantages that both approaches to the PSW could offer, we created a hybrid of sorts that demonstrates our attempts to balance between the extremes of freedom and control. The status quo had typically reserved space in the PSW for several sessions that tapped shared interests. In addition to those addressing institutional policies, procedures, and services, such sessions included, for example, introductions to strategies for designing and sequencing assignments and for responding to and assessing student writing. Obviously, the revised PSW would need to address such issues, but the relevant sessions could broach them with the standardized syllabus in mind.

More specifically, assignment sessions would focus not on how to craft a potential 101 assignment from scratch but, rather, on how to explicate, troubleshoot, and best prepare students for succeeding on the prescribed assignments. The response and evaluation sessions—rather than focusing on general strategies for assessing papers written in response to a 101 assignment that GTAs might or might not end up teaching—could engage GTAs with actual papers written in response to the assignments they *knew* that they *would* be teaching. Indeed, the revised PSW would be organized around units of the standardized curriculum, with two-hour sessions devoted to each. Further, the newly focused response and evaluation sessions would be supplemented with sessions on rubric construction that, again, could be tailored specifically to assignments on the common syllabus, compounding the potential for the new GTAs to begin internalizing criteria for assessing essays written in response to the prescribed assignments.

Not only would the standardized syllabus make for PSW sessions that offered the new GTAs more in-depth, immediately relevant training regarding assignments and response practices, but these benefits would be mirrored in the preparation of experienced GTAs who would be leading these particular sessions.[7] Before, experienced GTAs volunteered or were asked to cover PSW sessions with the expectation that they would depend largely on scholarship introduced in English 502 and their prior experience to independently generate content for their respective sessions. To be sure, these GTAs proved very capable and put together some impressive presentations and workshops. But with the PSW focused on components of the standardized syllabus, they would be able to capitalize on the intellectual benefits of collaborating on common, already generated material, and they would be assured of the continuity of their own sessions with those led by their peers, bolstering their confidence that they were offering the new GTAs information that would effectively serve them. Add to this the fact that the GTAs leading these core sessions on assignment and response had already taught English 101 by the standardized syllabus at least once (in contrast to a system in which everyone seemed to be teaching 101 differently, albeit with some overlap), and the quality of pre-service preparation would stand a good chance of increasing. The preparation to lead the training sessions under the new system would promote greater control over the content of large portions of the PSW, but just as the standardized syllabus units provided for some choice and flexibility, so would the plans for the collaboratively generated training sessions to teach these units. Moreover, on the subject of freedom, the revised PSW would still include a batch of concurrent sessions for which GTAs could propose topics relevant to first-year composition instruction that were in keeping with their individual interests.

For all the advantages of moving to a standardized syllabus anticipated from the vantage point of the writing studies office, it, too, was subject to tensions arising from the struggle between freedom and control. On top of dealing with some unrest resulting from the move toward standardization voiced by GTAs who had been in the program for a while, my office was forced to limit its pre-existing agenda to compensate for the heavy additional workload that accompanied standardization. After the time-intensive responsibility of conceiving a collection of assignments we could stand behind and a viable plan for sequencing them, we would need to devote considerable attention to drafting and revising the assignment handouts. Furthermore, we faced the challenge of vetting rhetorics and ancillary materials that would specifically support the syllabus—the considerable energy devoted to such committee responsibilities taking its toll on our available hours. But the controls on our time would not stop there; on the contrary, we quickly realized that the move to a standardized syllabus raised the temptation to plagiarize tenfold, as there would be numerous copies of essays written in response to the same 101 assignments lying around campus. Consequently, we would need to revise the assignments every semester (for up to at least four years, when the first class introduced to the standardized syllabus had presumably graduated) to reflect different topics while maintaining the carefully designed sequence of genres (e.g., a literacy narrative could focus on different types of literacy from semester to semester; an ad analysis could focus on different categories of products or services, etc.). In addition, certain assignments (summary/response, synthesis) would require us to change readings that would provide content for those essays every semester, and, therefore, we would have to adopt a custom reader that we could tailor to the changing topics.

While these ongoing revisions to the standardized syllabus promised to consume much of our energy, the writing studies staff projected that these additional controls on our time would be compensated in that the instructional challenges faced by the GTAs when teaching a pre-existing syllabus that had been the focus of their training would be considerably fewer, and, as a result, the demands on the staff for extra classroom observations or one-on-one consultations would be reduced. More importantly, a reduction in these types of instructional challenges would heighten the instructors' confidence and, by extension, students' confidence in and contentment with their instructors. Predictions such as this one, and many of those discussed earlier in this section, cannot be verified in any absolute sense for the complexity of the constructs and for lack of comparative data prior to implementation of the standardized syllabus. But after its having been in existence for almost four years, anecdotal support for the standardized syllabus is mounting, motivating us to continue with our approach.

REFLECTIONS ON THE MOVE TO STANDARDIZATION AT SIUC: WHAT WE KNOW, WHAT WE SENSE, AND WHAT WE IMAGINE

Despite the early resistance of experienced GTAs to the announcement that we were moving to a standardized English 101 syllabus, the staff of newly minted PhDs and advanced GTAs who have assisted me over the past few years in keeping track of the accomplishments, difficulties, and dispositions of our GTA population at large have reported that any overt resistance to the idea of standardization has weakened. One reason for this might be that many of the original naysayers have graduated; another reason might be that those who are still around typically teach advanced writing or core curriculum literature courses, and, therefore, their attentions are concentrated elsewhere. Whatever the case may be with regard to the old guard, the classes of incoming GTAs who have entered our program since adoption of the common syllabus don't seem to be troubled by it, or at least they are not professing their discontent openly.

Even if they are troubled by the idea of standardization or certain aspects of our syllabus in particular, these points of resistance seem to be outweighed by the appreciation they feel for having the course figured out for them ahead of time. Maybe, as Sarah Liggett and Betty P. Pytlik observe, they foresee that, though appealing on some levels, the act of "learning to teach by teaching can be an inefficient and frustrating method of professional development" (xv). To be sure, it doesn't take long for these new GTAs to pick up on the fact that there is considerable planning and stress involved in teaching, even with a standardized curriculum to guide them. In light of the common syllabus, because the assignments, sequencing, readings, and activities are already plotted for them, they have time to concentrate more fully on other aspects of instruction—such as lesson plan design, response, assessment, and classroom management. Furthermore, they have time to *reflect* on their teaching (Bamberg 151; Borrowman 4). Perhaps their anticipation of these benefits explains why, in recent years, when I've mentioned the history of the standardized syllabus while leading a PSW session or teaching English 502, I've commonly been met with exclamations of disbelief: "I can't imagine teaching 101 for the first time without having a standardized syllabus." Reportedly, members of my staff have heard echoes of this sentiment in various mentoring situations.

This positivity may be fueled by the knowledge that the syllabus has been revised in direct response to the feedback of GTAs who taught earlier versions of it—knowledge that we readily share with each new class of teaching assistants. More specifically, my staff and I switched the order of a couple of the assignments, changed the genre of one of the assignments to render the exercise more authentic, and reduced the number of subtasks involved in yet another of the assignments. In addition, we extensively revised the assignment handouts, editing context and toning down the register, so as to make them more reader-friendly. Though recent rounds of focus group conversations have not yielded any substantial suggestions for further improving the curriculum, we continue to invite feedback from the GTAs with the intent of strengthening their sense of ownership over English 101.

On the subject of PSW, though most of the incoming GTAs have nothing to compare it to, my staff members who attended or helped with the orientation before implementation of the common syllabus have reported over the last few years their appreciation for the tighter focus and depth of training that the new system enables. And because they have taught by the very same syllabus themselves, they feel especially self-assured in their abilities to introduce the units, help the incoming GTAs anticipate unique challenges posed by the various assignments, and share insights about activities and exercises that have proven helpful in their own 101 classrooms. Again, under the previous system, returning GTAs helping to initiate their new colleagues talked about first-year composition pedagogy theoretically or relevant to example materials, but not necessarily with the help of materials that those new colleagues would actually be using. Under the new system, returning GTAs are able to draw from their own experiences executing the syllabus as they tackle its various objectives in a way that the new GTAs view as directly and specifically relevant to what they'll be doing in the classroom. Of course, this similarity of experience can strengthen mentoring relationships, and, although such bonds were undoubtedly created under the status quo, before the standardized syllabus, the experiential distance between mentor and mentee was inevitably greater.

This similarity of GTA experience in English 101 across sections, and even years, presents various additional benefits that might be considered more practical or logistical in nature. Regarding one such benefit, though I have long been impressed with the willingness of our GTAs to cover classes for each other in the event of illness, conferences, family emergencies, etc., the continuity across English 101 sections clearly supports their altruistic tendencies. Now, a 101 instructor who is compelled to miss class can put out a call on the listserv used for writing studies communications, and, in many instances, within a moment's notice, can find several colleagues who are willing to step in immediately and cover the class meeting. And, importantly, since they are teaching or have already taught the standardized curriculum, they can do so knowing that they are well prepared to conduct a substantive class meeting, as opposed to simply monitoring a generic activity.

Not only can English GTAs readily stand in for each other on occasion, but the students themselves are in a better position, by means of the standardized syllabus, to help each other learn. Because all English 101 students are working on the same assignments, are covering mostly the same readings, and are on the same weekly schedules, there exists ample opportunity for cooperative learning beyond the classroom, even between students who are not assigned to the same section. Therefore, roommates, other dorm residents, sorority sisters and fraternity brothers, club or athletic team members—all who are enrolled in English 101 and, to some degree, who have already taken it—become potential resources of particularly germane information and assistance when instructors or 101 section-mates are not available.

The similarity in experiences across sections also holds advantages for English 101 students who seek assistance from the writing center and for the tutors who are working with those students. Since implementation of the common syllabus, training for writing center tutors has included familiarization with key components of the curriculum and practice conference sessions using papers written in response to that curriculum. Immersion in the specifics of the 101 syllabus and the fact that many of the students they are tutoring are studying the same genres and strategies at a given point in time ostensibly combine to deepen the tutors' knowledge base at a comparatively rapid pace and to reinforce their confidence. This situation promises to positively impact 101 tutoring sessions in ways that were not possible before there was a standard base of knowledge from which to draw.

Beyond the benefits for English 101 teachers, tutors, and students, the continuity in sections of this course can, as well, aid instructors and students of English 102 (the second course in our required first-year composition sequence)—and, indeed, of all subsequent composition courses—in capitalizing on the potential for backward-reaching knowledge transfer. For backward-reaching transfer to occur, teachers must be able to make explicit for students, and students must be able to discern, how a concept, strategy or skill they are currently teaching or learning is similar to one they've encountered in the past (Perkins and Salomon 26). Since, under the standardized syllabus, all writing studies instructors have taught virtually the same English 101 course and all students have taken virtually the same course, there are numerous common examples and experiences that English 102 teachers can highlight to promote learning. Theoretically, to the degree that teachers of writing intensive courses across the curriculum have familiarized themselves with the standardized English 101 curriculum,[8] they can capitalize on the same potential for backward-reaching transfer as teachers in general composition courses.

Of greater certainty at this point, however, than concerted attention to knowledge transfer is the general attitude of the university community toward standardization of English 101—a positive attitude grounded primarily in the realization that the writing studies office can articulate more definitively the specific genres, strategies, skills and readings that all English 101 students will be exposed to in that course. Prior to the standardized approach, lore that placed the course and its instructors in a questionable light proliferated. Of course, the standardized curriculum alone has not stopped the uncharitable chatter of those who don't understand or try to understand the research-based practices of our field or the nature of writing difficulties as matters that cannot be easily and forever "fixed." That being said, what the standardized syllabus does provide is a detailed accounting of what is supposed to be happening in every section on a weekly basis, and that detailed accounting renders it more difficult for critics to misconstrue or unjustly portray the nature of English 101 to those who are genuinely interested in learning about the course. To be sure, on more than one occasion when I was invited to speak with a given university constituency about the standardized curriculum, pieces of lore shared by the audience (e.g., "We don't understand why you don't teach grammar." "How come none of the writing assignments are academic in nature?") were readily dispelled by pointing to specific activities, readings, and assignments outlined on the common assignment handouts and unit schedules.

Also relevant to the attitudes of the university community at large is the issue of assessment. The Core Curriculum Executive Council regularly assesses English 101, and, during the early years of this process, the director of writing studies and the English department chair were presented with the same critique: "There is too much disparity across sections of this course." Obviously the Council's concern was rooted in the desire for a reliable assessment and, by extension, the desire that all English 101 students were receiving similar educational experiences in the interest of laying foundations for individual progress, retention, and knowledge transfer. Since implementing the standardized syllabus, response from the Council has been decidedly complimentary relevant to these factors. Admittedly, core assessment is largely impressionistic, based primarily on instructional materials provided by a sample of 101 teachers and on those teachers' analyses of their own performance and their students' work. As the university at large is in the process of ratcheting up its treatment of campus-wide assessment, English 101 stands ready to provide for a more valid and reliable assessment than was previously possible. Indeed, the continuity of content and structure across the many 101 sections and the greater control of pedagogical variables enabled by the common syllabus increase confidence in our capacity to evaluate students' performance (e.g., regarding attainment of objectives; any improvement, or lack thereof, in writing ability; transfer of knowledge to subsequent courses) relevant to particular constructs that are now more readily isolated for consideration.

It seems appropriate to end this profile by re-emphasizing projected benefits of the standardized syllabus that may contribute to the outcome most privileged by the movement for change—that is, improvement in the quality of first-year students' English 101 experience. As I noted earlier, there is no doubt that quality instruction was taking place in many 101 classrooms before the standardized syllabus was implemented. Many GTAs entered the halls of SIUC with keen instructional instincts, and the training our GTAs received in the PSW and English 502 strove to be thorough and current. Nonetheless, it seems clear that the turn toward standardization stands to more effectively support the numerous GTAs who begin the program with no experience and with undeveloped pedagogical inclinations.

Rather than having the new GTAs build a course in addition to managing all the other challenges of composition instruction (not to mention the challenges of the graduate courses they are taking), at least part of the work is already completed, and, thus, the threat of becoming overwhelmed or having to "shoot from the hip" in pulling the course together is lessened. Because 101 is plotted for the GTAs, we can be more certain that the assignments, activities, and readings that undergraduates encounter are collectively giving due attention to the entire set of 101 course objectives and that "best practices" are more consistently in play. In addition, pedagogical support from a community of peers who are teaching (or have taught) the exact same assignments, readings, and activities is abundant and substantive; furthermore, writing center tutors, having been trained on (or some even having taught) the same materials, are prepared to reinforce lessons at a very specific level. Though these and many other complex factors that impact the quality of instruction and student performance are difficult to isolate for cause/effect analysis, it is not much of a stretch to imagine that the nature of preparation and support made possible by a standardized syllabus would raise the potential for effective pedagogy, particularly in courses that are being taught by inexperienced and/or apprehensive teaching assistants.

Coda: Where We Are Now

Shortly after *Composition Forum* published my program profile, I decided to step away for a while from the writing studies directorship at SIUC so that I could fulfill expectations for a textbook contract I had just signed. Consequently, when I was invited to reflect on my profile for this collection, I knew I would be doing so from a remote perspective. Even so, I welcomed the opportunity to think further about the process of implementing our common FYC syllabus—especially in light of a discussion I had recently entered on the WPA-L discussion list regarding pros and cons of standardization ("Standard Syllabus Advice?" WPA-L Archives. ASU-East. Web. 25 Jan. 2013).

Although my profile addresses many issues germane to implementing a common syllabus, a point that may deserve more emphasis relevant to that WPA-L exchange (particularly charges that standardization is stifling, if not demoralizing), it is my sense that the decision to standardize is most productively viewed not as an all-or-nothing prospect but, rather, as enacting a continuum for teacher support. On one end of the continuum lies complete autonomy to build a course around program objectives. This end best serves programs that are staffed by adjunct instructors and/or faculty with considerable teaching experience and a solid knowledge base in the field. For such programs, autonomy commensurate with experience would seem essential to morale. At the other extreme lies prescriptions for reading and writing assignments, as well as daily activities. This end of the continuum best serves programs staffed by entry-level master's students who've had no teaching experience or exposure to best practices in writing instruction. In such cases, considerable guidance seems warranted, not only so the undergraduates in their classes will have access to these best practices, but also for the new teachers themselves, who can feel intimidated and overwhelmed during their first semester in the classroom and who may fear that actively reaching out for guidance is to admit inadequacy.

But as I've continually been reminded over the past few years while interacting with new instructors in my graduate seminar on teaching college composition, even in programs staffed largely by inexperienced graduate assistants, any manifestation of a common syllabus stands the best chance of succeeding if the teachers are integrally involved in ongoing evaluation of it and revisions to it. That is, the syllabus must fluctuate as new GTAs test it, as their students react to it, as the contexts for instruction shift over time and with new institutional initiatives. In other words, this ongoing adjustment of the syllabus must reflect the evolving ecology of the writing program, must adapt to the interplay of forces within and without.

Importantly, GTAs' increasing involvement in determining curriculum heightens their sense of investment, increases enthusiasm for instruction, and helps seal their status as contributing members of an instructional staff. These affective outcomes complement the more tangible benefits of their involvement, including improvements to major assignments, assignment sequences, preparatory activities, reading selections, and so forth—improvements rooted in their intimate familiarity with the global plan for the course and their close interaction with students working through that curriculum. Indeed, this collective evaluation and revision can't help but surpass what the WPA might accomplish alone, even if he or she teaches a section of FYC now and again. To be sure, a writing program ecology thrives on synergy between the WPA's leadership based in scholarship and experience over time and the instructional staff's up-to-the minute, multi-sectional immersion in the course and interaction with students.

In addition to seeking their input on curriculum, validating GTAs' growing expertise requires that any manifestation of a common syllabus will offer increasing autonomy as the semester unfolds. On the program level, the degree of autonomy should increase across courses, perhaps moving from standardized assignments and unit outlines, to standardized writing assignments alone, to menus of unit assignments, and, ultimately, to opportunities for designing a syllabus from scratch. By scaffolding development in this way, WPAs can effectively balance their concern for the quality of undergraduate education with their realization that inexperienced teachers need some freedom to stretch themselves and learn through experimentation, while establishing a relational agency within the writing program's ecology. What's more, GTAs receive the specific support they need as they are learning to manage their new roles as graduate students and educators.

Appendix

Overview of English 101 at SIUC

[These introductory notes are presented here as they are presented to students, with the exception of some submission and formatting guidelines (irrelevant to the essence of 101) that have been removed. This document was authored by Dr. C. L. Costello and Dr. Ronda Leathers Dively, with miscellaneous contributions from various other Writing Program staff members.]

ENGLISH 101 DESCRIPTIVE OVERVIEW

English 101 provides students with the rhetorical foundations that prepare them for the demands of academic and professional writing. In this course, students will learn and practice the strategies and processes that successful writers employ as they work to accomplish specific purposes. In college, these purposes include comprehension, instruction, entertainment, persuasion, investigation, problem-resolution, evaluation, explanation, and refutation. In addition to preparing students for academic communication, this Core-Curriculum course prepares students to use writing to realize professional and personal goals. Accordingly, class discussion and readings will address the function of rhetoric and of composing processes in a variety of contexts, with attention to various audiences. Throughout the course, while engaged in a diversity of composing endeavors, students will learn to respond constructively to their peers' texts and to use peer responses (along with extensive instructor feedback) to improve the quality of their own work.

COURSE GOALS

After taking English 101, students should be able to:

- generate effective compositions using various methods for critical thought, for the development of ideas, for the arrangement of those ideas to achieve a specific rhetorical goal, for the application of an appropriate style, and for revision and editing;
- demonstrate understanding of the ways that language and communication shape experience, construct meaning, and foster community;
- analyze and describe rhetorical contexts and use such descriptions to increase the efficacy of communicative acts;
- analyze and use the forms and conventions of academic writing, particularly the forms and conventions of argumentative and analytical writing;
- produce texts that demonstrate an understanding of how purpose, process, subject matter, form, style, tone, and diction are shaped by particular audiences and by specific communicative constraints and opportunities;
- understand the importance of research to writing, explain the kind of research required by different kinds of writing, and com-

pose effective texts by judiciously using field research, library resources, and sources retrieved from electronic media;

- employ critical reading and listening as forms of invention;
- efficiently compose reading and lecture notes that are concise and clear;
- synthesize different and divergent information, using the integration of information from multiple sources to engage in critical discourse;
- use Edited American English appropriately.

COURSE WORK

During the semester, your instructor will require you to write frequently—for a variety of purposes, for a variety of audiences, and in a variety of forms. Most of this work will provide direct or indirect contributions to the culminating project of English 101, the course portfolio (explained below). The portfolio will contain revised versions of your major assignments and an analysis of your writing and your communicative development during the semester.

Unit Projects

English 101 is divided into five units. By the end of each unit, you will produce a significant "formal" composition that is the equivalent of three to six double-spaced pages. For each unit, your instructor will distribute detailed assignment guidelines for the major composition associated with it.

Unit One—Literacy Narrative: For an audience of your 101 class, you will narrate and address the significance of an experience in which you learned the literate practices of a given field or community and, as a result, gained access to that field or community.

Unit Two—Advertisement Analysis: For a business audience, you will compose a report that evaluates the effectiveness of a given advertisement in the context of the magazine in which it appears.

Unit Three—Summary/Rhetorical Analysis: For an academic audience, you will summarize an article to be assigned by your instructor, as well as critique the rhetorical strategies employed by that article's author.

Unit Four—Literature Review: For an academic audience, you will synthesize information from various sources about a controversial or debatable issue as designated by your instructor.

Unit Five—Reflective Introduction: With attention to course readings and activities, as well as to the contents of your portfolio, you will compose an essay, targeted for readers in English 101, that discusses your development as a writer during English 101.

Each of these texts will emerge from a process approach to writing, in which you engage in invention activities, planning activities, drafting activities, and revision/editing activities (including peer review). The formal composition for each unit and the materials used to write the composition will be submitted in a "working folder," which is a folder that documents your work during a particular unit.

Submission of Working Folders: During each of the five units listed above, your instructor, on pre-determined due dates, will collect preliminary informal exercises (idea sheets, plans, drafts, peer comments) for purposes of providing you with some feedback, and he or she will keep track of your timely and engaged attention to these exercises in his or her grade book. At the end of the unit, your instructor will collect some or all of this material again as part of a "working folder," or a record of your effort and development during the unit; thus, it will be imperative that you retain all informal exercises produced in the context of the unit. Failure to submit your responses to such assignments in timely and thorough fashion relevant to their original due dates will result in a deduction from the unit grade.

The working folder for each unit will also contain a draft of the major assignment or essay associated with that unit. The entire working folder contents for a given unit, then, will be assigned a grade that ultimately will account for 10% of your course grade. In addition, your essay will be assigned an "advisory grade," or an indication of its quality at the time you submitted it. The advisory grades will not contribute to your final grade for the course since you will be able to revise most essays until the end of the term, but the unit drafts will be an integral part of the holistic working folder grade. Indeed, it will be impossible for you to receive higher than fifty percent of the points available for the working folder grade without having submitted a *substantial* draft of the unit essay in addition to the informal assignments required by your instructor. (Important note: Because you will need to consult the working folder contents for all units at the end of the semester as you are assembling your portfolio and composing your reflective introduction, you will need to keep all the working folder contents from previous units in a safe, readily accessible place as you embark on each subsequent unit.)

Informal Exercises

In some sense, each unit project will serve as a model for the portfolio that you will submit near the end of the semester. The working folder for each unit will be a collection of your work during that unit (the major unit assignment and smaller daily assignments). Each working folder that you compile should provide evidence of your growth as a writer during a specific unit (much as the course portfolio will provide evidence of your growth as a writer during the semester). During each unit, you will engage in work that will assist in preparing the text that you will submit for review at the end of the unit. Often, these small assignments will constitute stages in your own writing process for a particular major essay, but they might include other documents such as a peer review of a classmate's work or a detailed summary of a reading. In determining the grades for working folders at the end of each unit, the instructor will "weight" exercises in accordance with their length and complexity. Though this course does not have a specific participation grade, the informal exercises will indicate your level of effort and engagement.

Portfolio

This course has been designed to increase your ability to communicate, particularly in writing, by encouraging you to develop and then exercise a rhetorical sensitivity by which you identify the constraints and opportunities of any communicative challenge and respond appropriately. To improve this ability (which you already posses), this course is structured around a portfolio system, in which a large portion of your grade (fifty percent) is based on texts that you will be able to revise for much of the semester, drawing upon the rhetorical sensitivity that you develop, your instructor's comments, your peers' comments, and other resources that you might employ (for instance, the Writing Center). Near the end of the semester, you will submit your portfolio by gathering essays that you have completed during the semester and polished to "presentation quality" text. You will present this work to your instructor (in a two-pocket folder) as evidence of your ability to write and as evidence of your learning during the course of the semester. This collection of finished essays will be graded on the quality of the writing, not on effort. (Effort will be rewarded in the context of the working folder.)

Your instructor will judge the portfolio by engaging the collection of texts largely as an experienced reader (rather than an as educator). As he or she will have made regular comments on your writing (if you submit your rough drafts and visit him or her during the semester to discuss revision), your instructor will read your portfolio attentively but no longer with the kind of attention that supports formative commentary. Your instructor will read these texts against a rubric, based on the course guidelines, to see if your work is rhetorically effective and indicates that you have achieved the communicative goals set by the English 101 objectives. In the process of preparing your portfolio for presentation to your instructor, you will be asked to compose a Reflective Introduction (Unit 5 essay) that comments on your development as a writer as evidenced by the other formal essays that you've decided to submit.

Exam

In this class, you will be required to take a final exam *during the officially scheduled exam period*. The exam will ask you to generate an essay (employing strategies explicitly addressed in the context of English 101) on a subject matter to be announced near the end of the semester.

Percentages

Unit 1 working folder (including draft of Literacy Narrative) 10

Unit 2 working folder (including draft of Advertisement Analysis) 10

Unit 3 working folder (including draft of Summary/Response) 10

Unit 4 working folder (including draft of Literature Review) 10

Unit 5 portfolio (including Reflective Introduction) 50

Final Exam (in-class essay—form and subject matter TBA) 10

[The overview of English 101 presented above is distributed to students at the beginning of each semester in a document that also contains all course policies.]

Notes

1. English 101 classes are capped at twenty students, and, typically, new GTAs are assigned two 101 sections each.

2. "Core Curriculum" in this context refers to the office that administers SIUC's general education requirements, which call for twelve hours in "Foundation Skills" (including the first-year composition sequence), twenty-three hours in "Disciplinary Studies," and six hours in "Integrative Studies."

3. For an empirical investigation of the obstacles to knowledge transfer in this particular institutional context, see "Perceived Roadblocks to the Transfer of Composition Knowledge: A Pilot Study" (Nelms and Dively).

4. Of course, some programs manage to support and require completion of graduate courses in composition theory and pedagogy during the summer before GTAs begin teaching (e.g., Powell et al. 122, 124) and/or enjoy the luxury of introducing instructional responsibilities more slowly (e.g., Yancey 67).

5. The literature professor, Dr. Mark Amos, was assisted by Dr. C. L. Costello, who later became my assistant. Dr. Costello—now a Humanities instructor at Reading Area Community College in Reading, PA—had a large hand in developing SIUC's standardized 101 curriculum.

6. See http://compositionforum.com/issue/22/siuc-appx.php#appx2 for sample assignments.

7. I am assisted in planning and leading the PSW by a team of eight seasoned GTAs—two Administrative Assistants (PhD candidates in rhetoric and composition), two Instructional Assistants, and four small-group leaders who, based on their reputations as highly effective English 101 teachers, are asked to assume primary responsibility for introducing the new GTAs to the standardized syllabus.

8. The Communications Across the Curriculum (CAC) initiative at SIUC has ebbed and flowed along with inconsistent financial support. Soon after the standardized syllabus was implemented, I participated in CAC forums across campus to explain the reasons for the change in approach, the nature of the curriculum, and the projected benefits of standardization at that level.

Works Cited

Bamberg, Betty. "Creating a Culture of Reflective Practice: A Program for Continuing TA Preparation after the Practicum." *Preparing College Teachers of Writing: Histories, Theories, Programs, Practices.* Eds. Betty P. Pytlik and Sarah Liggett. New York: Oxford UP, 2002. 147–158. Print.

Borrowman, Shane. "First-Year Training for First-Year Composition: TA Training from the Inside." Conference of the National Council of Writing Program Administrators. Tucson. July 1999. ERIC. Web. 2 November 2009.

Carey, L. J., and Linda Flower. "Foundations for Creativity in the Writing Process: Rhetorical Representations of Ill-Defined Problems." Center for the Study of Writing, Technical Report, No. 32. June 1989. ERIC. ED 313699. Web. 15 June 2004.

Csikszentmihalyi, Mihaly, and Keith Sawyer. "Creative Insight: The Social Dimension of a Solitary Moment." *The Nature of Insight*. Eds. Robert J. Sternberg and Janet E. Davidson. Cambridge: MIT Press, 1995. 329–363. Print.

Dively, Ronda Leathers. *Preludes to Insight: Creativity, Incubation, and Expository Writing*. Cresskill, NJ: Hampton Press, Inc., 2006. Print.

Griffith, Kevin. "Readers, Writers, Teachers: The Process of Pedagogy." Conference on College Composition and Communication. Boston. Mar. 1991. ERIC. Web. 2 November 2009.

Kinney, Kelly, and Kristi Murray Costello. "Back to the Future: First-Year Writing in the Binghamton University Writing Initiative, State University of New York." *Composition Forum* 21 (Spring 2010): n. pag. Web. 21 May, 2010.

Liggett, Sarah, and Betty P. Pytlik. Preface. *Preparing College Teachers of Writing: Histories, Theories, Programs, Practices*. Eds. Betty P. Pytlik and Sarah Liggett. New York: Oxford UP, 2002. xv-xxii. Print.

Martin, Wanda, and Charles Paine. "Mentors, Models, and Agents of Change: Veteran TAs Preparing Teachers of Writing." *Preparing College Teachers of Writing: Histories, Theories, Programs, Practices*. Eds. Betty P. Pytlik and Sarah Liggett. New York: Oxford UP, 2002. 222–232. Print.

Nelms, Gerald, and Ronda Leathers Dively. "Perceived Roadblocks to the Transfer of Composition Knowledge: A Pilot Study." *Writing Program Administration* 31.1/2 (2007): 214–240. Print.

Nyquist, Jody D., and Donald H. Wulff. *Working Effectively with Graduate Assistants*. Thousand Oaks, CA: Sage Publications, Inc., 1996. Print.

Perkins, David N., and Gavriel Salomon. "Teaching for Transfer." *Educational Leadership* 46.1 (1998): 22–32. Print.

Powell, Katrina M., Peggy O'Neill, Cassandra Mach Phillips and Brian Huot. "Negotiating Resistance and Change: One Composition Program's Struggle Not to Convert." *Preparing College Teachers of Writing: Histories, Theories, Programs, Practices*. Eds. Betty P. Pytlik and Sarah Liggett. New York: Oxford UP, 2002. 121–132. Print.

Ward, Thomas B., Ronald A. Finke, and Steven M. Smith. *Creativity and the Mind: Discovering the Genius Within*. New York: Plenum Press, 1995. Print.

Welch, Nancy. "Resisting the Faith: Conversion, Resistance and the Training of Teachers." *College English* 55.4 (1993): 387–401. Print.

Yancey, Kathleen Blake. "The Professionalization of TA Development Programs: A Heuristic For Curriculum Design." *Preparing College Teachers of Writing: Histories, Theories, Programs, Practices*. Eds. Betty P. Pytlik and Sarah Liggett. New York: Oxford UP, 2002. 63–74. Print.

3 Taking the High Road: Teaching for Transfer in an FYC Program

Jenn Fishman and Mary Jo Reiff

> The bottom line for writing instruction may be this: We get what we teach for. And if we want to help students to transfer what they have learned, we must teach them how to do so. That is, we must find ways to help novices see the similarities between what they already know and what they might apply from that previously learned knowledge to other writing tasks.
>
> —David Smit, *The End of Composition Studies*

Current theoretical conversations in the field of rhetoric and composition, particularly conversations related to first-year curricular design, are increasingly concerned with the issue of "teaching for transfer," a trend evidenced by the recent flurry of books, articles, empirical studies, and professional discussions on the topic. Many writing teachers and scholars who discuss knowledge transfer draw on work by educational theorists D.N. Perkins and Gavriel Salomon, who together define two different types of transfer and two related pedagogies. Distinguishing between reflexive and reflective cognitive acts, Perkins and Salomon describe low road transfer as an automatic and highly routinized cognitive practice, while they characterize high road transfer as both mindful and analytic (25). Answering the question, "Can we teach for transfer?" in the affirmative, Perkins and Salomon offer two techniques: "hugging" and "bridging" (28). While the former "means teaching so as to better meet the resemblance conditions for low road transfer" by creating situations that trigger automatic responses, the latter, bridging, "'mediates' the needed processes of abstraction and connection-making" through activities that promote problem solving and generalizing across disparate examples (29). Both in theory and in practice, writing teachers generally prioritize high road transfer that entails application of writing knowledge to other courses and contexts, although David Smit and others believe that it is "rare" for general education writing courses to teach high road transfer effectively (134).

While developing successful transfer pedagogy is a challenging undertaking, one that may require writing instructors to revise both course content and teaching styles, there are good reasons for facing the challenge, starting with the changing face of college writing instruction. Over the past several years, both stand-alone first-year composition (FYC) programs and programs within English departments have become less isolated from the rest of the university, and it is not uncommon for them to be identified with (or even renamed) "University Writing Programs." In addition, FYC is often affiliated with writing across the curriculum (WAC) or writing in the disciplines (WID) programs, and FYC is frequently supported by writing centers, which serve college and university communities as true centers of local writing culture. In this context, first-year writing gains new responsibilities, becoming students' introduction not only to college composition, but also to writing transfer in other courses and contexts. Anne Beaufort addresses this shift in *College Writing and Beyond: a New Framework for University Writing Instruction,* where she explains the role that FYC has the potential to play in extending writing knowledge beyond the first year:

> Freshman writing, if taught with an eye toward transfer of learning and with an explicit acknowledgement of the context of freshman writing itself as a social practice, can set students on a course of life-long learning so that they know how to learn to become better and better writers in a variety of social contexts. (7)

For writing program administrators (WPAs) and FYC teachers who accept this challenge, the question they jointly face is how to develop a program that can help students acquire the rhetorical knowledge and skills vital to communicating effectively in multiple contexts. How, in other words, do we design a writing curriculum that creates the conditions for high road transfer? In response Smit provides a general suggestion: "If we want to promote the transfer of certain kinds of writing abilities from one class to another or one context to another, then we are going to have to find the means to institutionalize instruction in the similarities between the way writing is done in a variety of contexts" (120). Clarifying "the means" by which we might accomplish this feat, Beaufort and others have suggested that we should "teach those broad concepts (discourse community, genre, rhetorical tools, etc.) which will give writers the tools to analyze similarities and differences among writing situations they encounter" (149).

When we initiated program-wide FYC curriculum revisions at the University of Tennessee-Knoxville, we considered these issues closely in relation to the twinned goals of teaching students core writing strategies and encouraging them to communicate confidently in multiple situations. Regarding rhetoric as the best "mediator" between these two goals, we chose to center our new curriculum on rhetoric and the transferability of rhetorical knowledge across different situations, mediums and assignments. As a result, we reinforced and strengthened the rhetorical foundations of both English 101 and English 102, and we worked to ensure each course retained distinct content and focus. Ultimately, then, English 101 concentrates on analysis and argument, and we redesigned shared assignment sequences to emphasize the importance of learning to read and write critically and self-consciously about other people's texts. In English 102, our new priorities resulted in even greater changes: namely, the integration of multiple methods (including field research and historical research) into a course that previously focused on literary studies and included only traditional academic research methods.

Thinking especially about our desire to increase students' awareness of how writing can be used in different ways for different purposes, we also chose to integrate an expanded range of texts, including multi-media and digital texts, more strongly into our courses. As we began program-wide curriculum revision, we revised our outcomes to emphasize the ways in which emerging multimedia, multimodal, and multi-disciplinary FYC curricula inform the acquisition and transferability of academic literacy learning, a move necessitated by the changing nature of writing in the academy. As one of our new outcomes states, our revised courses are designed "to help students develop a variety of strategies for writing for multiple audiences and purposes; to develop their ability to create a wide range of texts (including multimedia and electronic texts) and communicate by means of multiple modes of communication" (UTK Composition Program Outcomes Statement). In making this change, we join those teachers and scholars of writing and literacy who have recently renewed their commitment to better understanding and promoting multimodal literacies, or literacy across different mediums and communication situations (Jewitt and Kress; Kress; Selber; Wysocki). As Kathleen Blake Yancey describes in "Made Not Only In Words: Composition in a New Key," multiple literacies play a leading part in "new curriculum for the 21st century" and programs of study designed to address how texts "move across contexts, between media, across time" (312). Attentive, thus, to the seemingly endless, past and present proliferation of rhetorical situations, genres, and communicative media, the pedagogical model Yancey describes is not only a guide to the new products of twenty-first century composition, but also an index to "the *content* of composition" for the millennials and those who teach them (308).

As first-year composition programs reassess their curricula and come to grips with the "content envy" that Richard Fulkerson identifies in his recent summary and critique of "Composition at the Turn of the Twenty-First Century," WPAs face critical questions about the transferability of learning in their courses. Such concerns invite us to think about how we can develop FYC programs that are both responsible and responsive to the writing needs of students beyond the first year. The program profile that follows focuses on how members of the FYC program at a Research-1 university, the University of Tennessee-Knoxville, came together over the course of three years to address these questions in curricular revisions. Examining the institutional context for curriculum review and the process of implementing new courses, we describe the impetus for change along with the goals and theories that informed each stage of decision-making. It is our hope that the local example we present here will provide insights into the challenges and opportunities that composition programs nationally face as they are increasingly called upon to address the question of transfer.[1]

THE LOCAL CONTEXT FOR CURRICULAR CHANGE

Our curriculum review began in Fall 2004 and was equally prompted and shaped by a combination of departmental, institutional, and programmatic factors. First of all, the departmental context: The English department at the University of Tennessee is a fairly traditional one comprised of three divisions: literature, criticism, and textual studies; creative writing; and rhetoric, writing and linguistics or RWL. While the RWL faculty has eight members and the department recently formalized a PhD track in the division, the majority of tenure-track positions are held by literature specialists and—with a few exceptions from the RWL group—none of the professorial faculty regularly teaches FYC. Instead, composition courses are taught by very qualified and talented lecturers and graduate TAs, although with a few exceptions, most of these instructors have disciplinary interests in literature or creative writing. This was the demographic and the situation that external reviewers addressed when, in 2003, our department underwent a 10-year review. At the end of that process, reviewers recommended strongly that FYC undergo "a thorough review of the requirements (led by the rhetoric and composition staff)," and they singled out English 102, a course focused on literature and composition, as requiring particular attention and reconsideration. The reviewers' final report offered this summary:

> English 102 was criticized inside and outside the department for being inconsistent over sections in requirements; some sections require much writing, while others (often taught by TAs, who probably see the course as preparation for teaching advanced literature) are taught as literature courses, with comparatively little writing. The program director should have the authority to examine book orders and syllabi to see that the focus in 102 stays on writing. (Final Report, Department of English, External Review, 2003)

While this report underscored the need to shift the focus in 102 from literary analysis to writing and rhetoric and to make the course more meaningful for students within varied fields of study, there were other institutional changes taking place that further emphasized the need to focus on broader, more transferable skills. In fall 2004, UT implemented new general education requirements, which grouped English 101 and 102 under "basic skills requirements" and added a third required writing course that students could take in one of 32 disciplines. These changes transformed FYC courses, turning them in effect, into prerequisites for writing in the disciplines. The UTK undergraduate catalog underscores this idea in the general education section under the heading "Communicating through Writing":

> Good writing skills enable students to create and share ideas, investigate and describe values, and record discoveries—all skills that are necessary not only for professional success but also for personal fulfillment in a world where communication increasingly takes place through electronic media. Students must be .

. . aware that different audiences and purposes call for different rhetorical responses. To satisfy this requirement, students take the first-year composition sequence and, upon completion of English 101 and 102 or their equivalent, take one other course designated as "writing-intensive" (WC), [requiring] formal and informal writing assignments that total 5,000 words.

Together, these institutional factors—both the outside review and new general education requirements—pushed us to reconsider how our curriculum is responsive to the writing needs of students beyond the first year.

These changes helped to create the context in which the composition committee began revising the first-year writing curriculum in the Fall of 2004. The committee includes the director of first-year composition (Mary Jo Reiff); the director of ESL; the director of the writing center; an RWL faculty representative (Jenn Fishman); a full-time lecturer, a part-time lecturer, and a graduate student. Working together, we turned to our knowledge of "best practices" in the field, and we looked for ways to reinforce the objectives described in the "WPA Outcomes Statement" (the development of rhetorical knowledge, writing processes, critical thinking and reading skills, and use of conventions); at the same time we worked to incorporate current research and pedagogical perspectives on transfer, such as Beaufort's knowledge domains (which include rhetorical knowledge, contextual knowledge, and writing process skills). We designed our new curricula to emphasize the idea that rhetorical actions are culturally situated; we formulated assignments to create better opportunities for students to learn rhetorical awareness through experiences with a diverse range of texts and genres, including multi-media texts; and we highlighted for students and teachers both the role of visual and multi-modal rhetorics in student engagement and learning—an additional outcome that Yancey calls for in her afterword to *The Outcomes Book* (220–21). Based on all of these factors, we identified three main curricular objectives:

1. Rhetoric: To increase students' rhetorical awareness of how their subject, purpose, audience and context for writing can shape their message, mode of inquiry, methods of research, and presentation of ideas.
2. Multiplicity: To expose students to a diverse range of texts (print, digital, multimedia), methods and perspectives and to teach

them to communicate via multiple modes of communication—written, spoken, and visual.
3. Transferability: By way of the first two goals, to further develop the rhetorical tools of inquiry, analysis, and research and experience with writing in varied situations and mediums that will transfer to writing situations outside of FYC.

In addition, we were concerned with more programmatic issues. First, we wanted to find ways to reestablish and then maintain a coherent program so that students across our 200 sections could see clearly they were working to reach the same goals and fulfill the same general requirements. Second, we wanted to find ways to improve the curriculum for faculty. Recognizing the expertise our FYC faculty bring to the classroom as well as the full range of responsibilities they carry within the department, we wanted to provide teachers with new ways to personalize their courses and teach to their strengths. Further, we wanted to provide them better support, taking into account the needs of both TAs, who are usually teaching two courses while also taking two courses, and lecturers, who generally teach across the department's undergraduate curriculum and carry a 4:4 teaching load.

In an effort to meet these goals, we took steps to engage teachers fully in the process of revision. To that end, we proceeded gradually over several semesters, and we solicited feedback at every stage, beginning with a faculty survey of current teaching practices.[2] While our questions covered every aspect of our FYC program, we were especially interested in teachers' comments about the organization and content of 102, and we received the following replies to the open-ended question, "What are the biggest challenges of teaching English 102?":

- Student motivation, apathy for literature;
- Keeping argument (vs. literary analysis) the focus of the course; getting students to see literature as an argument;
- Negotiating between teaching writing and research and teaching literature;
- Establishing relevance of literature courses to writing across the curriculum;
- Difficulty with teaching literary research and question of whether or not to expose students to literary criticism;
- Difficulty with teaching the generic research paper

Reading these responses, we were struck by their consistency with the external review and with our own sense of how 102 had become a course that limited high road transfer. Teachers confirmed this idea during an open forum, and their comments further reinforced our belief that 102, in particular, was ripe for reform. As teachers reported, the standing goals of 102 seemed disparate, even contradictory, and they required teachers to do—and to try to do well—the nearly impossible task of teaching both a fully-fledged introduction to college-level research and research writing *and* a comprehensive introduction to literary studies and analysis. Deciding that, at best, our curriculum promoted low road transfer, or students' relatively automatic application of previously acquired reading and writing knowledge, we began to articulate new program-wide goals. From this point onward, we turned our attention to identifying pedagogical practices we could build on and revise, while also singling out specific curricular areas with the greatest need for attention and improvement.

To continue making our revision process as collaborative as possible, the Composition Committee organized focus groups in Spring 2005, inviting teachers to meet with us in small groups and share their ideas about how the first-year writing curriculum could better serve its students and better support their writing needs beyond their first year.[3] From the focus groups, we learned that teachers were generally receptive to shifting the emphasis in 102 from literature to inquiry, and they were intrigued by the idea of defining their own inquiry subjects. At the same time, instructors were skeptical of their ability to increase attention to the rhetorical aspects of writing while also teaching diverse methods of inquiry (field research, historical research, and academic research). In addition, while there was general appreciation for a more structured and coherent curriculum, many teachers also expressed concerns about the additional work that would come with having the freedom to shape their courses and assignments. As all of us noticed, few of the inquiry methods and subjects we discussed are covered adequately (if at all) by the FYC textbooks and readers that crowd our bookshelves, and that observation prompted us to talked at length about how we might balance the intellectual and pedagogical opportunities presented by teaching without textbooks and the work involved in gathering or designing substitute resources, from readings and related reading guides to assignment sequences, rubrics, and prompts.

The courses we designed as a result of these deliberations speak equally to our desire for curricular coherence and faculty support, and we addressed these two goals in similar ways in both English 101 and English 102. Similarly, then, our new curricula are anchored by common assignment sequences, and both courses foreground broad knowledge domains (rhetorical learning, genre knowledge, discourse community knowledge). Additionally, they also feature transferable tools or bridging mechanisms that our program goals emphasize:

- To teach students a rhetorical awareness;
- To teach students to read rhetorically and to read as writers;
- To help students develop strategies for writing to multiple audiences for multiple purposes using multiple mediums and modes of expression;
- To teach students to produce complex arguments that matter (i.e., to teach students a sense of rhetorical exigence);
- To give students extended practice with composing processes;
- To demonstrate how to argue purposefully with sources;
- To teach students how to conduct purposeful primary and secondary research;
- To move students from analysis of texts (how texts work) to production of knowledge through inquiry and research.

Our new courses address these goals recursively, giving students opportunities to acquire new knowledge and skills cumulatively over the semesters, while also enabling students to read and write different kinds of texts consistently over the year. Our new curricula also establish new bridges between courses: with rhetoric as a foundation, the revised sequence moves from a first-semester focus on analysis to a second-semester focus on production. In this way, we hoped to move from "hugging" to "bridging" or from low-road to high-road transfer as we asked students first to develop their rhetorical, social, and genre knowledge and use it within the local contexts of the 101 classroom, and then, in English 102, to transfer that knowledge to the production of new knowledge in contexts extending beyond the English classroom.

English 101: (Re)Emphasizing Rhetoric

In our new curriculum, English 101 is the intellectual as well as the practical foundation for English 102, and we redesigned the former to teach students general rhetorical strategies that will, we hope, provoke "broad-spectrum practice" that reaches beyond discipline-specific subject matter (Perkins and Salomon 25). As Perkins and Salomon might say, we "shaped instruction to hug closer to the transfer desired" (30), and we developed the following sequence of assignments, which moves from reading rhetorically to arguing with multiple sources:

- Reading rhetorically: assignments/activities help students develop strategies for reading critically; they also introduce rhetorical terms and concepts (primarily as heuristic tools);
- Rhetorical analysis: assignments/activities emphasize ways writers use language and textual conventions effectively to communicate with readers;
- Contextual analysis: activities/assignments explore ways context shapes communication for both writers and readers;
- Argument paper (i.e., position paper): activities/assignments invite students to find common ground and take a stance (or more than one stance) on an issue that invites deliberation or debate;
- Source-based argument: activities/assignments go through the stages of developing a well-supported, focused argument using provided (largely) by the instructor.

As illustrated by these assignments, our aim when we begin teaching reading rhetorically is both to focus on reading critically and to introduce rhetorical concepts and terms, including the rhetorical triangle, the rhetorical situation, and rhetorical appeals. The next two assignments, rhetorical and contextual analyses, ask students to begin working self consciously with rhetoric as not only readers but also writers. The former asks students to read (or view or listen to) and describe other people's arguments, while the latter asks students to put such focused analysis into broader cultural and/or historical context. The next assignment, the position paper, emphasizes the importance of finding common ground for arguments at the same time that it compels students to take a stance or argue more than one position on a complex and possibly controversial topic. In sequence, these assignments are designed and ordered to build toward the final source-based essay, which invites students to work closely and deliberately with multiple sources in order to formulate and sustain an extended argument.

During fall 2005, Jenn Fishman and Stacey Pigg led a group of instructors in a writing study, the Embodied Literacies Project,[4] that piloted the new English 101. The course, titled "Self, Community, and Culture," had three main units, each of which focused on a different primary text paired with a set of related secondary readings, which researchers compiled in a custom-published reader. In addition, each unit sought ways to lead students from hugging to bridging by working from literary texts (and literary analysis students learned to do in high school) toward rhetorical analysis and synthesis of different genres, mediums, and types of argument. Teachers started the semester by asking students to work with *The Curious Incident of the Dog in the Night-time,* the common book chosen by the university for all incoming students to read. Rather than treating the book as an entrée into college-level literary studies, assignments asked students to read *Curious Incident* alongside texts that offered contrasting perspectives on the novel's subject, autism. As a result, students learned rhetorical terms and ways of reading while they compared perceptions of autism and arguments about its treatment in medical texts, print and online essays, and images, as well as their own writing.

The second unit included both the rhetorical and contextual analysis assignments, and it brought visual and verbal rhetorics to the fore by focusing on Marjane Satrapi's graphic autobiography *Persepolis*. Satrapi grew up in Iran during the Iranian Revolution, and students read her work alongside materials that helped put her story into historical and cultural context. Unit three turned to *The Laramie Project,* a play that explores reactions to the murder of Matthew Shepherd, a gay college student killed by two Laramie, Wyoming, residents in 1998. The play is based on hundreds of interviews with members of the Laramie community, and it vividly portrays the town's response to a brutal, hate-motivated murder. As an English 101 text, the play dramatized both unit assignments (taking a stance and arguing with sources) and the close relationship between research and community, which is emphasized in English 102. Establishing connections between FYC semesters, the final 101 unit was designed to help students work with multiple sources in order to take an informed stance and sustain an extended argument. This unit also challenged students to select and group texts drawn primarily from a custom-published course reader in order to insert themselves into ongoing arguments about topics ranging from masculinity and the West to defining and understanding hate crimes. In both of these units (according to the research plan that structured different course activities), some teachers integrated online activities, including blog writing and peer reviewing as well as Instant Messenger discussions and conferences. Further, all teachers incorporated both formal and informal oral presentations into several units, developing creating ways for students to use oral communication both as part of their writing process and as a final means of delivering well-prepared and well-supported arguments.

ENGLISH 102: FROM LITERATURE TO INQUIRY

While the revised English 101 course focuses on rhetorical analysis and argument, English 102 moved from its previous focus on literature (and low road transfer) to engaging students in rhetoric and research as part of different communities of inquiry (high road transfer). To this end, new 102 units and assignments—while focused on a subject or topic defined by the instructor—concentrate on identifying various expert communities; recognizing each community's primary research resources, questions, and strategies; and joining in expert conversations as reviewers, participant-observers, and contributors. According to our revised outcomes for English 102 students who complete this course should be able to:

- read critically to identify, define, and evaluate problems/complex issues, taking into account audience, as well as intercultural issues and multiple points of view;
- recognize how writing/research methods are used in different expert communities;
- enter and participate in different ongoing expert conversations using a range of written and visual texts;
- use multiple investigative methodologies (field research, historical research, academic research) to define and develop positions on issues/questions of their own choosing;
- construct effective arguments using appropriate material gained through "active scholarship";
- locate and evaluate information for specific research questions and audiences using a range of research sources;

- present research effectively in multi-modal formats, using genres and rhetorical appeals appropriate to audience and purpose.

With these goals, our 102 clearly emulates many WAC and WID curricula, although our courses also remain distinct. While we want English 102 to help students learn, first-hand, how different groups of people—both in and outside of the university—argue about issues that concern them, our courses aim to teach students general rhetorical concepts and moves that they can use as tools for communicating in different and different kinds of contexts. To that end, English 102 guides students into different communities in order to help them gain a sense of how different groups of people produce knowledge recognized as valid and authoritative. We also want students to see themselves as writers and rhetors who are capable of participating in varied scenes of writing, and we want students to develop "habits of self-monitoring" so that they can support their own learning for transfer after they leave our classrooms (Perkins and Salomon 29).

English 102 approaches these goals by focusing not only on rhetoric and writing, but also on research, and we place a great deal of emphasis on teaching students different research methods along with related argumentative strategies.[5] In this respect, our new curriculum differs substantially from our previous course, which followed the model offered by many FYC textbooks and treated research as a generalist activity. Falling into the trap that Gerald Graff criticizes in *Clueless in Academe*, overly general instruction fails to prepare students for authentic situated scholarly work because it fails to help them understand how resources as well as information gathering- and producing- activities differ from discipline to discipline and community to community. Our new curriculum responds to this problem by resituating research into a sequence of discrete course units, each focusing on the rhetoric and rationale of different research methods. As we describe in our formal overview of the curriculum, English 102 now spans three general types of research (field research, historical research, and academic research), divided into three units:

Unit 1: Being "Present" in Your Research: Field Research

In this unit, students learn hands-on research or "field research" methods, with the focus on the tools of observation, interviews, and surveys. They write field notes or descriptions of places and activities they observed, profiles of persons interviewed, or reports of data gathered in surveys. As an option, students with particular interests/majors might interview an individual working in their field or a professor or advanced student in that field (or might survey a group of student majors) and then write up their findings.

Unit 2: Answering Questions by Looking to the Past: Historical Research

In the second unit, students carry out historical research by examining artifacts, visiting museums, consulting archives, reviewing or (since methods overlap) interviewing someone with an historical perspective on the question being asked. Students might write oral histories or create visual essays or family trees (paired with a family history paper). As an option, students with particular interests/majors might explore the history of a discipline or area of study or analyze an artifact of the field or discipline and its significance.

Unit 3: Entering the Parlor: Academic Research

This unit introduces students to a range of traditional research methods (library sources, databases, Internet sources) and disciplinary methodologies. For their research project, students decide on methods and genres appropriate to audience and purpose (and may choose to write a website or pamphlet to a popular audience, for instance). Students may also choose to do a research project exploring how research is conducted in their field of interest; for instance, they could examine journals in the field to determine methods used, interview people, or observe disciplinary electronic listservs.

THE ENGLISH 102 PILOT PROJECT

Although our new curriculum was formally implemented in Spring 2007, we invited teachers to pilot the new 102 during the spring of 2006. The collaborations between WPAs, the Composition Committee, and teachers in the program were key to our revision process, and 16 teachers volunteered for this project. Together, we worked to devise pilot syllabi and assignments and to create a custom-published rhetoric for 102, which has been paired with an instructional website featuring materials developed by teachers.[6] The pilot project was invaluable, not least because it enabled us to give the revised 102 a test run, and it allowed us to discover objectives and methods that needed to be defined more clearly or supported more fully with better teaching materials. Even more importantly, the pilot teachers became a corps of experienced instructors who can share resources and co-lead faculty development workshops as we have begun implementing the new curriculum.[7]

At the beginning of the pilot process, interested teachers self-selected by answering a program-wide call for proposals. A mix of graduate students and lectures, these volunteer teachers submitted a course description, a list of texts, and writing assignment descriptions. The composition committee reviewed proposals and organized follow-up workshops for teachers whose proposals had been accepted. Some of the inquiry topics for our pilot sections included "Inquiry into Weblogs," "Inquiry into Travel," "Inquiry into Southern Appalachia," "Inquiry into Protest Writing," and "Inquiry into the Unreal." Once proposals were accepted (and in some cases, revised), we posted them to a 102 pilot Blackboard (course management) site, and we returned to the electronic space not only for posting similar documents (additional syllabus materials and assignments), but also for hosting discussion groups and for troubleshooting.

During the spring semester, while pilot courses were in session, we also held meetings with pilot instructors, and they shared successes, challenges, and concerns. Some of the problems teachers identified were good problems, like the overlap between research units and methods, which Perkins and Salomon have indicated can actually foster transfer. As they discuss in "Teaching for Transfer," points of intersection are highly teachable moments, which teachers use to help students make generalizations about inquiry rhetorics and practices across multiple subject areas and settings. Of course, other concerns were more challenging, especially the difficulty teachers had introducing students to unfamiliar research methods (like field research) and research in unfamiliar fields. Since students were encouraged to investigate subjects of personal interest, teachers also found themselves planning lessons for students who were researching and writing about very different texts and topics, and they found it difficult, at times, to maintain a unified focus in class. Perkins and Salomon suggest mediating this process of abstraction and connection by teaching transferable principles: general rhetorical strategies, research strategies, strategies for critical inquiry, and an awareness of investigative methodologies that can be applied in multiple contexts. Beaufort agrees with the idea of "teaching a set of tools for analyzing and learning writing standards and practices in multiple contexts" (11), with which our multidisciplinary investigative approaches in 102 are consistent.

When some of our 102 teachers reflected on their experiences at the 2006 National Council of Writing Program Administrators Conference (in Chattanooga), they focused on transfer, and they sought to articulate how our new curriculum fosters the transfer of writing and research knowledge across multiple contexts. Catherine Phillips, an instructor who taught "Inquiry into Urban Legends and Myths," describes the way her topic of inquiry guided students down the high road, creating a variety of opportunities for them to practice rhetorical analysis:

> Rhetoric and writing were always an explicit part of the course, and my students spent as much time working on voice, audience, and other rhetorical strategies in their own writing (in workshops, in in-class assignments, and so on) as on any other aspect of the course. But also, because my section's content moved from one discipline (folkloristics) to quite a few others (including media studies, sociology, psychology, and even cryptozoology), my students quickly became familiar with many different expert communities and the various research methods and methodologies therein. All of these papers demonstrated not only a fairly sophisticated grasp of the writing and rhetorical skills my students had been building since 101, but also a surprising willingness to think critically, to explore various disciplines, and to add to critical conversations—with quite a bit of confidence.

Catherine's reflections run parallel to Elizabeth Wardle's research findings. As Wardle argues, reflecting on her longitudinal study of FYC transfer, teaching rhetorical strategies and rhetorical analysis of texts can create a "meta-awareness" that cultivates transfer. While discipline-specific writing is best taught in the context of particular disciplines, "what FYC can do," Wardle argues, "is help students think about writing in the university, the varied conventions of different disciplines, and their own writing strategies in light of various assignments and expectations" (82).

Another 102 pilot teacher, Jessica Abernathy, reflects on how her course, "Inquiry into Madness," combined emphasis on different aspects of the topic and curricular objectives, while reinforcing, in miniature, the three primary goals of rhetoric, multiplicity, and transferability:

The "Representing Madness" unit was particularly geared towards increasing students' rhetorical awareness, to encourage them to concentrate on the mechanisms, effects, and successfulness of various texts (including their own), and to help them realize the crucial role of medium, context, and audience in shaping texts. In doing so, students were exposed to a variety of texts using different techniques and different media—print, film, art, music, and more. Students not only learned to analyze these texts in medium-specific ways, but also to communicate in different media themselves. Finally, though other course units and assignments were more specifically targeted at transferability (teaching students mutability, or "code-switching," across various fields), this unit encouraged flexibility, fidelity to research, and awareness of audience and context, all of which aids students in writing and communicating in the world outside of first-year composition.

IMPLEMENTATION OF THE REVISED ENGLISH 102 COURSE

Beginning in spring 2007, all 102 sections have followed the new inquiry model. During the fall semester before teaching 102, teachers are asked to submit a proposal describing the course topic, goals, assignments, and texts.[8]

Once proposals are reviewed by the Composition program staff, including the director, associate director (a lecturer), assistant director (a GTA), and the writing center director, teachers submit their own book orders. All 102 sections share the common custom-published text, *Rhetoric of Inquiry,* but teachers are free to choose their own supplementary texts. Following approval of the proposals, the English 102 topics are then listed in the university's course timetable, and short descriptions (written for a student audience) are also listed on our department website.[9] For us, this relatively straightforward process represents a significant administrative change that is worth noting. Prior to our curriculum revisions, sections of English 102 were listed in the timetable only by date, time, and instructor name, and course registration proceeded as though additional details were irrelevant. However, with the thematic focus of the new 102s, we felt it was important for students to have more information. For the first time, then, and after a significant revision of our enrollment procedures, our students can now read about the courses online, and they can make informed choices about what they choose to take.

Conclusion, or "If We Knew Then What We Know Now"

In order to figure out "What We Know Now," several of us in RWL have been conducting studies of our first-year program, making transfer our focus. The studies we have undertaken include an cross-institutional inquiry (with University of Washington) into college students' transfer of prior genre knowledge into FYC; the Embodied Literacies Project's two-year investigation into first- and second-year students' transfer of rhetorical knowledge across academic writing situations and media and a survey-based study of how students perceive the transferability of FYC research skills. In addition, during the 2007–2008 academic year, the writing center director, Kirsten Benson, led a study entitled "Perceptions and Performance of Research and Writing in English 102." Gathering survey and writing data, as well as teacher feedback, her study will produce comparative results from year two of the new curriculum.

While our research is ongoing and our assessment of the new curriculum is still in progress, some of our early findings are useful indicators of the general success of our curriculum revisions. In particular, results from surveys administered during the spring of 2006 and 2007 show high levels of student engagement, with a majority of students choosing their 102 section based on their interest in the topic.[10] In addition, preliminary findings from the Embodied Literacies Project suggest that students who take our courses are finding a road to transfer, carrying knowledge across different media and assignments (Year 1) and from course to course (Year 2). Although, to be sure, we have a great deal more work to do, we nonetheless find this data encouraging because it suggests the potential for fostering high road transfer in FYC. Thus, while some teachers and scholars associate teaching for transfer with the end of first-year composition, we see reason to remain committed to general writing courses. As Smit argues, novice writers "may need to be immersed in the discourses they need or want to learn as part of their own goals and ambitions" (159), and they "may need to be introduced to the critical frameworks necessary to understanding how groups function, so that they can develop a metacognitive sense of how writing functions in groups" (159). These types of activities take students and teachers alike far from the "generic" undergraduate research papers that Graff and others have criticized, and they leave behind the various "skill and drill" exercises and cookie-cutter, workbook-like assignments that are familiar to many of our students from high school language arts. And yet, as we are discovering, especially in our new English 102, when first-year writing involves genuine inquiry and research, and when it is supported by substantial rhetorical instruction, college writing has the potential to become a vastly different landscape.

As the teacher testimonials above illustrate, when course content varies, as it does from section to section of 102, students still gain the common ground of shared rhetorical concepts and writing experiences. Of course, this important benefit does not erase the problems that arise when writing courses have strong thematic elements, and both teachers and students have raised concerns about the potential for inquiry subjects to overshadow rhetoric and writing and thus limit the high road transfer of rhetorical knowledge and skills in 102. Perkins and Salomon note that "the occasional bridging question or reading carefully chosen to 'hug' a transfer target gets lost amid the overwhelming emphasis on subject matter-specific, topic-specific, fact-based questions and activities" (25). However, because the focus in 102 is inquiry—with goals emphasizing the investigation of a subject from multiple perspectives, methods, and methodologies—we encourage students to engage with subject matter self-reflectively. That is to say, we encourage them to explore content issues that take them outside the boundaries of the course, and we also ask them to be self-conscious about how and why they pursue the inquiries they choose. Such activities are "bridging" activities, which Perkins and Salomon describe as "broad-spectrum practices [capable of] reaching beyond subject matters" to establish knowledge of general, transferable principles and procedures (25). With the focus on course units that emphasize various methods of research and with the goal of engaging students in reading and writing a variety of different types of texts and writing for various situations, we hope that students leave English 101 and 102 better able to "take the high road" to transfer, and we hope they are able to continue developing their transfer abilities as they travel into new academic, professional, and personal rhetorical situations. Understanding that it is not only teachers who "get what we teach for," but also students, we firmly hope our students get it, and though conventional wisdom tells us "you can't take it with you," we hope in this case, they do.

Coda: Where We Are Now

At the conclusion of our 2008 program profile, we expressed our hope that first-year composition students at the University of Tennessee-Knoxville would "continue developing their transfer abilities as they travel into new academic, professional, and personal rhetorical situations." Little did we know that we would travel, too: Jenn to Marquette University and Mary Jo to the University of Kansas. While neither of us stepped immediately into WPA roles, our new institutional positions and the passage of time give us a unique perspective on the FYC sequence we designed, the ecology in which we designed it, and the complex system in which it continues to emerge.

From 2004 to 2007, we collaborated with program colleagues to invent and implement a transfer-focused curriculum we hoped would support a rich and sustainable teaching and learning ecology. Since that time, benchmarks of our success have included a 2011 CCCC Writing Program Certificate of Excellence award and ongoing, updated publication of the custom textbook Mary Jo launched in 2006. Of course, it is likely that there have also been challenges. As we discuss in "Taking It on the Road: Transferring Knowledge about Rhetoric and Writing across Curricula and Campuses" (2011), transfer curricula draw heavily on rhetoric and composition research, scholarship, and pedagogies, and as a result they can be difficult to maintain, especially in programs whose faculty have diverse disciplinary training and commitments.

Thus, while we consider FYC part of ecological systems based on individual campuses and set within local communities, the courses we developed affirmed the essential role disciplinary knowledge plays in writing instruction. This role is similarly reinforced by another contributor to this section, Elizabeth Wardle, whose curricular approach also explicitly connects FYC with rhetoric and composition knowledge. Distinct from the site-based, extra-programmatic forces that shape writing instruction, including the institutional staffing structures and state legislative mandates discussed by other contributors to this section, the research and scholarship we and Wardle imported into our programs is perhaps best understood as what ecologists term "introduced species" or "exotics": non-native species that we introduced with great care and cultivated for specific purposes. By placing priority on rhetorical knowledge, on multiple methods of inquiry, and on the movement from rhetorical analysis and reception to production, we hoped to build a sustainable curriculum, one that would invite successive classes of students to consider how different methods of inquiry offer writers distinct means of addressing rhetorical situations in dialogue—and even dispute—with different audiences.

Our desire for sustainability informed not only our turn to rhetoric and composition but also our focus on multiplicities, including multiple sources for "consumption" or reading, listening, viewing, and interacting; multiple methods for inquiry; and multiple media for "production" or writing writ large. If we found copious resources for expanding FYC "reading" lists to include everything from art and architecture to Zappos.com ads, we faced greater challenges—and greater rewards—diversifying methods of inquiry and production. Importantly, our own inquiries grounded our decisions. As WPA, Mary Jo conducted student and faculty surveys to establish context and priorities for revising the curriculum, and Jenn's leadership of the Embodied Literacies Project enabled us to document students' rhetorical knowledge transfer both within individual sections of the new English 101 and one year after students completed the course <http://el-utk.blogspot.com/>.

Looking back on the FYC courses we developed, what stands out to us now is not what has or has not proven sustainable but instead what is emergent. More than five years after the implementation of our two-course sequence, UTK is moving toward greater curricular standardization. Driven by many of the same motives Ronda Leathers Dively discusses with regard to FYC program developments at Southern Illinois University Carbondale, this choice is best charted "along a continuum for teachers support" that has complete autonomy at one end and, at the other, the professional mentorship and practical guidance provided by set readings and assignments. We also think about emergence in relation to our current institutions and the FYC programs we will be leading shortly. As we prepare for new roles and responsibilities, we pay particular attention to the patterns and evolving needs of current stakeholders, as Sara Webb-Sunderhaus and Stevens Amidon emphasize, and with Wardle's experiences in mind, we think about how not only staffing but also the credentialing of staff shapes our programs along with state mandates and changing interpretations of our institutions' missions. In deciding how best to respond, how to go with the flow of ever fluctuating ecologies, we hope our ongoing work can contribute to the collective endeavor signaled by this anthology: namely, replication of best program practices, aggregation of programs' top achievements, and production of new data to support the overall resilience of FYC.

NOTES

1. This article is based on a paper the authors presented at the 2006 National Council of Writing Program Administrator's Conference held July 13–16 in Chattanooga, TN. The session was entitled "Bringing Rhetoric Back on Board: Toward an Inquiry-Based FYC Curricula." Joining Jenn Fishman and Mary Jo Reiff were several instructors who piloted 102 co-presented and shared their experiences teaching the new curriculum: Jessica Abernathy, Casie Fedukovich, Bill Hardwig, Christopher Kilgore, Misty Krueger, and Catherine Phillips.

2. See faculty survey at http://compositionforum.com/issue/18/tennessee-appx.php#appx1.

3. See invitation memo at http://compositionforum.com/issue/18/tennessee-appx.php#appx2.

4. The Embodied Literacies Project began as a semester-long study of first-year writing designed to examine whether and how embodying literacy through oral and digital performance helps college students develop rhetorical strategies associated with transferability, especially in academic writing contexts. Together, co-principal investigators Jenn Fishman and Stacey Pigg developed and led the first phase of the project, which involved a team of co-teachers and researchers, including Miya Abbott, Devon Asdell, Bill Doyle, Casie Fedukovich, Nina Nell Haeckel, Jerod Hollyfield, Mary Jo Reiff, Hiie Saumaa, and Amanda Watkins. In the second year, Jenn Fishman and Mary Jo Reiff co-lead follow-up research, working with a sample from the original research subjects (N=204), and their preliminary results can be found online: <http://el-utk.blogspot.com>. See the syllabus for the pilot 101 project at http://compositionforum.com/issue/18/tennessee-appx.php#appx3.

5. See sample 102 syllabus at http://compositionforum.com/issue/18/tennessee-appx.php#appx4.

6. See Table of Contents at http://compositionforum.com/issue/18/tennessee-appx.php#appx5.

7. Along with our pilot teachers, colleagues from the university libraries and from different technology and computing centers on campus have helped us realize the curricula we designed. We owe particular thanks to Kawanna Bright and Kristin Bullard from Hodges Main Library; Nick Wyman from Hoskins Special Collections Library; Michelle Brannen and the staff of The Studio; and Chris Hodge and his Sunsite colleagues.

8. See CFP at: http://compositionforum.com/issue/18/tennessee-appx.php#appx6; a few sample descriptions of courses and research units/assignments included as part of the proposal can be found at: http://compositionforum.com/issue/18/tennessee-appx.php#appx7.

9. See website at http://web.utk.edu/~english/courses/f_102desc.shtml.

10. For a sample of the 2007 survey, see: http://compositionforum.com/issue/18/tennessee-appx.php#appx8.

Works Cited

Beaufort, Anne. *College Writing and Beyond: A New Framework for University Writing Instruction.* Logan, Utah: Utah State UP, 2007. Print.

Fishman, Jenn and Mary Jo Reiff. "Taking it on the Road: Transferring Knowledge about Rhetoric and Writing across Curricula and Campuses." *Composition Studies* 39.2 (2011): 121–44. Web.

Fulkerson, Richard. "Composition at the Turn of the Twenty-First Century." *College Composition and Communication* 56.4 (2005): 654–87. Print.

Graff, Gerald. *Clueless in Academe: How Schooling Obscures the Life of the Mind.* New Haven: Yale UP, 2004. Print.

Haddon, Mark. *The Curious Incident of the Dog in the Night-Time.* New York: Random House, 2003. Print.

Jewitt, Carey and Gunther Kress. *Multimodal Literacy.* New York: Lang, 2003. Print.

Kaufman, Moises. *The Laramie Project.* New York: Random House, 2001. Print.

Kress, Gunther. *Literacy in the New Media Age.* London: Routledge, 2003. Print.

Perkins, D.N. and Gavriel Salomon. "Teaching for Transfer." *Educational Leadership* 46.1 (Sept. 1988): 22–32. Print.

Satrapi, Marjane. *Persepolis: The Story of a Childhood.* New York: Random House, 2003. Print.

Selber, Stuart. *Multiliteracies for a Digital Age.* Carbondale: Southern Illinois UP, 2004. Print.

Smit, David. *The End of Composition Studies.* Carbondale: Southern Illinois UP, 2004. Print.

Wardle, Elizabeth. "Understanding Transfer from FYC: Preliminary Results of a Longitudinal Study." *Writing Program Administration* 31.1.2 (Fall/Winter 2007): 65–85. Print.

Wysocki, Anne Frances, Johndan Johnson-Eilola, Cynthia L. Selfe, and Geoffrey Sirc. *Writing New Media: Theory and Applications for Expanding the Teaching of Composition.* Logan: Utah State University Press, 2004. Print.

Yancey, Kathleen Blake. "Bowling Together: Developing, Distributing, and using the WPA Outcomes Statement and Making Cultural Change." In *The Outcomes Book: Debate and Consensus after the WPA Outcomes Statement.* Eds. Susanmarie Harrington, Keith Rhodes, Ruth Overman Fischer, and Rita Malenczyk. Logan: Utah State UP, 2005. Print.

—. "Made Not Only in Words: Composition in a New Key." *College Composition and Communication* 56.2 (Dec. 2004): 297–328. Print.

4 Intractable Writing Program Problems, *Kairos,* and Writing-about-Writing: A Profile of the University of Central Florida's First-Year Composition Program

Elizabeth Wardle

First-year composition programs, both large and small (but particularly large), have long faced a seemingly intractable set of problems. In my experience as both a writing program administrator (WPA) and writing program researcher, I have seen those problems play out this way: Macro-level knowledge and resolutions from the larger field of writing studies are frequently unable to inform the micro-level of individual composition classes, largely because of our field's infamous labor problems. In other words, composition curricula and programs often struggle to act out of the knowledge of the field—not because we don't know how to do so, but because we are often caught in a cycle of having to hire part-time instructors at the last minute for very little pay and asking those teachers (who often don't have degrees in rhetoric and composition) to begin teaching a course within a week or two. Often these courses are far larger than the class size suggested by NCTE, likely because of the high cost of lowering class size and of widespread misconceptions about what writing is (a "basic skill") and what writing classes do ("fix" writing problems). In addition, composition courses continue to be housed largely in English departments, where they tend to get the least attention and funding of all the low-funded English programs and where sometimes faculty with little interest in or training to teach writing are nonetheless required to do so. Sometimes entire composition programs are staffed with brand new graduate students, many if not most of whom are graduate students in fields other than rhetoric and composition, and who have taken, at most, one graduate course in how to teach writing before walking into a classroom.

Contrast this situation with any course taught in any other field, and the difference is stark: no administrator would ever send untrained faculty members or graduate students from another discipline to staff an entire segment of courses in, say, biology or history or mathematics or economics. Yet this happens every day in composition programs. Because of these and other entrenched practices, locations, and labor conditions,[1] and despite our field's advances in how best to teach writing, we can still find composition classrooms where the students are learning modes or grammar or literature in formalistic ways, or are learning popular culture with little to no attention to writing itself, in courses sometimes if not frequently taught by faculty or graduate students with little to no training (or even interest) in teaching writing.

The fact that research has suggested for many decades now that students in composition courses often do not reach desired course outcomes or improve as writers in measurable ways in one or two composition courses is not an unrelated problem. It seems reasonable to assume that if we staffed any set of courses in any discipline with teachers who had little training or interest in teaching them, we would likely see a problem in student achievement. Of course, lack of student achievement in writing courses can exist for a variety of reasons beyond labor and training that others have outlined quite comprehensively; for example, writing development is not linear or necessarily fast, tools used for measuring improvement are not necessarily reliable or valid, and so on. In addition, research clearly tells us that no single writing class can serve as a writing inoculation, and thus composition courses are more effective when they are seen as entry points to writing in the university, the beginning (not the end) of an ongoing rhetorical education. Such vertical and comprehensive writing structures do exist but are not the national norm; where they exist, it can be difficult, time-consuming, and expensive to soundly assess writing development longitudinally across and even outside of courses (see Wardle and Roozen).

The fact that composition courses often do not seem to achieve desired outcomes is made more complex because our field does not necessarily agree on what appropriate outcomes are or should be for first-year composition. Despite the valiant and important efforts of those who worked (and continue to work) on the WPA Outcomes Statement, beliefs about what outcomes should be for composition still seem to vary widely. Should composition courses help prepare students for what they will write later? If so, what counts as "later"? School settings? Which school settings? Work settings? Personal settings? If transferable knowledge and skills are not the desired outcome, then what do we focus on instead? Self awareness? Cultural awareness? Artistic and creative enjoyment of writing?

So we have a tangled set of longstanding problems and questions surrounding first-year composition: What should it do? How can it do that? How can well-prepared teachers be appropriately employed to undertake this work? And how can we assess our efforts there? The historical circumstances surrounding this set of questions have led to and perpetuated systems where writing program administrators often have no choice but to hire (sometimes at the last minute) teachers who know little about our field's research, to teach classes that are often too large, and to achieve outcomes that can vary wildly from classroom to classroom, program to program, state to state. In turn, these teachers are often treated as expendable, and institutions tend to invest very little in either their remuneration or professional development and advancement.

Many sorts of resolutions have been attempted to address this set of problems; eliminating first-year composition altogether has been the most dramatic solution posed. Most of the proposed solutions are less dramatic and more the responses of WPAs who face practical problems that need to be solved and reported to stakeholders tomorrow if not sooner. For example, because labor is unstable, some programs attempt to ensure programmatic consistency by giving part-time teachers and graduate students (some of whom teach even their first semester as MA students) program syllabi and specific and fairly rigid assignments to teach. Many programs make efforts to provide ongoing professional development for adjunct instructors and graduate students, but these supports are in constant tension with material conditions related to pay and time constraints, including the fact that such underpaid adjunct instructors are often teaching numerous sections at multiple institutions, leaving them little time to participate in the life of any one department.

While solutions like consistent program syllabi do result in similar assignments across sections, they do not necessarily resolve or even address the underlying reasons why such solutions are necessary in the first place: that many composition teachers in many composition programs have little to no background in writing studies or training in the teaching of writing beyond one TA training class, and that the labor conditions and pay for those teachers often preclude them from gaining further expertise and preclude programs from employing teachers who already have that expertise.

This set of problems can be paralyzing, preventing composition courses and programs from moving forward and acting on the knowledge of our field in both their curricula and their employment practices. How can we act on the knowledge of our field in our composition curricula, particularly when that knowledge suggests multiple paths forward, and when so many of those actually in composition classrooms are not necessarily familiar with any of it? How can we work against entrenched labor practices and material conditions in order to make changes? The problems appear to be unsolvable, yet we know that they must be solved. Until all composition teachers have relevant theoretical and research-based knowledge about writing and teaching writing, and are treated as expert professionals by their institutions, any attempts at programmatic consistency seem bound to be reductionist. In other words, until composition faculty themselves have enough knowledge about writing research and theory to make their own informed choices about curricula, and to make informed arguments for changed material conditions, how can we move beyond a managerial mode in composition programs? There may not be one model for teaching composition that we can agree on across the country, but models that are informed by research and theory and that have been created, assessed, and revised by expert teachers themselves rather than director/managers (and driven by textbooks) would feel like a big step forward from where we have historically found ourselves.

If teachers are passive recipients of curricula they didn't help shape and philosophies they don't share, it seems likely that they can only enact them in a formulaic fashion, if they enact them at all. Such formulaic teaching (which our legislative bodies seem intent on pushing us even further toward) simply reinscribes all of the problems I have been outlining above: composition teachers are not seen as professionals with specialized disciplinary knowledge, and stakeholders assume that anyone can teach composition; and, thus, anyone can be hired to do so at the last minute, since there must not be much to learn or prepare for in teaching a composition class. The teachers most willing to teach composition for $2,000/course and no benefits are often (but not always) least involved in the field's discussions about writing and writing pedagogy; in turn, the composition courses they teach may not be informed by the knowledge of the field, and students are then less likely to achieve desired course outcomes, all of which set composition courses and programs up to be viewed as anything but academic or scholarly. And the cycle continues.

I have outlined this set of problems, as I understand them, not to be discouraging or overly pessimistic, but because I have researched writing programs and served in some administrative capacity in those programs since I was a graduate student. At three different institutions, public and private, in varying roles, I have found the very particular problem of how to inform micro-level classroom practices with macro-level disciplinary knowledge to be centrally important to our field's development and our students' learning—and singularly difficult to overcome. In this program profile, I will talk about the ways we have worked (and are still working) to overcome this problem at the University of Central Florida, and how we have succeeded in limited and often surprising ways, surprising most of all to me, given my general cynicism about the problems associated with first-year composition (see "Mutt Genres" and "Can Cross-Disciplinary Links," for example).

"Writing-About-Writing" Claims for Addressing These Problems

Doug Downs and I have been among the growing number of compositionists who are arguing that composition courses need to directly embrace and enact some of the research and theory about writing by:

1. Teaching students about writing in ways that can enable them to be more successful later, and
2. Explicitly and publicly making the case that composition courses can only serve as entry points to writing in the university and the larger world and cannot serve as inoculations.

This view of composition has led to an approach that has come to be known as "writing about writing" (or "WAW"). Writing about writing pedagogies vary widely, but their basic philosophy is that composition courses should teach both declarative and procedural knowledge about writing (declarative knowledge, for example, might include considering how writing works, how revision happens, how communities influence genres and conventions, and so on; procedural knowledge might include knowing how to revise for audience concerns or how to effectively use a semi-colon) and place a strong emphasis on metacognition (Charlton "Forgetting" and "Seeing"; Robertson; Taczak).[2]

Writing about writing advocates argue that this approach to teaching composition is not only theoretically sound, but that it also brings into clear and unavoidable focus the labor issues long associated with composition, thus potentially forcing a resolution to them. The arguments for such an approach go something like this:

- To teach a writing class informed by writing studies research, teachers must be or become familiar with relevant research in composition studies and then enact this knowledge in their classrooms.
- In gaining and enacting this expertise, those teachers enact a professional identity with disciplinary standing, which in turn might raise their status and the status of composition courses and programs themselves.
- If teachers must know the research of the field in order to teach composition classes, large groups of adjuncts can't be hired at the last minute and treated as expendable; rather, potential teachers must have some training (whether formal or informal) in rhetoric and composition.
- When composition teachers have this sort of disciplinary knowledge, they can teach to informed outcomes without being forced to a prescriptive and reductionist consistency, and they can be engaged and rewarded as expert colleagues, rather than "labor" to be "managed."
- This *should* result in better achievement of student outcomes related to writing. And better student outcomes with professionalized teachers *should* raise the status of composition courses and programs themselves.

In sum, teaching declarative concepts about writing requires *knowing* declarative concepts about writing, which requires some familiarity with the research of writing studies. There are two ways to assemble a faculty with such familiarity: hire all rhet/comp specialists (an expensive and difficult proposition) or implement sustained, scaffolded support for composition teachers from all backgrounds so that they can gain familiarity with some composition research. Doing the latter requires changing some material conditions for teachers, so that they have time to engage in professional development and are rewarded appropriately for doing so. Debra Dew has written about her successful attempts in this regard at University of Colorado, Colorado Springs, explaining their shift from a "writing with no content in particular (W-WNCP)" model to a "writing with specific content (W-WCS)" model that made their ENGL 131 class a content course with "rhetoric and writing studies . . . as content matter" and "reconstituted their instructional labor by aligning ENGL 131 with other content courses across the disciplines" (88). In this profile I will outline our own attempts to make such changes at the University of Central Florida (UCF).

At UCF we encountered and were able to take advantage of a kairotic moment to use a writing-about-writing approach to address the set of problems I outlined earlier. Our experience demonstrates how a programmatic writing-about-writing approach with timed implementation and training improved professionalization, informed micro-level classrooms with macro-level disciplinary knowledge, and, through both of these, improved student outcomes. For these changes to occur, particular institutional supports had to be in place, and an advocate in upper administration needed to serve as the catalyst to ensure the attempted changes came to fruition. Our experience at UCF demonstrates how deep cultural shifts and changed material conditions can be effected through a combination of kairos, piloting and assessment, advocacy, and laying bare our practices so that they are visible to stakeholders.

Our Institutional Context

The University of Central Florida (UCF) is the second largest university in the country, serving a diverse population of more than 45,000 undergraduate students and 60,000 students overall. Because a large number of our students are transfers from local community colleges, we serve only about 4,000 first-time-in-college (FTIC) students in composition each semester. When I was hired as the composition director in Fall 2008, the composition program had problems fairly endemic to most large programs, including a large number of adjunct instructors, large class sizes, and no recent incarnation of a professional development program for teachers. However, I was hired in with tenure, a positive sign that the nature of WPA work was understood as labor intensive and potentially politically fraught.

Housed in the Department of English where composition had not been a priority, the composition program had seen three different tenured or tenure-track directors over a ten-year period. These directors worked with one or two instructor-line (permanent, non-tenure track) composition coordinators and an administrative assistant. The writing program employed eighty teachers each semester to teach around 140 sections. Composition sections were capped at twenty-seven students, and the composition teaching faculty consisted of approximately 15 percent permanent, non-tenure-line instructors, 19 percent visiting instructors (full-time, limited contract non-renewable after four years), 19 percent graduate student instructors (mostly MA students), and 47 percent adjunct instructors (part-time, one-semester contract with no benefits). When tenure-line faculty taught composition, it was usually honors composition (to fulfill a longstanding and complicated salary debt that the English department owed to the Honors College).

ENC 1101 (part of a required two-semester sequence for all Florida public college and university students) at UCF had been taught for many years as a process-focused and assignment-based course, with a focus on revision and developing critical reading and writing skills. The four papers typically taught in the course were a memoir, a political or social commentary, a review, and a public argument. Students in this curriculum read models of the texts they were being asked to write.

The composition program had long experienced a chronic lack of resources for compensation, professional development, and hiring. The overwork of the directors and coordinators, along with the reliance on a large and ever-changing adjunct labor pool, had a predictable impact on how the curriculum came to be implemented. Although the curriculum was designed by rhet/comp faculty to focus on outcomes, in practice over time, the daily focus came to be more on teaching the four papers as rigid modes for which students supplied the written content. To be clear, this emphasis came about not because of a lack of awareness on the part of the directors, but because of the working conditions of the many teachers who moved in and out of the program each semester (adjunct instructors who might be teaching up to 8 classes a semester at three different schools). The composition coordinators told me they spent most of their training time helping new teachers teach the specific papers (and also responding to the sorts of crises that arise when so many people are teaching a course for so little compensation and with no guarantee of future employment). Teachers were provided with descriptions of the four major paper assignments and given rubrics with which to grade them. The coordinators reported that as the years progressed, a great deal of the program administrators' time came to be spent helping teachers follow guidelines in order to try to attain some programmatic consistency in the face of few resources and a large, ever-changing staff.

When I came to UCF in fall 2008, I asked for volunteers willing to try a writing-about-writing curriculum. I hoped that engaging some interested teachers in such an approach would open a door to shared reading and discussion and consideration of alternatives to the "core" papers that had become so ingrained in the program. Several teachers took up the challenge in spring 2009. This was a very loose pilot staffed by willing teachers who "figured it out" as they went. The curriculum they tried was based on the belief that writers need both declarative and procedural knowledge about writing, including a deep understanding of writing-related concepts such as rhetorical situation and genre. For example, they talked with students about discourse communities and their impact on how writing comes to be and is understood, and about writing processes and how they differ across situations and genres. The approach did not specify genres or "modes" that must be written, or even particular assignments, but rather focused on declarative course outcomes and encouraged teachers to draw on their own expertise to teach toward those outcomes. In this very early stage, the piloting consisted of a lot of informal conversation, suggested reading, and syllabus sharing. The teachers who "tried it out" that first year looked at my syllabi and built on those and then talked with each other. My primary goal at that time was simply to engage teaches who were looking for a change and begin to encourage and facilitate a way for them to become familiar with rhetoric and composition scholarship.

KAIROS

During my first year at UCF, while this initial loose pilot was going on, our vice provost and dean of undergraduate studies, at the request of the UCF president, turned her attention to composition, the writing center, and the college algebra course as sites that might make demonstrable differences in undergraduate student learning outcomes. In her position, she was charged with helping UCF meet one of its strategic goals, to provide the best undergraduate education. Although she had always had this charge, a new tuition increase provided some resources with which to innovate. Because the state of Florida had created specific guidelines about how the money ("tuition differential") should be used (primarily for undergraduate education and advising), composition and algebra were natural sites for early innovation. The dean of undergraduate studies' role as an advocate for changes in the institutional structures around the writing program (as well as the math program, which is a story in itself) cannot be overestimated. Knowing that some new funding was going to be available through a tuition increase, she made a proposal to the president for reducing composition class size from twenty-seven to twenty-five and conducting a three-year study of smaller class size, providing comparison groups of nineteen. She also argued for six new full-time instructor positions, four in 2009–2010 and two more in 2010–2011. The president agreed to what she proposed, launching the President's Class Size Initiative (PCSI), with the understanding that everything we did would be audited, assessed, and presented to stakeholders at any time. Our dean of undergraduate studies understood how funding worked, what funding might be available, and she had access to one of the few stakeholders who could effect structural change immediately. Without her advocacy for improved writing instruction, and without the interest in and support of change by our president, the opportunity to change the writing structures might not have come—at least not so dramatically or suddenly.

The vice provost's actions initiated a moment, *kairos,* what Carolyn Miller has described as a critical occasion for decision or actions. Miller notes that, although "[t]he ancient Greek term is most often translated in temporal terms, as 'the right time' or 'timeliness,'" it can also be understood as a "spatial metaphor, that of a critical opening. The earliest Greek uses of the term, in both archery and weaving, referred to a 'penetrable opening, an aperture,' through which an arrow or a shuttle must pass (Onians, 1951/1973, 345)" (84). Eric Charles White shares this view, understanding *kairos* as "a passing instant when an opening appears which must be driven through with force if success is to be achieved" (13). Rhetoricians like Miller note that these "passing instants" can be both constructed and discovered (Miller 84), and *kairos* can be understood as either continuous (linked to a history wherein a rhetor can always create an opportunity) or as discontinuous and thus appearing as revolutionary, breaking with the past (Miller 84). Our vice provost's intervention illustrates how *kairos* can be both continuous and discontinuous simultaneously; her efforts built on years of efforts by writing faculty at UCF, but also created an opportunity to make headway where previous efforts had stalled out. Previous arguments had been made that relying on part-time faculty was unethical and full-time faculty should be hired for that reason; this argument, despite its merit, was not effective. Previous committees had been formed to outline and push for best practices, few of which were supported outside the department. Yet previous victories had also been won: the college of arts and humanities dean had previously held off an attempt to raise composition class sizes beyond twenty-seven. Thus, efforts to improve the material conditions of the first-year composition program had been continuing and ongoing, though not widely successful. New efforts could not necessarily build on what had previously been attempted unsuccessfully, but they certainly continued those efforts.

Our opportunity with the dean of undergraduate studies and the tuition increase created a kairotic opportunity but did not define it or ensure that it would be acted upon fruitfully. Instead, her intervention helped create a moment in which our writing program could test claims made for writing-about-writing with the possibility of changing the constellation of problems facing our composition program; it was a moment when we had an advocate and our president's attention. The president wanted to make changes that impacted student outcomes, so it seemed that if we could show results of some kind, the president would be listening, and open to action.

The first unofficial mini pilot (spring 2009) had engaged some teachers and garnered interest with other teachers. The composition coordinator (with a degree in creative writing) had piloted a writing-about-writing approach with what she felt were positive results—students were engaged, the curriculum was challenging, and she felt a direct focus on writing enabled her to act out of her own expertise in ways the previous curriculum had not. Given these results, we were ready to hire the first four presidential instructors to be the official guinea pigs for the changes we wanted to investigate, starting in fall 2009. We hired the smartest, most enthusiastic, most flexible people we could find, none of whom had rhet/comp degrees. Two were recent MFA graduates, one was a regular adjunct with a literature MA, and the fourth was ABD in literature. The current and previous composition coordinators (the latter of whom had a graduate degree in library science) agreed to *officially* pilot with this group in fall 2009. Their involvement was central to the success of the official pilot, since they had longstanding credibility with the teachers that I, as a new hire, did not. So we all took a deep breath and ran into the opening with force, as White suggests must be done for *kairos* to result in change.

Preparing, Piloting, & Assessing

We had hired smart, enthusiastic, willing, good teachers without graduate training in rhetoric and composition. Here was an opportunity to find a way around what Doug Downs and I called "the elephant in the room" in our 2007 *CCC* article about describing our writing-about-writing approach. There we had argued that only faculty with rhet/comp training could teach a writing-focused composition course. Yet, here as a WPA of a large program, I was faced with the impossibility of such staffing conditions. If we wanted to move forward with a writing-about-writing pilot, we had to find another way. So what to do?

As we prepared the pilot curriculum for ENC 1101, I pulled together some seminal articles that provided a relevant rationale for a WAW approach and gave the teachers a very rough draft of the *Writing about Writing* textbook that Doug and I were at that time in the process of writing. That summer the new teachers plus the current and previous composition coordinator read all the materials, and we together discussed outcomes and possible units and assignments. The coordinator and I brainstormed some sample assignments and took others from the developing textbook, but we encouraged the teachers to design their own units and assignments according to their expertise and instincts.[3] The units these teachers ended up teaching addressed concepts such as how texts are constructed, how writers write, how communities shape writing, and so on. Assignments included genre analysis, rhetorical analysis, autoethnography, writing memoir, discourse community ethnography, and analysis of science accommodation.[4] The declarative concepts covered were those that seemed relevant to all of us and also teachable for people from different backgrounds. It seemed to us that the genres assigned should help achieve the goals of the assignments and help students learn the declarative concepts under consideration. For example, as students studied their own writing processes, they might write about what they learned in the form of a research report, or they might write about it somewhat reflectively. Both types of texts could help students learn the declarative concepts about writing under consideration. The teachers met regularly to plan and share goals and sometimes chose to teach similar assignments. However, no assignments or genres were prescribed, and teachers made changes to plans and assignments as the semester progressed.

These teachers, in addition to piloting a WAW approach, were also the President's Class Size Initiative (PCSI) study guinea pigs, each teaching two sections of nineteen and two of twenty-five (all four sections were taught as Writing about Writing). They were observed frequently, gave their students pre and post surveys, and turned in all of their students' portfolios at the end of the semester. We gathered all sorts of data from their students, including grades. These teachers read extensively, tried new things, critiqued their own practices, and took suggestions willingly. They laid bare their teaching practices for the possibility of structural improvement, and their willingness to do so can also (like the dean of undergraduate studies' advocacy) not be overestimated. Had they been less flexible, less open, or less willing to change practices, our experiment could have ended as quickly as it began.

At the same time, the composition coordinator agreed to also train the new Teaching Assistants to pilot the WAW curriculum, all of whom had previously taken ENC 5705: Composition Theory and Pedagogy.[5] The new resources being invested in the program allowed us to provide much more professional development for all teachers in the program than was previously possible. Instead of providing teachers with a common rubric, for example, we sponsored workshops on responding to writing and grading, in order to educate teachers in the research of rhetoric and composition so that they were empowered to create their own grading rubrics (or not). Regardless of whether we ultimately moved to a writing-about-writing curriculum, I wanted to start to move toward a program in which all teachers had the expertise to participate in decision-making about curriculum and assessment. For that to happen, we had to start making cultural shifts, such as providing teachers direct access to knowledge and resources in the field; setting an expectation that conversations would be about research, theory, and innovation rather than complaints about students; and demonstrating that questions about pedagogy would always come back to program outcomes (see Appendix) rather than to rigid discussions regarding how exactly to teach a specific mode. These changes happened in stages and primarily through peer example. As the pilot teachers became engaged in research and theory, and as they tried new ideas, they shared their experiences with their colleagues, some of whom then became interested in reading and exploring new ideas themselves.

This was a stressful but rewarding year. As a result of their experience reading and implementing the scholarship of the field, the new instructors engaged in careful reflective practice, adjusting their assignments and activities as they progressed, and they indicated that they found the curriculum and the reading rewarding. For example, they enjoyed talking about ideas and research with one another, rather than relying on a textbook or a set curriculum to show them what their class would look like. They indicated that the new curriculum challenged their students, and thus made their own experiences in the classroom more engaging.[6] Yet the changes in the pilot were making ripples in the larger program; while some teachers became interested in new ideas, others became concerned about the possible changes in curriculum, culture, and criteria for evaluation. Some teachers had been teaching the same way for many years, with little oversight and little expectation that they would be involved in a community of teachers/scholars. Some were uncomfortable at what they perceived to be higher expectations and more scrutiny of classroom practices. For example, there had long been a cultural norm in the program that teachers did not share assignments with one another because they were concerned that doing so would encourage similar assignments and increase plagiarism. The developing new expectation that teachers would talk about and share ideas (and that they could do so without creating the same assignment) challenged many parts of this long-standing practice (including the apparent belief that preventing plagiarism is the most important aspect of assignment creation or that students would surely plagiarize if given the opportunity). In addition, teachers had not made a practice of observing one another's classrooms, and we were encouraging this as a basic way to get feedback on practices and to learn new ideas from colleagues. Some teachers saw their classroom as their domain and did not welcome the prospect of having their colleagues ask to attend. We were asking teachers to read research, assess their practices, share their ideas with others, and be observed—all new expectations. We were expecting that professional teachers would want to engage in ongoing professional development. And we were doing this with very little change in material conditions, with only the hope that we would be able to change material conditions down the road. Some teachers were invigorated by these changes; others were unhappy enough about them to leave our program.[7]

When the pilot began, we immediately began assessing student work. Portfolio assessment had been conducted before, but not in recent years. We instituted portfolio assessment in fall 2009 to learn how the program was achieving outcomes, overall, and how the pilot sections were achieving outcomes as compared to the "regular" sections. We created a portfolio assessment team of full-time, part-time, and graduate student teachers, and we trained them to assess portfolios. We read sample portfolios together and created a rubric based on our current program outcomes, those shared across both pilot and existing curricula. Two raters read each portfolio independently, and a third rater read if the first two disagreed by more than one point. Portfolio assessment of the first pilot was encouraging.[8] On every item measured, the WAW sections of nineteen had higher mean scores than the other groups, with the WAW sections of twenty-five scoring second highest on most measures. The findings suggested that the WAW curriculum (regardless of class size) had a positive and significantly different impact than the traditional curriculum on teaching students about writing as a process, rhetorical analysis, and integration of outside sources. On two other measures, higher order thinking and correct in-text citation, class size may have either played more of a role than the WAW curriculum or enabled the WAW curriculum to have a greater impact.

Taking It to the President

In March 2010 (not even a full year into the new initiatives), the dean of undergraduate studies asked me to present the portfolio assessment findings to the president and his advisory board. Again, her assertive advocacy for improved writing instruction enabled access to decision-makers we may not have had otherwise.

Although we were undertaking a three-year study, this sort of access to powerful stakeholders was unusual and unlikely to happen again. In the presentation I began by focusing broadly on the composition course itself—what it is, why it exists, and why it is not sufficient to meet the needs of students. I did this in order to try to situate our current pilot efforts into a larger context and to emphasize that, no matter how successful our composition pilot might be, the composition course by itself would never be enough to "improve student writing" in the ways that stakeholders desire. Toward this end, I described how rhetorical training historically evolved into the composition course and why this one class was not sufficient to meet the needs of students. I suggested some reasons why Harvard's nineteenth century composition design was ineffective and how that design became institutionalized nationally despite this lack of effectiveness. But, I noted, the model had started to change at UCF, thanks to the president's funding, which enabled the work of our new instructors and led to the results of our portfolio assessment. I argued that the key to meaningful change in writing instruction, based on early results, seemed to be smaller class sizes, plus new curriculum, plus well trained and supported teachers. However, these changes, while important, would not be sufficient because one or two writing courses will never be enough to show measurable change in student writing. To provide a truly distinctive and sound writing education for students, UCF as an institution needed to embrace a comprehensive, vertical model of writing education, with rhetorical training across all four years. Our view of our new writing-focused approach to first-year composition was that it would provide a useful entry point into writing in the university, which other writing courses across the university could build on. In creating such a model, UCF would be at the forefront of national best practices. At the end of the talk, the advisory panel asked numerous questions and brainstormed strategies for more effective writing structures.

Within two months, funding had been set aside to convert all of our adjunct lines to permanent instructor lines (eighteen instructor lines over three years), fully fund our long-ignored writing center and hire a tenured writing center director, and begin and fund a new writing across the curriculum program with a tenured director. The decision was made to invest in these writing initiatives outside of English, in a new department of writing and rhetoric that would also include undergraduate and graduate writing degrees. This new department and its initiatives came into existence on July 1, 2010. I cannot speak to the reasons why we were moved out of English. I can only speak to the material changes that such a move effected.

Fully Implementing the WAW Curriculum

In the 2010 to 2011 year, we began fully moving to a writing-about-writing curriculum across our program. Teachers who had already piloted the curriculum led small reading and discussion groups with willing faculty,[9] and we provided small stipends for teachers who completed this training. Because we were hiring full-time instructors as a result of the president's investment, there was fairly widespread motivation for people to participate in this training. By the end of the training, they had begun to establish some familiarity with writing studies research, had seen examples of the WAW curriculum, and had created their own syllabus.

By the 2011 to 2012 school year, the program looked very different than it had only a few years before. We employed only a few adjunct (part-time) and visiting (full-time but limited term) instructors. Most of our teaching staff were permanent instructors, tenure-line faculty in rhetoric and composition, or carefully trained graduate students (second-year MA students or PhD students, all of whom completed a graduate course in composition theory and pedagogy before being awarded an assistantship). By this point, we were hosting six to eight workshops or reading groups each semester, many led by instructors and graduate students.[10] Portfolio assessment had become a normal part of our yearly activities and attracted a large number of volunteers (12–14), many of whom said they saw portfolio assessment as a way to get good ideas from other teachers, rather than an onerous rating task. Instructors had begun to attend and present at national conferences. One instructor was the founding editor of our in-house, peer-reviewed journal of first-year writing, while another edited an in-house quarterly newsletter. We had begun hosting a large annual celebration of the best student writing from composition, which our president and provost attended. UCF president John Hitt read all the student pieces published in our journal and began awarding a book scholarship to the best piece.[11]

All of these changes meant that, when the time came to begin looking at the second of our composition courses, ENC 1102, we could engage in this work very differently than we historically engaged in curricular change. ENC 1101 had been our primary writing-about-writing course, teaching what we considered to be foundational declarative concepts about writing. ENC 1102 was historically a research class. Although most of our teachers had begun adapting their 1102 practices in light of what they were doing in 1101, we had not undertaken a formal examination of 1102.

The 1102 revision committee convened early in Fall 2011 and consisted of instructors, visiting instructors, tenured faculty, and tenure track faculty. The discussions were both theoretical and practical, and wide-ranging. Every person on the committee shared ideas, submitted articles for thought and discussion, invented new syllabus ideas, and pushed us in new directions. We discussed several difficult questions, including which kinds of declarative writing knowledge students needed to work on further after 1101 and what the implications of genre theory are for designing and teaching assignments. In spring 2012 we facilitated a series of lectures on rhetorical theory and genre theory, coupled with reading groups and hands-on workshops, so that teachers across the department could further explore what 1102 can do. At the end of the year, we had generally agreed on principles and outcomes for 1102 (see Appendix) but had not agreed on what specific practices should be in the 1102 class. As a result, we decided to solicit pilot proposals for fall 2012. In December 2012 teachers who have piloted new approaches in 1102 will share them at an open discussion. We are not in a hurry to make radical changes, but instead feel that the reading, discussing, and piloting are themselves important parts of improving our practices and ensuring that we continue dialogue about research and theory across classrooms. The discussion and planning can be open and collaborative in this way instead of managed from the top down (Strickland) because of the changed material conditions of our teachers. Our teaching faculty members have permanent jobs with security and benefits; they know they'll be here next semester to take up the discussion where we left off and to help pilot any initiatives that make it to the next stage. They also have expertise in writing studies because they have been reading the research and teaching it to their own students.

Current State of the Program

The material, intellectual, and teaching conditions of our lives have changed dramatically over the past few years. In fall 2012 we employed only seven adjunct instructors to teach composition (three of whom have full-time jobs but like teaching a night class for us, and two of whom are new MA graduates who are preparing for national job searches or doctoral programs), down from the thirty-three adjuncts we employed in fall 2008. We have hired fifteen new, full-time permanent, benefitted and unionized instructors. We also hired tenured writing center and WAC directors, both of whom participate actively in the activities of our first-year composition program. So far we have hired four tenure-track assistant professors, all of whom teach composition each year and participate in various aspects of maintaining the composition program (including course revision and assessment), in addition to their duties with the upper-level and MA programs. This gives us a composition program with courses taught by ten tenured or tenure track rhet/comp faculty; twenty-nine full-time, benefited instructors or visiting instructors; seven adjunct instructors; and twelve or so graduate teaching assistants (none of whom teach until they have eighteen hours of relevant graduate credit). Many of the instructors and visiting instructors received small equity raises when we became an independent department.

Our program also feels very different than it did in 2008, in ways that are much harder to quantify. The conversations in the hallways and copy room are about ideas, assignments, and texts recently read. When teachers come to see me, it is usually to discuss and get feedback about new ideas. Most telling is what teachers are **not** coming to see me about: they don't come by to complain about students or working conditions or their colleagues. I receive very few grade appeals or teacher complaints of any kind. We have new problems: there are more teachers who have ideas for leading workshops than there are workshop slots to be filled; there is no time in the day when everyone's schedule is free to attend a workshop or reading group, and people don't like being left out; there are more people who want to pilot new technologies or attend conferences than there is money. These are much better problems to have.

It is not hyperbolic to argue that reading and teaching the knowledge of the field changed the culture of our composition program. However, reading composition theory and research did not *by itself* magically bring about these changes. This was not a "moment of miracle," as Miller argues that *kairos* understood as discontinuous can sometimes appear to be (83). Rather, this was a result of continuous work resulting in cyclical, positive, and self-perpetuating changes. Teachers who read and enacted composition theory and research changed their classrooms, and the measurable results of *that* change, in turn, quite literally changed the teachers' professional positioning and our entire program by helping us convince our administration to essentially eliminate contingent labor. Having full-time permanent jobs and being treated as members of the field changed the material conditions and intellectual orientations of our teachers even further. The composition program now feels like—and *is*—a professionalized group of expert writing teachers and scholars instead of what it had always previously felt like, Harry Potter's cupboard under the stairs.

The Lessons

I think there are at least two, but likely three, lessons to be learned from our experiences.

First, our experience illustrates that sometimes there are moments when change is more possible than usual, and as rhetoricians and writing program administrators, we can and must be prepared to take advantage of them. We might fail, and the passing opening might close. But it is possible to leverage our field's knowledge and narrative to work with our good teaching faculty and make changes. Often our field's narratives about composition programs are about the forces at work that keep change from happening. But change is possible, and structures are created, destroyed, and recreated by human beings. And we, as rhetoricians, can engage with our stakeholders in ways that can effect structural change. Here, our attempts to engage in positive ways usually entail starting a conversation by talking about exciting ideas for change and considering ways these new initiatives can help achieve current university goals or solve problems shared across programs, departments, and colleges.

Second, our experience also illustrates the necessity of advocates. Without advocates in positions to bridge or broker boundaries, we may never know about potential resources or make our way in front of the presidents and provosts and deans who have the power to enable structural change. Cultivating relationships with potential bridge builders and boundary brokers seems to be an essential part of our role as writing program administrators. One way to cultivate such advocates is to consider administrators as partners rather than adversaries, and to consider what solutions we might be able to offer to problems they are facing from their stakeholders (legislators and parents, for example). By drawing on our own expertise to help find solutions to shared problems, rather than always playing the role of supplicant and asking for resources, we are more likely to be seen as colleagues who are wanted at the table and to be included in conversations.

Finally, I think our experience illustrates that welcoming our part-time faculty into our field—and engaging with them around the knowledge of our field—is a possible way to break the logjam I described when I began this program profile. Reading and teaching the knowledge of the field can change the culture of a composition program, affect its measurable outcomes, and impact the ways that writing teachers view themselves and the ways they are viewed (and compensated) by others.

Would we have done anything differently in retrospect? The factors we were able to control—piloting, assessment, implementation over time to facilitate buy in—all worked well. Several of the factors we could not control made life difficult at times: the fast pace at which we were asked to pilot, assess, and present was stressful; the suddenness of our separation from the English department, and the fact that we received resources for new programs at a time when resources were limited, strained some relationships. Despite the difficulties of being in the spotlight for a sudden and high stakes change, the benefits for students and teachers continue to outweigh the drawbacks. Our students are better served now than they were four years ago, and our teachers have a better quality of life, both materially and intellectually.

More important than what we might have done differently is what we must continue to do in order to protect our program and ensure that it continues to thrive. Even as our own institution has provided increasing support for writing instruction, our work is threatened at the state level by a governor and legislature who are unfriendly to a liberal arts education and to a meaningful general education component. The general education requirement in Florida has recently been cut by six hours, the consequences of which are still to be determined for writing courses. Our university saw a budget cut of $52 million this year alone, with more cuts expected. Despite rising numbers of enrolled students, actual credit hour production is flat (likely because more students are forced to attend part time). Due to increasing emphasis on helping students "test out" of composition and other gen ed requirements, enrollments in our composition courses are decreasing. All of these changes, combined with our governor's tendency to look to Texas for models of how to influence higher education, suggest that we have many battles ahead as we fight for the ability to provide an excellent, research-based writing education to our students.

The experiences of the past few years have reminded me repeatedly that we are a discipline with a body of knowledge. At UCF we have learned that when we can find ways to inform micro-level classroom practices with this macro-level disciplinary knowledge, we can change the structures that have for so long controlled us and our programs to the detriment of our students and teachers. As rhetoricians, we do have the tools for empowerment. With the help of our advocates and stakeholders, we can pick up those tools and use them to construct, discover, and respond to kairotic openings.[12]

Coda: Where We Are Now

Since my program profile was published, I find that two issues have arisen—or continued to arise. The first is the issue of expertise, and the second is the issue of external forces that exert strong influences on our programs at unexpected times. Both of these issues demonstrate the ecological nature of writing programs: they are complex and interconnected not just internally but externally; in addition, no matter how visible and prioritized a writing program may be at any time, its identity and resources are always fluctuating. Changes are always occurring, whether we seek them out or not, and in order for writing programs to survive and thrive, our responses to changes must be flexible and are always negotiated in a complex relationship with others, both those within and outside of the writing programs.

Expertise

While our efforts to invite teachers without rhet/comp training into a conversation about our field's research and to convert adjunct lines into instructor lines have meant we gained a cadre of enthusiastic teachers willing to read and teach the research of the field, these instructors are not on tenure-track lines. They are permanent employees, unionized and benefitted, certainly, but they make about half the starting salary that a new tenure-track faculty member makes, they have a heavy teaching and service load, and research is not part of their annual assignment. Some of them assist tenured program directors in leading composition, WAC, and writing center initiatives, but they are not expected (or allowed) to lead these programs themselves. The university recently implemented a promotion track for instructors, enabling them to apply for associate and then full instructor status after a number of years of service. But it's clear there is a two-tier system in place and that such a system has real limitations.

Our department offers a number of upper-level and graduate writing courses. What "credential" does a faculty member need to teach these? What constitutes "expertise"? Certainly, a graduate degree in rhet/comp is one form of credentialing, but not all the tenure-line rhet/comp faculty members are qualified to teach the variety of specialized courses we have created. How do we encourage our instructors to take a central role in the life of our programs and to teach a variety of courses if they are qualified to do so (and even to conduct research if they are so inclined) when the two-tiered university structure does not afford these opportunities and when there is no clear way to determine what qualifies a person to teach various classes?

At UCF we were able to eliminate our reliance on adjunct labor, but many questions about reliance on instructors—even when instructors don't *replace* tenure lines—still loom large, not just in rhet/comp departments but in institutions of higher learning nationally. Ours is not the only independent writing department that will have to find answers to these questions over the next few years.

External Factors

I ended the program profile by describing recent legislative attacks and other external factors that could seriously undermine our program's advances. None of those issues has been resolved; they simply continue to fluctuate and change, sometimes within weeks or months. Watching them continue to play out has reinforced the need of our faculty and administrators to be aware of threats to a sound writing education; our ability to function effectively as a writing program is dependent upon the decisions and priorities of stakeholders who are unlikely to recognize how closely our jobs and their decisions are interrelated. The decision to cut six hours from general education was reversed earlier this year though what administrators called a "glitch bill"; had this cut not reversed, we likely would have lost one of the two courses in our composition sequence and thus potentially have needed to lay off up to ten instructors. Although our program made it unscathed through this potential change, recent budget cuts and legislative interference have made it plain that there will be continued attempts to control higher education, and the next attack is already rounding the corner. Furthermore, proposed cuts, such as to Florida's state-funded scholarship program, Bright Futures, will likely mean that fewer low-income and minority students will be able to attend college, and those who are will need to take as many courses as possible at schools with the lowest tuition rates. This economic pressure will likely mean that fewer and fewer students take composition at UCF, while more of them take composition at local community colleges that rely on both a curriculum and a labor structure very much like the ones we worked so hard to eliminate. Writing program ecologies are interconnected. The budget cuts continue to have other unintended but certainly interrelated consequences. As administrators have looked for additional ways to raise funds, they are turning to international students.

One recently proposed plan would entail anywhere from 400 to 1,200 international students per year gaining admission to UCF—and taking composition. This population would dramatically change our writing program's staffing as well as the training that our staff would need. It seems we are trading the potential for cuts and layoffs for the potential for astronomical growth and an entirely new population of students, students for whom our curriculum may not be appropriate in its current form.

Our experiences in the writing program at UCF, then, demonstrate how deeply ecological writing programs are: connected with and dependent on many other variables, requiring constant change and adaptability, resulting in a program whose identity is always emerging and never static.

Appendix

Current program outcomes for ENC 1101 and 1102

Outcomes Statement for 1101

1. Students should demonstrate an understanding of writing processes and how writing processes change depending on the writing contexts by . . .

- using acquired vocabulary for talking about writing processes and themselves as writers, including terms like incubation, recursiveness, and revision.
- creating drafts that demonstrate substantial and successful revision.
- responding to substantive issues raised by instructor and peer feedback.

2. Students should demonstrate an awareness of rhetorical situation and acquire strategies for writing in different writing contexts by . . .

- using acquired vocabulary for talking about rhetorical situations, including terms like
- audience, exigence, and constraints.
- employing style, tone, and conventions appropriate to the demands of at least one particular genre and situation.

- demonstrating the ability to write for different audiences and contexts, both inside and outside the university.
- articulating and assessing the effects of theirwriting choices.

3. Students should improve as readers of complex texts by . . .

- identifying and explaining the "moves" common to academic, scholarly texts (e.g. CARS, references to prior research, explanation of methodology).
- using college-level texts in strategic, focused ways (e.g. summarized, cited, applied, challenged, re-contextualized) to support writing goals.

4. Students should demonstrate an awareness of the relationship discourse conventions, lexis, genres, and their related communities by . . .

- using acquired vocabulary for analyzing how language mediates a community's actions, including terms like discourse community, genre, lexis, authority, and literacy.
- identifying and analyzing discourses, communities, and conventions.
- demonstrating an ability to respond to varied discourse conventions and genres in different situations, including different classes.
- demonstrating a responsible and appropriate use of intertextuality within discourse communities (direct and indirect citation practices).

Outcomes Statement 1102

1. Students should demonstrate an awareness of the dynamic relationship between rhetorical situation, discourse community, genre, and inquiry by . . .

- using acquired vocabulary for reflecting on their own writing processes and writing situations, including terms like genre, discourse conventions, and rhetorical situation.
- articulating and assessing their inquiry choices and the inquiry choices of others.
- demonstrating a clear understanding of their audience, with various aspects of the writing (mode of inquiry, content, structure, appeals, tone, sentences, and word choice)
- being addressed and strategically pitched to that audience.

- demonstrating an awareness that genres are not idiosyncratic, but responses to a community's inquiry processes and discourse conventions.

2. Students should engage in a meaningful, dynamic, and inquiry-based research process by . . .

- demonstrating an ability to work flexibly and iteratively with primary and secondary research.
- demonstrating an understanding of how to frame and pose a research question or problem
- utilizing multiple kinds of evidence gathered from various sources (primary and secondary—e.g. library research, interviews, questionnaires, observations, cultural artifacts) in order to support writing goals.

3. Students should read, analyze, and synthesize complex texts and incorporate multiple kinds of evidence purposefully in order to generate and support their writing by . . .

- using course texts in strategic, focused ways to both enter into and respond to on-going inquiry.
- writing intertextually, meaning that a "conversation" between texts and ideas is created in support of the writer's goals.
- demonstrating responsible use of genre conventions, including formatting, document design, and system of documenting sources (e.g. MLA).

4. Students should produce complex, analytic, persuasive arguments that matter in academic contexts by . . .

- producing at least one argument that is appropriately complex, based in a claim that emerges from and explores a line of inquiry.
- persuasively articulating the stakes of at least one argument (why what is being argued matters).
- producing at least one argument that involves analysis, which is the close scrutiny and examination of evidence and assumptions in support of a larger set of ideas.
- producing at least one argument that is persuasive, taking into consideration counterclaims and multiple points of view as the writer generates his or her own perspective and position.

- producing at least one argument that utilizes a clear organizational strategy and effective transitions that develop a line of inquiry.

5. Students should develop flexible strategies for revising, editing, and proofreading writing by . . .

- creating drafts that demonstrate substantial and successful revision.
- responding to substantive issues raised by instructor and peer feedback.
- proofreading and editing to ensure that errors of grammar, punctuation, and mechanics do not interfere with reading and understanding the writing.

Notes

1. Of course, there is a rich intellectual tradition in our field discussing labor issues, which there is not time to review here. Interested readers might look to Bousquet, Scott, and Parascondola; Robertson and Slevin; Schell; Scott; Sledd; Slevin; and Strickland, among others, for more on this topic.

2. For more detailed information about how such curricula are enacted, see Wardle and Downs, "Reimagining" and Downs and Wardle "Teaching."

3. Matt Bryan and Scott Launier were two of the first four PCSI instructor hires; Laura Martinez was an MA GTA who taught the curriculum early in its piloting, studied it for her thesis project, and is now an instructor in our program. For current sample syllabi, including assignments and calendars from these instructors, see: http://compositionforum.com/issue/27/ucf-appendix1-bryan-launier.php and: http://compositionforum.com/issue/27/ucf-appendix1-martinez.pdf.

4. This accommodation assignment drew on Jeanne Fahenstock's work considering how descriptions of scientific studies change depending on the genre, audience, and media outlet. Our accommodation assignment (now viewable in the *Writing about Writing* textbook) is largely a rhetorical analysis.

5. ENC 5705 provides a history of the field, an overview of approaches to teaching composition, and some deeper exploration of some of the field's underlying knowledge and beliefs about writing and the teaching of writing. At the end of the class, students create a teaching portfolio that includes a syllabus, assignments, and rationale for their approach to teaching composition.

6. For a response from one of our teachers, Adele Richardson, see: http://compositionforum.com/issue/27/ucf-appendix3.php.

7. For a response from a teacher, Scott Launier, see: http://compositionforum.com/issue/27/ucf-appendix4.php.

8. For full results of the pilot portfolio assessment, see http://compositionforum.com/issue/27/ucf-appendix5.php.

9. For the teacher training curriculum, see: http://compositionforum.com/issue/27/ucf-appendix6.php.

10. For last year's workshop and reading group schedule, see: http://compositionforum.com/issue/27/ucf-appendix7.php.

11. For a description of the program, see http://writingandrhetoric.cah.ucf.edu/composition.php.

12. This article, and the changes it describes, would not have been possible without the hard work of numerous UCF faculty members and administrators. I want to particularly thank Deborah Weaver, current composition coordinator, and Lindee Owens, past composition coordinator and current writing across the curriculum coordinator, for their leadership in the pilot initiative. The first four PCSI instructors are largely responsible for the assessment results that so impressed our upper administration: Adele Richardson, Scott Launier, Laurie Uttich, and Matthew Bryan. Our former vice provost and dean of undergraduate studies, Alison Morrison-Shetlar, is largely responsible for putting this chain of events in motion. Our president, Dr. John Hitt's, willingness to invest in the writing initiatives that serve our undergraduate students has changed the face of writing instruction in Central Florida. And our dean, José Fernandez, has worked to support, protect, and extend the changes to writing education. Truly, changes like those we have experienced really do "take a village."

Works Cited

Bousquet, Marc, Tony Scott, and Leo Parascondola, Eds. *Tenured Bosses and Disposable Teachers: Writing Instruction in the Managed University*. Carbondale, IL: Southern Illinois University Press 2004. Print.

Charlton, Colin. "Forgetting Developmental English: Re-Reading College Reading Curricula." *Basic Writing eJournal* 8/9 (2009/10). Web.

Charlton, Jonnika. "Seeing is Believing: Writing Studies with 'Basic Writing' Students." *Basic Writing eJournal* 8/9 (2009/10). Web.

Dew, Debra. "Language Matters: Rhetoric and Writing I as Content Course." *WPA: Writing Program Administration* 26.3 (2003); 87–104. Print.

Downs, Douglas, and Elizabeth Wardle. "Teaching about Writing, Righting Misconceptions: (Re)envisioning First-Year Composition as Intro to Writing Studies." *College Composition and Communication* 58.4 (2007): 552–84. Print.

—. "Re-Imagining the Nature of FYC: Trends in Writing-about-Writing Pedagogies." *Exploring Composition Studies*. Eds. Kelly Ritter and Paul Kei Matsuda. Utah State University Press, 2012. 123–44. Print.

Miller, Carolyn. "Opportunity, Opportunism, and Progress: *Kairos* in the Rhetoric of Technology." *Argumentation* 8 (1994): 81–96. Print.

Robertson, Liane. *The Significance of Course Content in the Transfer of Writing Knowledge from First-Year Composition to Other Academic Writing Contexts.* Diss. Florida State U, 2011. Print.

Robertson, Linda, and James Slevin. "The Status of Composition Faculty: Resolving Reforms." *Rhetoric Review* 5.2 (1987): 190–194. Print.

Scott, Tony. *Dangerous Writing: Understanding the Political Economy of Composition.* Logan, UT: Utah State University Press, 2009. Print.

Schell, Eileen. "Part-Time/Adjunct Issues: Working Toward Change." *The Writing Program Administrator's Resource: A Guide to Reflective Institutional Practice.* Eds. Stuart C. Brown, Theresa Enos, and Catherine Chaput. Mahwah, NJ: Erlbaum, 2002. 181–202. Print.

Schell, Eileen, and Patricia L. Stock, eds. *Moving a Mountain: Transforming the Role of Contingent Faculty in Composition Studies and Higher Education.* Urbana, IL: NCTE, 2000. Print.

—. *Gypsy Academics and Mother-Teachers: Gender, Contingent Labor, and Writing Instruction.* Portsmouth, NH: Boynton/Cook Publishers, 1998. Print.

Sledd, James. "On Buying In and Selling Out: A Note for Bosses Old and New." *College Composition and Communication* 53.1 (2001): 146–9. Print.

Strickland, Donna. *The Managerial Unconscious in the History of Composition Studies.* Carbondale: Southern Illinois UP, 2011. Print.

Taczak, Kara. *Connecting the Dots: Does Reflection Foster Transfer?* Diss. Florida State University, 2011. Print.

Wardle, Elizabeth. "Can Cross-Disciplinary Links Help us Teach 'Academic Discourse' in FYC?" *Across the Disciplines* 1 (2004): Web.

—. "'Mutt Genres' and the Goal of FYC: How Can We Help Students Write the Genres of the University?" *College Composition and Communication* 60.4 (2009): 765–88. Print.

Wardle, Elizabeth, and Doug Downs. *Writing About Writing: A College Reader.* Boston: Bedford/St. Martin's, 2011. Print.

Wardle, Elizabeth, and Kevin Roozen. "Addressing Multiple Dimensions of Writing Development: Toward an Ecological Model of Assessment." *Assessing Writing* 17.2 (2012): 106–19. Print.

White, Eric Charles. *Kaironomia: On the Will-to-Invent.* Ithaca: Cornell UP, 1987. Print.

Part II. Remapping Interdisciplinary Ecologies: WAC and WID Programs

The profiles in this section highlight a variety of programs that aim to integrate writing instruction across and within disciplinary contexts, acknowledging that no one writing course can accomplish the curricular or pedagogical goal of teaching students "how to write"; they, therefore, showcase in pointed ways the complex ecologies that writing in the disciplines or writing across the curriculum programs negotiate, illustrating the key ecological principle of interconnectedness and the networks of dynamic interaction within writing program ecologies. That is, such programs must negotiate competing demands of students, faculty, departments, administration, and state legislators, and typically work in concert with other, existing writing initiatives (with their perhaps competing administrative goals, curricular structures, and guiding principles), such as first-year composition requirements and other "capstone" expectations in individual departments. WAC/WID programs often find themselves working in the interstices of existing, mandated writing programs or find themselves challenged with inventing new programs that coordinate or reconceive writing requirements—emergent structures that evolve and develop within complex writing program ecologies. And in all instances such programs face severe budgetary restraints, lack of training of faculty across the campus, potential labor exploitation of graduate student assistants or lecturers, and, as one of our authors noted, the "snail's pace" of institutional change—demonstrating the range of variables (temporal, spatial, material) that account for the complexity of writing program ecologies.

In the three featured profiles of this section, each professes the importance of interconnectedness and of contextual, disciplinary learning and writing, while acknowledging that no one course can prepare students sufficiently to perform either in a first-year writing course or an upper-division majors course. The first, Michelle Ballif's "The Writing Intensive Program at the University of Georgia" (published in 2006), describes a college of arts and sciences initiative that builds on the required first-year composition experience by offering "writing intensive" courses in the disciplines, taught by faculty in the disciplines and supported by trained "writing coaches" also in the discipline. The aim is to help students understand the "ways of knowing" and communicating in any given discipline, and to provide additional opportunities to write and to receive constructive feedback on their writing. Although funded by the college, the Writing Intensive Program (WIP) is not actually housed anywhere, and the WIP course offerings vary from semester to semester. This program represents a "value-added" program rather than an institutionally mandated curriculum, reflecting the interdependent network of affiliations that shape writing programs. That is, the program works to complement other on-campus writing initiatives, including the successful first-year composition program (a required, two-course sequence in the first year) and the Writing Certificate Program (students earn certificates after taking so many "writing intensive" courses and a capstone portfolio course). Working in the "interstices" of writing instruction on campus, the WIP clearly reflects the ecological attribute of interconnectedness by aiming to provide writing experiences and instruction throughout a student's curriculum. Yet, as the "Where We Are Now" coda articulates, the program, although hailed by faculty, students, and administrators, continues to succeed despite a lack of resources, funding, and administrative support, demonstrating how the network of affiliations within writing program ecologies can shape and reshape programs in both constructive and detrimental ways.

In the other two programs featured here, the programs reflect the ecological trait of emergence as they self-organize in complex institutional systems, gaining autonomous and university-wide status and integrating more programmatically and administratively into the curricular requirements. The Binghamton University Writing Initiative, as profiled in 2010 by Kelly Kinney and Kristi Murray Costello in "Back to the Future: First-Year Writing in the Binghamton University Writing Initiative, State University of New York," exists as an "autonomous campus unit devoted exclusively to literacy instruction." The success of this WAC program is its "counterintuitive" return to a first-year writing sequence that fosters writing transfer by teaching students—before they have declared a major—"to foster a cohesive learning community that helps students develop an understanding of how writing conventions *vary according to context*." Ecologies are seen as emergent when the actors or objects within them form more complex behaviors as a collective, and this attribute of emergence is clearly reflected in this writing program ecology. The ecological context for this program's development is a less-than-successful writing across the curriculum program that was developed—in theoretical and pedagogical good faith—to replace the required first-year composition requirement. Administrators of the new program found that—without a first-year requirement—students delayed taking any writing courses until they were anticipating graduation, effectively undermining the scaffolding hopes of the instantiated writing program. So, going "back to the future," a new initiative emerged—a first-year writing requirement that emphasized genre awareness that would, it was hoped, foster transfer into more discipline-specific writing requirements in advanced courses. As their "Where We Are Now" coda describes, they are pleased with their programmatic success, but most especially with the improvement in the working conditions of the faculty who work in the writing program. Once again, the Binghamton University Writing Initiative, as do so many other writing initiatives, demonstrates the delicate, ecological balance that program administrators must manage: to provide stellar writing instruction and course offerings, while negotiating complex and fluctuating constraints, such as top-down administrative demands that these efforts be done "cheaply," as an institutional service to the "real" academic work of the university.

In "Imagining a Writing and Rhetoric Program Based on Principles of Knowledge 'Transfer': Dartmouth's Institute for Writing and Rhetoric" (published in 2012), Stephanie Boone, Sara Biggs Chaney, Josh Compton, Christiane Donahue, and Karen Gocsik describe the emergence of Dartmouth's Institute for Writing and Rhetoric as "the largest academic enterprise on Dartmouth's campus, offering 125 courses in writing and speech per year." Specifically, the institute oversees and administers first-year writing courses (a two-term, required sequence of courses), as well as first-year seminar courses, public speaking courses, and optional upper-level speech and writing courses, covering a variety of topics—from speechwriting to writing in the workplace. The success of the institute is attributed to overcoming a "lack of administrative structure"—a veritable "hydra-headed" administration of a variety of writing and communication courses. The program that developed and evolved, constituted not by its individual parts but by the complex behaviors of actors as a collective, illustrates the ecological feature of emergence and the "self-organization that arises globally in networks of simple components connected to each other and acting locally" (Syverson 183). In 2004, when the various writing initiatives were conjoined, the institute, its programmatic offerings, and its success flourished, suggesting—as the "Where We Are Now" coda emphasizes—that a synergized ecology goes a long way in ensuring faculty buy-in into not only the course offerings themselves, but also into the greater "transfer" of skills goals that the program has set for itself. The healthy combination of writing and speech represents a unique program, especially since on many campuses the two fields remain in separate departments. This program, in particular, aims to understand "how students transfer their knowledge and know-how about writing and speaking from one course to another," while "determining how to create a coherent curriculum in which the possibility of transfer is fostered." All three programs profiled in this section reveal how WAC/WID courses not only negotiate their own ecological contexts—institutionally as well as publicly—but also teach students to negotiate their discursive worlds.

Works Cited

Syverson, Margaret. *The Wealth of Reality: An Ecology of Composition*. Carbondale: SIU Press, 1999. Print.

5 The Writing Intensive Program at the University of Georgia

Michelle Ballif

General Description

The University of Georgia's Writing Intensive Program began—as do many college-wide writing initiatives—with faculty concern about the quality of student writing at the university. Acknowledging that the responsibility for this quality—or lack thereof—belongs to all faculty, in all disciplines, not just to first-year composition instructors typically housed in the English department, the Writing Intensive Program (WIP) was founded in 1997 to strengthen student writing skills specifically in the context of disciplinary demands.

In so doing, The Writing Intensive Program works against several time-honored misperceptions about student writing. The first: faculty perception of a lack of quality student writing is based on the misperception that "good writing" transcends disciplinary differences and is clear, concise, and grammatically correct. A second misperception is that there exists a proverbial set of "writing skills" that students should have "picked up" sometime during their K-12 years, but if not, then surely during their first-year composition experience. Both these misperceptions, argues Lee Ann Carroll in her longitudinal study of college writers, are a function of a "fantasy" that "students should already know how to write for situations they have not yet encountered" (xvi). Carroll argues that faculty reports of "poor" or "fair" student writing skills obscures the reality that college students must developmentally mature as writers—not as necessarily "better" writers, but as writers who must write "differently" as they are required from year to year, and from discipline to discipline, "to produce new, more complicated forms addressing challenging topics with greater depth, complexity, and rhetorical situation. What are often called 'writing assignments' in college are, in fact, complex 'literacy tasks' calling for high-level reading, research, and critical analysis" (xiv). "When professors assign 'writing,' and students are unsuccessful," Carroll argues, "professors may assume that students don't know 'how to write'" (129–30), when in fact, students are being asked to complete increasingly complex and disciplinary-specific "literacy tasks," which professors may be unable or unwilling to articulate.

I'm preaching to the choir, to be sure, but The Writing Intensive Program presumes that student writing "skills"—what Carroll calls more accurately "literacy tasks"—are inseparable from what Judith Langer terms the "ways of knowing" of a particular discipline, and therefore, "[w]riting (and the thinking that accompanies it) [is] a primary and necessary vehicle for practicing the ways of organizing and presenting ideas that are most appropriate to a particular subject area" (71). Further (and Langer would concur), writing (and the thinking that accompanies it) is the process by which disciplinary knowledge is constituted, the process by which one "comes to know" knowledge, and the process by which that knowledge is vetted. In short, writing *is* the academic dialogue that we, as educators, aim to introduce to our students. Hence, to teach writing is to teach the "ways of knowing" unique to any discipline: the methodology of inquiry, the conventions of evidence, the mode of presentation. Such a pedagogical goal, then, assumes that the most effective way to improve student writing is to do so within the context of disciplinary demands under the tutelage of committed faculty across the campus, who are willing and able to "articulate" those conventions.

Recognizing that such a pedagogical goal requires new responsibilities for and time demands on participating faculty, the program trains discipline-specific graduate students to serve as teaching assistants to support Writing Intensive Program courses. That is, each Writing Intensive Program course, competitively selected, is provided with a specially trained TA to support participating faculty—not as a "grader" but as a "writing coach."

Currently only a college of arts and sciences program, but recommended for university-wide expansion by the *Report of the Task Force on General Education and Student Learning* at UGA, the Writing Intensive Program serves from 1000 to 1500 students in approximately forty-five diverse courses across the college, ranging from art history, biology, classics, geology, mathematics, music, religion, sociology, and women's studies. Most of these courses are regular-enrollment sections, but one or two courses each semester are large-enrollment classes. Most regular-enrollment sections cap at 30 to 35 students; large-enrollment courses range from 75 to 300 students. In the regular-enrollment courses, one TA is assigned to assist; in the large-enrollment courses, we have employed alternative arrangements: one alternative is to have several "break out" sections of the entire course, each assigned a WIP TA, that are designated as "writing-intensive"; students are lured into these sections with guarantees that exams will be replaced by writing assignments, for example. That is, in this instance, only a portion of the students enrolled in a large-enrollment course participate in the writing-intensive aspect of the course. Another option is to have the assigned, trained Writing Intensive Program teaching assistant serve as an oversight TA for the other TAs assigned to assist the course, creating appropriate assignments and instructing the other TAs how to best (and most efficiently) respond to student writing. One experimental large-enrollment introductory biology course used a great deal of peer-review (guided by a peer-review-calibrated computer program) to handle the student writing. Open to new ideas and approaches, The Writing Intensive Program does not stipulate what kinds/amounts of writing are to be assigned, and faculty have tried various options. Although the program has some baseline criteria (see Appendix for WIP course guidelines and criteria), I like to argue that the Writing Intensive Program is not "writing exhaustive," but rather it aims to provide for students an "intense" engagement with writing—its processes and its conventions—and participating faculty are inventive in their methods of doing so.

A compelling majority of students enrolled in these courses consistently report that the pedagogical aim of the program has been realized. Student evaluations collected over a nine-year period report that their experience with the Writing Intensive Program introduced them to and gave them experience with writing within their discipline; additionally, as a result of the program's emphasis on revision, students report that their writing processes improve along with their writing. Not surprisingly, they also report that their Writing Intensive Program courses force them to keep up with the readings and to be better prepared for class discussions—in short, to be more engaged in the learning process. As well, they report that their critical and disciplinary thinking has improved. Here are some survey results: when asked if the WIP course strengthened their writing, a majority of students responded "yes" (ranging from 70 percent to 78 percent), with students commenting: "The multiple drafts allowed me to take time and revise my paper several times"; "It helped me to see different *possibilities* to use in the organizing of a paper, and I realized there is not necessarily one way that is *most* correct"; and "I did not truly know how to write an in-depth Art History paper before this class"; and "Once we get to these upper-level courses, it is expected that we can write, and this is not always the case. Having a TA to meet with specifically about writing a paper is very beneficial" (Fall 04 Student Responses). When asked if the writing assignments stimulated their critical thinking, a majority of students answered "yes" (ranging from 75 percent to 89 percent), with responses such as "I have to think as to *why* to solve a [mathematical] problem instead of plugging in formulas": "The writing assignments forced me to read the texts with a degree of critical thinking far greater than I normally would when reading for my classes" (Fall 04-Spring 05).

Faculty survey responses show that their Writing Intensive Program courses increase student engagement and improve student writing; additionally, faculty report that their participation in the Writing Intensive Program has strengthened their own teaching practices. Here is a sample of faculty responses: "WIP courses force you to ask, 'How do you design an intellectual experience?' They give you touch points, make you teach more slowly, allow students to incubate, and strengthen the connections they make." Another responds, "WIP courses force you to plan more from the beginning"; and another: "The WIP allows me to provide a level of contact with students that, frankly, I would have been unable to provide. Before WIP, it was sink or swim" (Spring 04).

Theory Informing the Program

Before designing the Writing Intensive Program, an arts and sciences senate ad-hoc committee on writing researched several other writing across the curriculum (WAC) and writing in the disciplines (WID) programs. As chair of the committee, I was most taken with Cornell's Knight ("Writing in the Majors") Program, now named the John S. Knight Institute for Writing in the Disciplines, because of its emphasis on the *disciplinary* nature of writing. Jonathan Moore, director of the program, argues:

Although WAC and WID are sometimes used synonymously or interchangeably, and both terms usefully suggest the importance of writing in all fields, these two approaches have very different implications for the role of writing instruction in higher education. While WAC emphasizes the commonality, portability, and communicability of writing practices, WID emphasizes disciplinary differences, diversity, and heterogeneity. That is WID emphasizes what remains incommensurable and irreducible in writing practices both within academic fields and from one field to the next. ("Writing and the Disciplines" 4)

Not only a big fan of the idea of the heterogeneity and incommensurability of language games, I also was drawn to the assumption that writing is inseparable from the production of knowledge—in a way more complex than the "writing-to-learn" adage. Writing, then, as I articulated above, is the conversation of scholars. As Monroe states it, "Once they have begun college-level work in writing, students have also begun, in earnest, the work of the university" ("Writing" 5). The work of the university, then, is the work of scholars, producing knowledge. (I realize the privilege of being at a research institution, where students—of privilege, typically—come to participate in the academic dialogue, some to continue on to become academicians themselves, others to document their privilege with a degree.)

In any event, at an institution such as The University of Georgia, "The writing issues our students confront, from entering students to advanced undergraduates, to graduate students, to the most distinguished scholars, remain in fundamental respects the same issues, including especially the process of socialization or acculturation into a particular field that may have recognizable beginnings . . . but has no end in sight for as long as one continues to be committed to the production of knowledge in that field" (Monroe, "Introduction" 8). And, hence, the Writing Intensive Program was founded: to improve student writing skills by providing them with multiple opportunities to acculturate themselves into a discipline and to learn the "ways of knowing" unique to their field and to engage in both through writing.

Description of the Program

Funded by the offices of the dean of the college of arts and sciences, the provost, and the vice president for academic affairs, the program typically supports from fifteen to eighteen graduate teaching assistants each year. The budget consists entirely of TA-lines. Early each spring semester, I distribute a call for course proposals to faculty in the college, which details Writing Intensive Program course criteria and guidelines (see Appendix). Courses are then competitively selected, and graduate student teaching assistants from the appropriate departments are chosen to support the selected course offerings. (In rare, but appropriate circumstances, a WIP TA might be assigned to a course outside of his/her immediate discipline: a comparative literature graduate student might be assigned to a cross-cultural speech communications course, for example—but only if the participating faculty agrees; in this instance, in fact, the faculty wanted a non-disciplinary TA.) The graduate teaching assistants are nominated by department heads in consultation with graduate coordinators (some departments use the Writing Intensive Program teaching assistantship as a recruitment tool). I interview the nominated graduate student to ascertain the TA's suitability for the role and explain the program and the TA's responsibilities. Typically, each TA supports four courses—two each semester, or alternately, each TA supports three courses—one for one semester and two for the next or vice versa. For example, four art history professors submit successful proposals for regular-enrollment (thirty-five students or so) courses, and I then hire an art history graduate student to support these four courses during the academic year. As another example, school of music faculty support several successful course proposals, two of which are for large-enrollment history of music classes. I will then hire and assign several teaching assistants to support the large-enrollment courses as a team, but each will also support, individually, a regular-enrollment music course. The process is complex but is meant to allow flexibility in course design and offerings.

Once all the faculty, courses, and teaching assistants are selected, I begin the process of introducing the participants to the principles of the Writing Intensive Program and training them in the most effective ways to assign and to respond to student writing. Participating faculty, a good number of whom have been involved in the program since its inception, are invited to a series of orientation sessions. Faculty who are new to the program attend a session at the beginning of the fall semester; veteran Writing Intensive Program faculty join the new faculty for an informal discussion about how they structure their courses, how they divide responding responsibilities with their supporting TA, what kinds of writing assignments they use, etc. I also host a couple of colloquia each year, featuring experienced faculty presenting their best practices.

The selected teaching assistants attend a week-long pre-fall semester orientation session, as well as enroll in, during fall semester, WIPP 7001, "Pedagogy of Writing in the Disciplines," which I teach, where they are further introduced to the research on and best practices of conferencing with students, responding to student writing, guiding the writing process, etc. During spring semester, we meet less formally, about every other week as a group to problem solve: What do I do if a student says, "I don't *do* drafts?" What do I do if a faculty member changed a grade I gave without consulting me? (Each WIP faculty negotiates with his/her WIP TA as to the grading responsibility of the TA. In some instances, the WIP TA handles all of the grading; in other instances, although the TA responds to all the work, the faculty member assigns grades. I have witnessed a range of successful arrangements. The key to success, however, is to have the faculty member and the TA work out before hand the roles and responsibilities of each, and to stand by this agreement, not allowing a "good cop" and "bad cop" scenario to emerge.)

I must admit that my favorite part of directing this program is working with these teaching assistants. Although inevitably the only time that is free in all 18 TAs' schedules is late Friday afternoons, I never feel put out by spending the last part of the work week or delaying happy hour to spend the time with them. I always leave inspired, energized, and recommitted. Although the program was constructed to help undergraduate students—and it does, it has had the happy side effect of revitalizing me each Friday afternoon. The TAs benefit, too.

They not only receive intense instruction in the most effective ways to aid student writing practices, but they also receive what amounts to the *only* training they will receive in their graduate careers in effective teaching. As many of our TAs are interested in pursuing faculty positions elsewhere, they value this instruction in basic pedagogical strategies. The TAs receive not only pedagogical training but experience, as well, as they work with faculty to instruct students in the course content as well as in the writing conventions of each discipline. The Writing Intensive Program TAs report an additional benefit; their *own* writing practices and writing efforts improve. A significant number of TAs have reported that their experiences as a Writing Intensive Program TA have enabled them to compose successful grant proposals, for example. Other TAs report that their experience as a Writing Intensive Program TA has supported their job searches, as many institutions (academic and otherwise) have noted and appreciated their WIP expertise. Additionally, the program affords the TAs an opportunity to interact with TAs from across the college, forming an intense camaraderie in the service of student learning.

Institutional/Budgetary Restraints Guiding or Controlling the Program's Growth and Development

During the past several years, the University of Georgia has, as have most institutions, experienced very serious, slashing budget cuts—indeed, it is a testament to the success of the program that our budget has been protected and has even witnessed a slight growth. Indeed, in response to a not-so-flattering survey on student engagement at UGA, a task force convened to study the undergraduate experience at UGA recently recommended that the program expand. The upper administration has indicated an interest to do so, but has not (yet?) allocated funds for such a university-wide expansion. I have not sought, to date, external funding, although I have been encouraged to do so by the associate dean of the college, as I do not want the program's growth to depend on "soft" money that may disappear and never be an institutional commitment. The most basic institutional restraint is simply a lack of funds, as I am committed to providing support to those faculty who participate, and I do not want the mandate, as the task force report suggested, "incorporate more writing into the curriculum" to become an unfunded mandate. Further, I do not want faculty to just assign more writing without the guidance of principled practices, based on solid research in rhetoric and composition scholarship.

What I've Learned, or If I Knew Then What I Know Now

From the very beginning, I would have argued for, even insisted on, a line for an academic professional to serve as an assistant to the director. I do receive a line to hire a TA to help me with administrating the program, but the stipend (and no benefits) is in no way proper recompense for the work and professionalism required by the position, which requires one to collaborate with the director on faculty and teaching assistant recruitment, training, and retention, as well as to create program materials—from teaching messages to assessment surveys to reports, including the program's website. (Indeed, the extraordinarily qualified assistant I had for some time, Parker Middleton, was lured away, understandably, by a more lucrative and secure position.)

Also, in terms of wished-for funding, I would have argued for faculty development funds, given the need to educate, support, and inspire participating faculty with visiting speakers and specialists from other universities. In the program's early years, I did receive funding to host visiting writing scholars, such as Pat Bizzell, and John and Tilly Warnock and Anne-Marie Hall for university-wide workshops, which drew a large number of faculty and sparked the original enthusiasm for the program. But, due to budget cuts, such funding quickly dried up.

But I suppose these are present and future issues.

As for the past? I would have named the program differently. The "intensive" in the title continues to frighten students, even though I insist that the program is "intensive" not because it requires an overwhelming amount of writing, but because it offers an *intense engagement* with the writing process. It's a tough sell to students. It doesn't help that our program's acronym, as a long-time faculty supporter in the school of music pointed out to me, is pronounced "whip." An intense whip—sounds like a whuping with writing with or without black leather.

In retrospect, I should have written and published about the program. I have almost ten years of data, stories, and hindsight—clearly a rich resource for publishable material. My primary research interests lie millennia away (in classical Athens) and are theoretically incommensurable (poststructuralism/postmodernism). However, there may still be an article or a book in the offing.

In Conclusion

Despite its limited and controlled funding, the Writing Intensive Program—with its devoted faculty and teaching assistants—has delivered on its initial challenge to "do something about student writing." And it has done much, much more: it has created an atmosphere of commitment to student writing in all disciplines. As has Jonathan Monroe, committed WIP faculty have aimed to create a "sense of shared responsibility for the teaching of writing, and above all for the enhancement of learning through writing, across all disciplines and at all levels of the curriculum" ("Introduction" 11). I wish to acknowledge the commitment and dedication of my colleagues at UGA—across the disciplines—who have shared this vision and responsibility, and I wish to thank associate dean Hugh Ruppersburg for his continued administrative support.

CODA: WHERE WE ARE NOW

Since the original program profile publication in 2006, the Writing Intensive Program has deepened and expanded its reach. Continuing to dedicate itself to its original program goals and expectations to articulate and to teach the disciplinary conventions of "good writing" in the various disciplines (ranging from chemistry to music to sociology), the WIP's effectiveness has improved as demonstrated by recent semester-end surveys. In their responses, the students praised their WIP courses, reporting that their experiences provided them with an increased confidence to speak and to write about the course content, as the writing assignments enhanced their understanding of the material and their engagement with the course. Specifically, this past academic year, 88 percent of students reported that their experience in the course strengthened their writing; 93 percent reported that the writing assignments enhanced their learning; 86 percent reported that the writing assignments encouraged them to be actively involved in the course; 90 percent reported that the writing assignments helped them understand the disciplinary writing conventions; and 89 percent reported that the coaching they received from their writing coach was effective. The faculty and teaching assistants participating in the program, likewise, laud the program as "fantastic" and "phenomenal," arguing "that the quality of student thinking, writing, and learning was better than it was in non-WIP classes."

Despite lean budgetary times, the WIP has been able to expand its influence in the following ways:

1. Training all of the graduate lab assistants who teach the introductory biology labs, via multiple sections of the WIP's preparation course, "Pedagogy of Writing in the Disciplines," so that *all* of the introductory labs on campus (approximately 100 sections each semester) are now taught in accordance with the Writing Intensive Program principles. This expands the reach of the program vastly.
2. Providing specially trained WIP TAs to serve in the writing centers on campus. With funding provided by the vice president for instruction, the campus writing centers have been able to staff tutors who have been trained by the WIP; this means that

students can work with writing center tutors who have training in writing instruction as well as disciplinary knowledge of the sciences or social sciences as per the teaching assistants' area of study.
3. Increasing the number of WIP courses. Just this year, the senior associate dean has approved ten more units of WIP TA support—which represents a 20 percent increase in the number of WIP courses we will support during the 2013–2014 academic year.

Although the WIP has, thus, seen a modest increase in size during the past six years, such an increase in no way answers to the university's continued call for its campus-wide expansion. As a recent internal review of the program reports: "We note that various committees and reports have over the past decade, with what might be described as monotonous regularity, called for WIP to be given the funding and staffing necessary to make it a campus-wide program." The director is hopeful that the program, which the recent review called "underfunded, understaffed, and underappreciated," will grow in size and influence, and remains hopeful that she will receive an assistant, trained in rhetoric and composition who could aid in the professionalization and growth of the program.[1]

Viewed from a specifically ecological perspective, the Writing Intensive Program's history and performance foregrounds and exemplifies issues of interconnectedness, fluctuation, complexity, and emergence. Developed, originally, in response to faculty concerns about the amount and quality of student writing, the program has—as a result of faculty, student, graduate student, and administrators' interactions—emerged as much more than simply a program focused on undergraduate writing, as it has supported faculty development by encouraging faculty to view attention to writing as a hallmark of exceptional pedagogical practices. Many WIP faculty tell me now that they teach all of their courses in harmony with advocated WIP principles, whether they have the benefit of the WIP-trained TA or not. The WIP-trained TAs report the same thing: when they have the opportunity to teach their own courses, they will continue to teach them in a writing-intensive way. Thus, one of the emergent products of the program is the development of better teachers across the campus.

Relatedly, the program is interconnected to other writing initiatives on campus, including the Writing Fellows program that provides intense training in and support of writing instruction; this program echoes WIP principles and encourages faculty to offer their own writing-intensive courses not necessarily under the aegis of the Writing Intensive Program in a coordinated effort to have as many writing-intensive courses offered on campus as possible, acknowledging the limited budget of the WIP. Additionally, to encourage students to seek out writing-intensive courses, the WIP works in conjunction with the Writing Certificate Program, which rewards students who have taken so many writing-intensive courses with a notation on their transcript; and finally, specifically in light of the rich, recent work on transfer, the WIP is exploring institutional ways to encourage the transfer of writing skills from UGA's award-winning First-Year Composition Program, by continuing to work to make messages about writing consistent between the various programs and to work to make visible the problem of transfer for faculty and the participating graduate students, so that they can encourage students to see the interconnectedness of their writing instruction and experiences through the curriculum and their degree programs.

Appendix

The Franklin College Writing Intensive Program Course Guidelines and Criteria

(Portions of this document were originally composed by Parker Middleton; additions and amendments made by Michelle Ballif.)

> "Because many students rely on pulling an 'all-nighter' when a paper is due, college courses serve an important role when they provide opportunities and methods for developing more orderly, reflective writing habits."
>
> —*Bedford Guide to Teaching Writing in the Disciplines* 16

As the case in any course that asks students to write, strong syllabi are especially important in writing-intensive courses because they model effective writing, establish the purpose and role of writing in the course, and give vital information about it. Furthermore, in a WIP course, syllabi teach the writing process by charting writing assignments, their stages, and due dates. Moreover, syllabi, both paper-based and online versions, are perhaps the course document that students are most likely to keep up with and to review most often. This makes the syllabus an important teaching tool—for students as well as for faculty. Previous

WIP faculty have reported that writing-enriched classes lead them to articulate more clearly their goals for a course and to state objectives and criteria for grading more explicitly in their syllabi and materials.

Some features of effective syllabi for Writing Intensive Program courses are listed below.

- Make writing in the disciplines philosophy, as well as the role and benefits of writing in the course, clear. WIP syllabi should integrate writing assignments fully into course learning. In some situations, this many entail more selling than in others: students in lecture-based courses may not expect to have to write and may be resistant because, as they put it, "It isn't an English class." This resistance may be met by emphasizing the benefits of writing to learn the course content, as well as by emphasizing writing as a critical competency in an information culture. It can also be countered with truly innovative writing assignments that students see as clearly relevant to their university and post-university careers and that boost their performance in the course.

- Model the writing process as a learning process. By providing students with a process for writing, WIP courses teach important process management skills. Many students simply have no experience with managing time and work, routinely making final-hour efforts when work is due. This lack of process skills becomes clear when we ask them to write. By modeling a process for getting work done in syllabi and other materials, "college courses serve an important role": they provide both "opportunities and methods for developing more orderly, reflective writing habits" (*Bedford* 16).

- Stage assignments, breaking them into manageable parts with opportunities for guidance, feedback, and revision. A course calendar or process timeline can be useful.

- Sequence assignments so that students can build on and elaborate early learning.

- Make effective use of WIP teaching assistants to help give more attention to student writing than would be possible in a non-writing intensive course (see below).

- Consider logistics of prompt feedback, conferences, and revision.

- Offer clear guidelines and criteria by which work will be generated and evaluated. Along with articulating the features of effective writing in a course and specifying the most important writing conventions, course materials should let students know what part disciplinary styles and formats such as MLA, APA, or CBE will play, as well as the role that editing errors involving grammar, punctuation, and spelling will play in finished work.

- Teach discipline-specific models, formats, conventions. Along with teaching the writing process, teaching students to write means teaching the writing conventions of a discipline: how to think, how to argue, how to write as scientists, sociologists, or music historians. Judith A. Langer's "Speaking of Knowing" emphasizes that "writing (and the thinking that accompanies it) then becomes a primary and a necessary vehicle for practicing the ways of organizing and presenting ideas that are most appropriate to a particular subject area " (71). This requires faculty to state explicitly the often tacitly held "rules of argument and evidence that represent the ways of thinking unique to each discipline" (83).

- Avoid the tendency to attempt too much. Two sequenced writing assignments with guidance and feedback in planning, drafting, and revising, for example, give students greater learning benefits than turning in a 20-page paper at the end of a course. Teaching with writing requires working smarter, not necessary harder. Strategic adjustments can bring big gains.

Working with a Writing Intensive Program Teaching Assistant: Talking Points for WIP Faculty and Teaching Assistants

WIP teaching assistants help faculty to provide the enhancements that a writing-intensive course offers. In general, the Writing Intensive Program selects discipline-specific teaching assistants who believe that writing is important, who demonstrate effective communication skills, who have a more sophisticated than usual understanding of the writing process, who show promise in working with others and leadership in problem solving.

WIP teaching assistants receive extensive training in teaching writing in the disciplines skills. Each August, those selected for the program attend a pre-seminar designed to introduce writing intensive pedagogy and the logistics of supporting writing intensive courses. During Fall Semester, WIP TAs meet weekly for WIPP 7001, a graduate course in the pedagogy of writing in the disciplines; they also meet with the Director during Spring Semester to further discuss pedagogical issues.

To make the best use of these specially trained "writing coaches," you will need to work to harmonize your course goals with the principles of the Writing Intensive Program. Hence, prior to the beginning of the semester, the participating faculty member and his/her assigned WIP teaching assistant should address the following in order to be clear on the logistics of the course, the roles and responsibilities of both the faculty member and the WIP TA, and, most especially, to guarantee that the WIP TA will be utilized as the WIP principles and best practices prescribe.

- How will the Writing Intensive Program and the Writing Intensive Program TA be introduced to the class? Will there be a statement on the syllabus that explains the program and explains the role and responsibility of the WIP teaching assistant? What will these statements be? Make sure students know the TA's purpose and role, and how they are to work with the TA, for example, whether meeting with the TA is required to get back drafts or papers.

- What kind(s) of writing will be assigned? Low-stakes (informal, writing-to-learn activities, such as journal entries) or high-stakes (formal essays, term papers, etc.)? Who will compose each actual assignment (the document that specifies the writing task, purpose, requirements, and evaluative criteria)? Will the specifics of each assignment be included in the syllabus, or will the assignment document be a separate handout? Will the faculty or the TA compose the assignment?

- How will the logistics of the course guarantee that the WIP TA is actively involved in the writing processes of the students and is not being used just as a grader? The WIP prescribes two, key course components: mandatory conferences and opportunities for revision. How will the syllabus provide for at least one, required (individual or group) student-TA conference? How will the syllabus (and the writing assignments) provide opportunities for revision? Will the TA have class time to discuss writing assignments or effective writing, in general?

- Related to above, if the faculty member will be assigning high-stakes writing assignments, how will the assignment be staged, so that the TA can intervene several times in the students' writing processes? How will the logistics of such be managed on the syllabus (e.g., establishing due dates for drafts)? How will low-stakes writing assignments be configured to allow TA-student interaction?

- How will the written assignments be evaluated? The Writing Intensive Program urges faculty to provide for students clear guidelines for effective writing and grading standards in your course. Criteria may be spelled out in a rubric that can be adapted for reviewing drafts and grading final products. Who will create these rubrics—the faculty member or the teaching assistant?

- Who will be responsible for an assignment's final grade? Or will both the faculty and the TA weigh in on the grade? However this is decided, make sure the students understand, and make sure that the TA has some authority. A brief, norming session may be helpful between the faculty member and the TA to help ensure that rubrics are used effectively and uniformly (in short: to make sure that a paper that the faculty member considers to be an "A" paper is the same as what the TA considers to be an "A" paper).

- Talk about the time and logistics involved in getting work back to students promptly, as well as how feedback strategies will be communicated to students.

- Talk about how grade disputes and disgruntled students will be handled.

- Talk about how to best communicate with each other (e.g., e-mail, during office hours). Establish whether the TA is expected to attend the class every class period. Discuss any other expectations that you have for the course and each other.

Note

1. At the time of production of this volume, the senior associate dean granted permission for the hire of an academic professional to support the administrative demands of the program.

WORKS CITED

Carroll, Lee Ann. *Rehearsing New Roles: How College Students Develop as Writers.* Carbondale, IL: Southern Illinois UP, 2002. Print.
Langer, Judith A. "Speaking of Knowing: Conceptions of Understanding in Academic Disciplines." *Writing, Teaching, and Learning in the Disciplines.* Ed. Anne Herrington and Charles Moran. New York: MLA, 1992. 69–85. Print.
Monroe, Jonathan. "Introduction: The Shapes of Fields." *Writing and Revising the Disciplines.* Ed. Jonathan Monroe. Ithaca and London: Cornell UP, 2002. 1–12. Print.
—. "Writing and the Disciplines." *Peer Review* 6.1 (Fall 2003): 4–7. Print.
Report of the Task Force on General Education and Student Learning. The University of Georgia. August 2005.

6 Back to the Future: First-Year Writing in the Binghamton University Writing Initiative, State University of New York

Kelly Kinney and Kristi Murray Costello

Co-founded in 2008–2009 by Kelly Kinney and Rebecca Moore Howard, the Binghamton University Writing Initiative is an autonomous campus unit devoted exclusively to literacy instruction. A central component within the Writing Initiative,[1] first-year writing was developed in response to the perception that the institution's general education composition offerings were not meeting the needs of students or the desires of faculty. While the State University of New York at Binghamton had endorsed the writing across the curriculum (WAC) model it adopted in the 1990s and had subsequently sustained a variety of WAC courses across campus, when Kinney accepted her position in Fall 2007, her home department of English, General Literature and Rhetoric was still offering the lion's share of composition courses and had a scarcity of faculty specialists in writing studies to support them. This essay seeks to explain the history that led to the establishment of first-year writing at Binghamton, a set of electives that seeks to complement discipline-specific and writing-to-learn courses while also providing first-year students a common experience in and comprehensive introduction to college writing. True to the genre of the "program profile," we also provide a description of the theories informing the program, an overview of its courses and institutional constraints, and a synthesis of the lessons learned during the inaugural year of first-year writing at Binghamton.

A History of WAC at Binghamton, and What Led Us (Back) to First-Year Writing

It may seem counterintuitive that, after initiating a WAC program, our institution chose to establish first-year writing (FYW); after all, many see WAC programs, particularly those that stress writing in the disciplines, as a progressive alternative to "generic" first-year composition (Smit 146). At face value, Binghamton's WAC model remains both theoretically sound and pedagogically rich. It was built on the principles that teaching writing is the responsibility of all faculty; that writing development is always "ongoing and lifelong"; that "writing about a subject is a powerful means of learning about a subject"; and that "writing to learn is as important as writing to communicate" (Gay and Tricomi 6). Consistent with Sharon Crowley's proposal to abolish universal first-year requirements (241), another attractive component of Binghamton's WAC model was that it did not force students to take a required first-year course or a mandated series of courses: instead, it proposed a range of composition offerings in a variety of disciplinary contexts, allowing students to pick and choose among electives that complement their educational goals.

But while Binghamton faculty had high hopes for the WAC program given its sound pedagogical promise and diverse curricular potential, these goals were not altogether realized. One of the WAC program's strengths in theory—namely, that it did not compel students to enroll in a required sequence of courses—became a notable weakness in practice. Despite good intentions, lack of sequencing made it possible for students to delay enrollment indefinitely. While Binghamton students do have options in completing their general education composition component,[2] nothing in the WAC model compels them to enroll in writing courses early in their college careers, leaving some ill-equipped to take on higher-order literacy tasks as they progress in the majors. Although various campus units developed a discrete range of well-designed writing courses at different levels, prerequisites and curricular sequencing were virtually non-existent. Thus, juniors and seniors routinely flooded 100-level composition courses both inside and outside their majors, leaving few seats for the first-years and sophomores for which these courses were designed and intended. This vicious cycle prompted the institution to reconsider the value of FYW—and to look to a future that would include it.

Although instituting a required sequence for composition courses already in place—rather than developing a new set of electives—may have been the simplest solution to the problem, there was a second, perhaps more pressing issue: because instructional resources were not significantly increased or reallocated across campus when the WAC program was initiated, the English department continued to offer the majority of 100-level composition courses. As a result, it was hit-or-miss whether students received the content-based or writing-to-learn instruction the initial WAC model had aspired to offer. Faculty across the disciplines recognized the flaws in the system, complaining that even their most promising seniors were often inexperienced and thus mediocre writers in the major. And while it might be reasonable to attribute some faculty perceptions to the pervasive cultural mantra that "Kids just can't write these days," it is also reasonable to attribute these perceptions to the faculty's knowledge of students' potential. Binghamton faculty are appropriately proud both of our students' vast diversity *and* outstanding academic potential: 33 percent of our undergraduates are students of color, 10 percent are international students, and all have competed in a highly selective process to gain entry into Binghamton, dubbed by *Greene's Guide* as a "Public Ivy" and by *Fiske Guide* as "the premiere public university in the Northeast."[3] It became clear that a lack of course sequencing and a dearth of diverse WAC courses resulted in too few students gaining the level of sophistication and facility in their writing that faculty rightly recognize as within their reach. All told, and as we return to below, Binghamton needed a set of first-year courses that would complement discipline-based instruction *across* departments *throughout* the institution.

From our perspective as writing program administrators, absence of faculty development and teaching assistant preparation also influenced undergraduates' writing performances. Although faculty development across the disciplines is not within the purview of our positions,[4] our experiences interacting with ladder faculty, instructors, and graduate assistants alike convinces us that too few writing teachers on our campus have been introduced to basic tenets of contemporary composition pedagogy. Not only has there been no concerted faculty development since the previous WAC Director stepped down several years ago,[5] but the English department suspended formal teaching assistant preparation when the previous director of composition left the university in 2005–2006, the same academic year she was hired. Because there has never been a critical mass of literacy-specialists on campus to support WAC, rather than relying on what Anne Herrington and Charles Moran describe as the two central approaches to WAC instruction—that is, the "writing-in-the-disciplines" approach that emphasizes the habits of mind and rhetorical conventions particular to a discipline, and the "writing-to-learn" approach that engages students in a range of activities in order to master course content (7–9)—many Binghamton instructors relied on disparate or ill-informed conceptions of what writing courses should do.

But even given these problems, when Kinney took over the English department's composition program in the fall of 2007, her first instinct was to work to strengthen and modify the 100-level composition courses already in place, courses taught almost exclusively by graduate students in English. At the time, English offered more than 90 percent of its composition courses in two formats: English 115, a course intended to cultivate students' critical literacy practices in non-literary contexts; and English 117, a theme-based course intended to teach students discipline-based conventions in writing about literature. Students in any major took either or both of these courses to help fulfill their general education composition component, and there was no connection or scaffolding between the two courses or among them and other, higher-level writing courses offered by English or other campus units. We came to recognize that this lack of course sequencing and coherency, coupled with few opportunities for teaching assistant preparation, made it next to impossible for instructors to cogently distinguish between the aims of English 115 and English 117. Most graduate assistants, our classroom observations and syllabus reviews made clear, substituted the primary curricular aims of either course—teaching writing to promote critical literacy (English 115) or teaching the conventions of literature-based writing (English 117)—for the goal of literature coverage. Frankly, this was no surprise; there was nothing in most of our instructors' pedagogical development that would have prompted them to do otherwise.

To work to reform the system in place, Kinney offered up a three-year plan for strengthening first-year writing instruction in the English department. It included developing a graduate seminar in rhetoric and composition theory, instituting monthly pedagogy workshops and a multi-day pre-semester orientation for composition instructors, and piloting a new, genre-based 100-level course, seeking feedback from the instructors who chose to take part in the pilot project.

But while Kinney had plans for incremental reform, the university administration was poised to move forward—and fast. Prompted by less than stellar results on the state of writing instruction reported in the National Survey of Student Engagement, the administration hired consultants to make recommendations for change. Enter Rebecca Moore Howard of Syracuse's Program in Writing and Rhetoric and Paul Sawyer of Cornell's Knight Institute of Writing. On two snowy days in December 2007, these nationally recognized WPAs drove hazardous upstate highways to interview Binghamton students and faculty, dialogue with deans and department heads, and listen to Kinney and other program directors describe their institutional observations, pedagogical perspectives, and plans for the future. Through their investigation, Moore Howard and Sawyer chronicled a range of deficiencies in the institution's support of writing instruction, including the abuse of graduate student labor, the lack of full-time and tenure track specialists in writing studies, the scarcity of faculty development on campus, and the absence of budgetary and hiring autonomy for the writing program.

By the beginning of classes the following semester, Moore Howard and Sawyer had completed a fifteen-page report of recommendations: perhaps its most far-reaching recommendation was to move the composition program out of the English department and to establish what would become the Binghamton University Writing Initiative,[6] an autonomous academic unit comprised of first-year writing, the writing center, and the English as a second language program. The goal of bringing the three programs together was to reorient the foundations of writing instruction on campus and to ground the university's teaching culture in the scholarship of writing studies and language-learner literacy. By uniting the missions of these programs and giving the new academic unit control over curriculum goals and hiring practices, Binghamton was positioning itself not only to overcome its sequencing problems, but—as we detail more thoroughly in the following sections—to develop a FYW program that is: 1) consistent with general education outcomes, 2) committed to teacher preparation, and 3) complementary to the WAC model in place.

Intrigued by Binghamton's newfound commitment to writing, Moore Howard was persuaded to take a year's leave from Syracuse to collaborate with the established directors and—perhaps most importantly—create coherent programmatic connections among WAC, FYW, the writing center, and ESL. In an effort to do so, Kinney, Moore Howard, and Kristi Murray Costello refined the common syllabus Kinney had piloted the previous year and built enough sections to offer to all incoming first-year students. Together Kinney and Murray Costello also created a second, co-enrollment course to support first generation college students and English language learners. As a result, in 2008–2009, the Writing Initiative offered two new courses for a combined total of 133 sections of 100-level composition. For all intents and purposes, Writing 111 and Writing 100 replaced English 117 and 115. Like the previous first-year courses offered in English, neither Writing 111 nor 100 would be required,[7] but unlike their predecessors, they would both be offered exclusively to first-year students. Both 111 and 100 would also focus on genre-awareness across contexts, a new curricular emphasis at Binghamton and, perhaps to a lesser degree, across the nation. Although discipline-specific composition courses would continue to be offered in various departments across campus, the foundation for writing instruction at Binghamton would now be built on first-year writing. By preserving the best of WAC and FYW, we had arrived "back" to the future.

THEORIES INFORMING FYW AT BINGHAMTON

Let us be clear, however, to distinguish key differences in our courses from those described in the scholarship above. We are well aware of criticisms waged against what many have come to call "generalist" writing courses; because our 100-level courses are better categorized as generalist than discipline-specific, they do not coincide neatly with the programs that Beaufort and Fishman and Reiff describe. Neither, however, do they undermine discipline-based instruction. While our courses are offered exclusively to first-year students—that is, "pre-majors" who have yet to confirm their disciplinary homes—they are also meant to foster a cohesive learning community that helps students develop an understanding of how writing conventions *vary according to context*. Thus, our courses are consistent with both FYW and WAC instructional models. Far from retrograde, through our reconfigured commitment to first-year writing, Binghamton University is moving forward in its efforts to improve writing instruction.

Accordingly, we embrace a model of first-year writing that is purposefully aligned with genre theory, as well as with social-expressivist and civic-rhetorical pedagogical traditions. Agreeing with Doug Downs' and Elizabeth Wardle's perception that writing courses should introduce students to the traditions of rhetoric and writing studies, but cognizant of the problems our campus has experienced by asking underprepared teachers to develop discipline-based composition courses, we instead forged our curriculum within interconnected traditions that embody the pedagogical (rather than strictly scholarly) trajectories of writing studies.

Of the many theories of writing instruction influencing FYW at Binghamton, genre theory is the most influential. Built on the notion that no writing class—no matter how well-designed—can teach students all there is to know about writing conventions, our courses help students analyze and negotiate genres. Consistent with Amy Devitt's *Writing Genres* and Anis Bawarshi's *Genre and the Invention of the Writer*, we see genre awareness as central to the development of mature writers—that is, increasingly independent learners who are able to recognize patterns and see similarities and differences across genres and, finally, to articulate their ideas within and against these genres for specific purposes. In keeping with the notion of genre awareness, however, our courses are not predicated on the *mastery* of the genres we introduce, per se, but on the negotiation of convention and audience across contexts. Devitt explains:

> Even if explicit teaching did help students acquire genres, even if teachers could know and articulate every nuanced detail of every genre, no writing class could possibly teach students all the genres they will need to succeed even in school, much less the workplace or in their civic lives. Hence the value of teaching genre awareness rather than acquisition of particular genres. The choice of genres to use as exemplars and assignments then must derive from the place of the course in its institution and the teacher's goals within that institution. (205)

Because we hope to foster the independent learning that comes from teaching genre awareness, and because the students enrolling in our FYW courses will move into unpredictable disciplinary contexts with different (and sometimes conflicting) writing conventions, we believe that teaching genre awareness lends itself well to any writing course, but particularly first-year writing courses. And we believe this not just because we recognize the diversity of genre conventions within the university, but because we hope to prepare students for writing both *in* academic contexts and *beyond:* that is, we hope to prepare writers to move into the always unpredictable future.

Of course, by asking students to examine salient political issues central to citizenship in participatory democracies, our courses are also fixed squarely within the classical Greek tradition. Grounded in a bi-fold conception of the "polis," we see politics not simply as the sphere of elected officials, but the sphere of all citizens, including college students—people compelled to work together for the benefit of their various communities and, by extension, the greater body politic. Focusing on a broad and diverse sphere of personal, civic, and academic discourse, we see first-year writing as a practice-ground for broad humanistic expression, a place where students can and should take on increasingly active roles as critical citizens in their various communities. By doing so, we hope to prepare students to be good writers *and* critical citizens, that is, socially-sensitive and politically-engaged adults capable of navigating in increasingly diverse discourse communities.

Indeed, we believe FYW courses should help students "come to voice" not simply as members of a confined academic community, but also as citizens of broader social and political domains. Influenced by Sherrie Gradin's *Romancing Rhetorics: Social Expressivist Perspectives on the Teaching of Writing,* our courses ask students to reflect on contemporary issues central to their roles as citizens in participatory democracies but also stress "writing to discover self and voice, and development of power and authority [in] one's own writing." As Gradin articulates,

> In order to be effective citizens and effective rhetorical beings, students must first learn how to carry out the negotiation between self and world. A first step in this negotiation must be to develop a clear sense of one's own beliefs as well as a clear sense of how one's own value system intersects or not with others, and how, finally, to communicate effectively. (xv)

Rather than embracing what we believe to be too narrow conceptions of either social-constructivist or Romantic-expressivist pedagogical traditions, our courses enact a social-expressivism consistent with both genre theory and feminist-critical pedagogy, one that "encourages students to work at coming to voice" even in public atmospheres and unfamiliar genres where they may not feel comfortable or may "see themselves at risk" (hooks 53). To this end, first-year writing at Binghamton engages students in what Kurt Spellmeyer and Richard Miller call "the most pressing problems of our times" (para. 12): our aim is to give students practice grappling with complex public problems (world health crises, global climate change, and religious, racial, and sexual violence, for example) in ever-changing literacy contexts (that is, contexts increasingly saturated with new varieties of textual and digital communication). In the ways that it can bridge civic participation with literacy acquisition, we see genre theory as an important extension of writing studies' turn toward the social.

Description of First-Year Writing Courses: Writing 111 and Writing 100

Our signature course, Writing 111—Coming to Voice: Writing Personal, Civic and Academic Arguments—asks students to write in different genres for a range of audiences, engage in intensive revision, and practice critical thinking through genre-diverse writing assignments. These assignments include a personal essay that connects private experience with a salient public issue; a rhetorical analysis that leads to a close reading of a single, complex text; an op-ed piece that responds to a local, national, or international controversy; and a researched argument that synthesizes a variety of perspectives on a salient social issue and integrates the writer's perspectives alongside and against published research. Through its emphasis on genre negotiation, audience awareness, and civic and scholarly participation, Writing 111 reinforces the notion that conventions differ according to their rhetorical situations. While Writing 111 asks students to apply genre awareness through the production of texts, the course also engages students in critical discussion of rich, challenging readings.[8] In sum, the course aims to help students develop the habits of mind necessary to succeed in future course environments where genre and audience expectations will differ, and where decreasing levels of instructor feedback will likely be the norm. By focusing on salient social issues and asking students to both research others' and articulate their own perspectives, Writing 111 is also in keeping with one of rhetoric and writing studies' most time-honored traditions: that is, to nurture in writers a sense of social responsibility as active agents in ideologically diverse discourse communities.

To provide a common first-year experience for Binghamton University students and—perhaps just as importantly—to help students recognize that the expectations of college writing are often quite different from those they practiced in high school, Writing 111 uses a shared syllabus, a portfolio that constitutes seventy percent of the course grade, and a *multi-reader portfolio grading system that seeks to articulate clear community standards for the four basic letter grades* (as the syllabus notes, our portfolio evaluation system is modeled after one fine-tuned at Grand Valley State University). Because of our commitment to process pedagogies, the course uses a deferred grading system: students turn in early and polished drafts for peer review and teacher commentary, but do not receive formal grades until the end of the semester, when they submit a final portfolio to a small group of instructors who come to consensus on an appropriate letter grade. By using deferred but common grading practices, our portfolio system not only ensures that students have plenty of time to get feedback on, re-imagine, revise, and polish their writing; it also reinforces the expectations of the first-year writing program for sophisticated content, attentive structure, and purposeful style. In this way, while necessarily modest in scope, Writing 111 seeks to give students a solid introduction to writing and revision processes and to the analysis and negotiation of genre, emphases that discipline-specific composition courses may not accentuate to as large a degree. In other words, the hope is that we are providing students with the practice and reflection necessary to move into discipline-specific writing courses in which they will need to engage in analyses of genre conventions, audience expectations, and revision practices in increasingly independent ways. What's more, because we successfully argued that our course capacity should be limited to sixteen students per section (in an effort to offer both high-quality instruction and reverse the labor exploitation of graduate students noted in Moore Howard and Sawyer's consultant report), Writing 111 teachers are able to give students the kind of sustained feedback and guidance that previous 100-level incarnations of composition at Binghamton were not.

Our second course—Writing 100: The Academic Writing Workshop—is a one-credit co-enrollment course designed to support students who appreciate additional tutoring in Writing 111. In this way, Writing 100 is the nexus of our collective efforts within the Writing Initiative, bringing together theories from first-year composition, the writing center, and the ESL program to support Binghamton's diverse student population. Despite their academic propensities in an array of sophisticated disciplines, we recognize that not all of our undergraduates come from educational or linguistic backgrounds that prepare them to write successfully without sustained support. Still, because we are committed to normalizing our instructors' attitudes towards language and educational diversity, as well as our student body's conceptions of what it means to live and learn in an increasingly diverse world, we believe in the merits of mainstreaming. As such, through Writing 100, we offer students who wish to receive additional writing support weekly seventy-five minute workshops with a tutor-teacher and no more than five students. These student-centered workshops are held in computer labs, where students receive mentored help while working on Writing 111 assignments.[9] Because students receive independent course credit and participation-based grades in Writing 100, and because we are sensitive to some students' desires to keep tutoring relationships confidential, tutor-teachers do not communicate formally with Writing 111 instructors about students' performance. Instead, Writing 100 tutor-teachers serve as readers of student writing who understand the assignments and goals of Writing 111, but who evaluate students according to their engagement and participation in Writing 100 alone. We believe this is appropriate not only to ensure student privacy, but because students are more apt to develop productive learning relationships with tutor-teachers if they see them as coaches rather than evaluators of their work.

Despite various institutional forces that may have persuaded us to the contrary, and following scholarship in the vein of Mike Rose's *Lives on the Boundary* and Ryuko Kubota and Kimberly Abels' "Improving Institutional ESL/EAP Support for International Student," we also made a purposeful decision *not* to describe Writing 100 as grammar-intensive, basic, or in any way remedial. Indeed, the course is *optional*, not *mandated*. Although some students choose to engage their self-identified grammar concerns in Writing 100, the primary emphases of the course include the development of sophisticated content and coherent structure, and the appropriate evaluation, use, and citation of sources. By characterizing the course as an opportunity to take control of one's learning rather than as a pre-disciplinary enterprise, we hope that students see Writing 100 as a proactive step in their academic enculturation. In the process, we have in large part avoided the stigma and controversy sometimes associated with "developmental" writing.

INSTITUTIONAL AND BUDGETARY CONSTRAINTS AFFECTING PROGRAM DEVELOPMENT

Perhaps the most controversial aspect of our program, however, has not been the development of Writing 100, but our decision to abandon literature-based composition courses and require teachers to use a shared syllabus in Writing 111. This decision was influenced by interconnected constraints, including the institution's desire to develop a cohesive learning community for first-year students, the problems associated with offering literature-based composition as the default composition course across campus, and FYW's commitment to ensuring that graduate assistants are not burdened with more work than they are prepared—or paid—to take on. In the old system, because of the labor-intensive work of designing their own courses, the higher course capacities, and the lack of formal teacher preparation, we believe that graduate students were working too many hours and that many first-year students were not getting the kind of genre-based immersion in the writing process that every undergraduate can benefit from. Although the movement away from literature-based instruction and toward a shared syllabus was upsetting to some, we assert it was the best decision given our commitment to strengthening both undergraduate and graduate education.

But while we acknowledge the complications involved in initiating a shared syllabus, particularly because the majority of our instructors—at least for the time being—will continue to come from a literature-centric English department,[10] we disagree with assertions that a shared syllabus creates a "rigid" environment where instructors are inappropriately robbed of professional autonomy (Marshall 416). On the contrary, we feel that a shared syllabus is beneficial not only in helping to develop a cohesive learning community for first-year students, but that it is a responsible choice for a writing program staffed primarily by graduate students with limited exposure to rhetoric and writing studies and little experience in teaching process-based, genre-focused composition. Said another way, our shared syllabus unifies our curriculum, providing first-year students opportunities to support each other *as it* provides our graduate assistants the support they need to balance the demands of their graduate coursework with the demands of writing instruction. Although our largely graduate-student staff is an intelligent, capable, and highly motivated group of teachers, because graduate assistantships should not require the same amount of labor expected of full-time faculty experts, we believe that the shared syllabus is not only entirely appropriate, but that it is essential if we are to enact the goals of first-year writing, support WAC, and flourish within the limited budgetary and institutional confines we have inherited.

In our efforts to support this common curriculum and the teachers who implement it, we have developed a threefold approach to teacher preparation that is both comprehensive and innovative. First, before graduate assistants enter our classrooms, they are now required to take a four-credit graduate seminar, "Rhetoric and Composition Theory and Practice." Appropriately, this seminar has a pedagogical emphasis, asking graduate students to develop a teaching philosophy, research current instructional trends, create a series of classroom-based activities, and observe their graduate student colleagues currently teaching first-year writing. But the seminar also reaches beyond the immediate praxis of the classroom, giving graduate students a thorough introduction to the major theories and intellectual traditions that inform scholarship and pedagogy in the field. Following Betty Pytlik and Sarah Liggett's *Preparing College Teachers of Writing*, we believe that theoretically grounded teacher preparation must expose new teachers not just to program practice, but also to the histories and theories that influence work in the discipline.

Second, we now require all new instructors to participate in a multi-day pre-semester orientation. Again mixing practical applications with disciplinary theories, the orientation introduces new teachers to our major assignments, demonstrates heuristics to teach core concepts, and compels new and returning teachers to collaboratively articulate, examine, and justify our evaluation practices. We also see our orientation as an important opportunity for teachers to meet, form bonds, and forge the supportive professional relationships we believe are vital to a strong culture of teaching. In the spirit of Bill Hendrick's "Working Alone Together: Labor Agency and Professional Exchange in the Teaching of Composition," we work to offer writing teachers frequent opportunities for "ongoing professional development [. . . and a] willingness to question old habits and test out new ideas" (236).

Third, and to this same end, all of our teachers gather in small groups for weekly meetings led by experienced full-time faculty or graduate student administrators. Sharing goals similar to Margaret Marshall's "Teaching Circles," our pedagogy groups instill a sense of community and support for instructors through candid discussions among beginning and experienced teachers (414). In sum, pedagogy groups allow us to discuss classroom dynamics, share teaching activities, examine scholarly trends and—most centrally—practice effective strategies for responding to and grading student writing. At the end of the semester, these same small groups come together to collaboratively grade student portfolios, an important opportunity for teachers to reflect on how successfully they have internalized the grading criteria embraced by the program as a whole. More than just weekly meetings, however, we view these small groups as learning communities for our instructors. Reflecting the same kind of social and educational benefits that inform our conception of a first-year learning community, pedagogy groups have been an important emphasis in the formation of first-year writing at Binghamton. Just as undergraduate "development is enhanced when there is extensive interaction among students who study together, tutor one another, and discuss issues that matter" (Smith et al. 142), faculty development is enhanced through shared work and ongoing pedagogical conversation. These pedagogy groups also allow us to better refine and articulate the common learning outcomes of our newly established first-year writing courses.

IF WE KNEW THEN WHAT WE KNOW NOW

While we are proud of our work during 2008–2009, looking back, there are certainly things that we would do differently. We would now, for instance, vet all our teachers more thoroughly. We would ask them candid questions about their attitudes toward and willingness to support the changes taking place. We would make a deliberate, concerted effort to hire only teachers who want to teach writing, who value composition, who enjoy working with students, and who are willing to embrace the goals we have developed. While we wish to note that all but a small fraction of teachers who taught with us the initial year were supportive of these goals, and that several under-vetted instructors became among the strongest teachers in our program, there were still a few who worked intentionally to undermine the program. As we move into our second year, we have had the time and space to carefully vet all of our teachers, and thus have the benefit of welcoming into our program committed, well-prepared instructors who understand and are supportive of first-year writing.

In future years, we also hope to do a better job of introducing new instructors to the materials we have created to support our courses, and of reinforcing the notion that these materials are not meant to close down opportunities for classroom innovation. To be sure, we enthusiastically encourage teachers to augment our tentative schedule of activities according to their personal teaching styles and students' learning needs. In hindsight, we realize that after our first orientation some teachers—having designed their courses independently in the past—saw the shared syllabus, tentative schedule of activities, and other teaching artifacts as confining rather than supportive. In the future, we plan to do a better job of addressing their anxieties. To this end, in fall 2009, we created a more nuanced presentation of the goals of our shared syllabus, as well as a clearer explanation of the pedagogical requirements and classroom innovation we hope to foster as first-year writing evolves and changes.

While there will always be some teachers who resist a composition curriculum that does not embrace literature as its raison d'être, since our first orientation, we have also made important strides to welcome all of our teachers—regardless of their intellectual concentrations—into the scholarly community of rhetoric and writing studies; this is perhaps our most important piece of advice for program directors working with graduate students. Indeed, it is not until graduate students have taken a course in writing studies that they can begin to see the ways in which rhetoric and composition's theories and pedagogies support their own conceptions of good teaching and their own experiences of learning to write. It is also through graduate coursework that new writing teachers begin to see the ways in which the intellectual tradition of rhetoric and composition coincides with—but necessarily deviates from—their disciplinary specializations in literature, creative writing, or other fields in the academy.

Likewise, by offering graduate students the opportunity to earn teaching certification in college composition,[11] in our second year we are giving teachers incentives to attend professional development forums that can enhance their knowledge of writing pedagogy and rhetoric and writing studies beyond coursework and orientation. To support the certificate as well as to grow a stronger community of teacher-scholars, in 2009–2010, we are offering both a pedagogy workshop series and a vibrant visiting speaker series.[12]

Finally, perhaps the most successful way we have found to enhance graduate students' professional engagement in the field is to devote budget resources to support attendance to regional and national conferences. In 2009, with the proceeds from the inaugural publication of *Binghamton Writes: A Journal of Undergraduate Composition*,[13] we were able to give full travel and accommodation support to seven graduate students presenting papers at a regional conference—the State University of New York Council on Writing at Buffalo State—and to eight different graduate students presenting papers at the Conference on College Composition and Communication in San Francisco. We plan to expand this support in 2010. By giving teachers ample exposure to the scholarly discourses and professional opportunities of writing studies, we continue to grow a stronger cohort of graduate student writing teachers—that is, a critical mass of well-prepared writing instructors eager to engage in the best practices of the field. Participation in such opportunities not only increases morale within the Writing Initiative, it forecasts how much Binghamton University's commitment to first-year writing, we hope, will continue to grow in the future.

Coda: Where We Are Now

As we look back over the first six years of first-year writing at Binghamton, the fluctuation and interconnectedness of writing program work resonate with our experiences and, perhaps more significantly, suggest an ecology of writing programs that is more than site-specific.

The fluctuations include the peaks and valleys our staffing structure has experienced over time. In general, the working conditions for the majority of writing program faculty have improved, a result of our reduction of course capacities, an increase in teaching assistant preparation, and an almost doubling of the per course wage our small number of adjuncts earn. Prior to the establishment of the Writing Initiative, not only were teaching assistants underprepared and overworked, but adjuncts were only making $2,500 per course section, and only one full-time faculty member was supporting the program. In 2014, first-year writing is supported by seven full-time faculty, five of whom are in tenure-line positions, and all appointed on the basis of their credentials and experience in the teaching of writing. Now, instead of relying on dozens of underpaid adjuncts, we have only two (well-vetted and thoroughly prepared) three-quarter-time instructors, both on an annual contract with full benefits, and making no less $4,500 per course.

In the four years that have passed since "Back to the Future" debuted in *Composition Forum*, our program has grown in substantial ways, particularly through our participation in our campus's residential learning community program, and in the expansion of our collaborations with the Educational Opportunity Program (EOP). In conjunction with the 2012 presidential election, we spearheaded eight election-themed sections of Writing 111 that were taught in four of Binghamton's residential learning communities, a collaboration that culminated in an election-night celebration where over 300 first-year students watched the returns, enjoyed a pancake dinner, and experienced the political theatre that revolves around election night coverage.

We also put in place a new first-year course developed to meet the needs of Binghamton University EOP students, Writing 110: Seeing and Writing the World. Its initial iteration was a collaborative effort among Kinney, Costello, and writing center director Paul Shovlin, and the course has since evolved to include sections taught by all of the Writing Initiative's full-time faculty. Adopting Donald and Christine McQuade's groundbreaking textbook, *Seeing and Writing*, the course is designed to help students gain confidence and fluidity in their writing, and it has been praised by students for helping them learn the habits of mind that separate high school from university writing conventions.

But alongside these successes, we have also experienced setbacks, including the snail's pace at which the university has moved to confer our program department status, and the derailment of plans to build a graduate program in writing studies. Although the current dean of our college of arts and sciences has assembled a writing advisory committee to make recommendations on our future status, and while faculty in three departments on campus have informally discussed collaborating with us on an MA in writing studies, we have also lost a spate of visiting professors, outstanding colleagues who just couldn't jeopardize their professional futures in the hope that we *might* convert their lines to permanent tenure-track positions, that we *might* develop a graduate program in writing studies.

This year we will also bid adieu to Costello, who began her work at Binghamton as a graduate teaching assistant and then transitioned into the role of full-time faculty member and associate director of first-year writing, but who is now assistant professor and director of the Writing Program at Arkansas State University. Costello is optimistic about the future of writing instruction at Arkansas State, to a large degree because of the lessons she's learned at Binghamton. Together with campus administration and a newly formed composition committee, she is making plans to create learning objectives for first-year writing courses, a credit-bearing seminar course for teaching assistants, a co-enrollment tutorial course for at-risk students, and a paid pre-semester faculty orientation. Costello is prioritizing these efforts because she learned from her experiences at Binghamton how professional and financial support of writing instructors can translate into better working conditions for faculty, and better learning conditions for students. She also saw how the Writing Initiative's successful partnerships with the residential learning communities and the EOP improved the educational experiences of first-generation college students and provided spaces for faculty from various programs to collaborate, teach, and support each other. Based on the successes of these relationships, Costello is building similar associations with Arkansas State's First-Year Studies and Living Learning Communities initiatives.

As Arkansas State's Writing Program is building and shaping its own identity, it is easy to see Binghamton's footprint on the budding program, just as it is easy to see how writing programs at Grand Valley State and Ohio University—where Kinney had her first job and did her graduate work—have had influence on the Writing Initiative at Binghamton. Given the strong culture of writing our program has helped promote at Binghamton, we'll end this updated version of our program profile with a bit more horn blowing: we are proud to have been awarded the Conference on College Composition and Communication's Certificate of Writing Program Excellence, an honor we were given in 2011. But because of the interrelatedness of WPA work, we see this honor as one we share with those programs and colleagues that have helped shape our pedagogies and practices, as well as those that will continue to shape us into the future.

Notes

1. As we explore in more detail below, in addition to being the home of first-year writing, the Writing Initiative also houses the writing center and the English as a second language program.

2. For a description of these general education composition components, see: http://compositionforum.com/issue/21/binghamton.php#appx1.

3. See Binghampton University's "About" website at: http://www.binghamton.edu/about/at-a-glance.html.

4. As director and assistant director of first-year writing, our work is focused on teaching assistant preparation in the Writing Initiative, not faculty development across the entire campus. Given our respective ranks as tenure-track assistant professor and PhD student, the scope of our administrative responsibilities is appropriately limited.

5. We wish to note that the past WAC Director, who remains on the faculty at Binghamton, also remains supportive of WAC and other campus writing efforts, including FYW. That said, the institution has not moved to hire a new WAC Director, and thus there is a large hole in the WAC infrastructure.

6. The term "initiative" was chosen as the unit's initial name—rather than "program" or "department"—because Binghamton's faculty governance bodies have not yet authorized our status as an official academic unit. Currently, we are moving to seek program status as we expand our undergraduate course offerings beyond first-year writing. Should we receive the institutional support to grow a graduate program and/or an undergraduate major, we may seek department status.

7. There are at least two practical reasons we have not moved quickly toward a first-year writing requirement, and a third theoretical one. The first is that Binghamton University has a long tradition of honoring student choice and thus *not* requiring courses: we knew that, appropriately, our student body would ruffle should the requirement be mandated without careful planning and consultation. The second reason, again (see footnote above), is that there are lengthy campus governing procedures that must be followed prior to establishing a requirement: while we were eager to offer first-year writing, we were also keenly aware that governance bodies would be hesitant to embrace a requirement without extensive data to support its value. In order to make the best case for first-year writing, we need time to collect this data, as well as to consider the pros and cons of establishing a requirement. This leads us to our third reason for moving cautiously: as scholars sympathetic to Crowley's abolitionist argument, we feel no need to rush. We have chosen to work slowly and with discipline in order to gauge the degree to which students enrolling in first-year writing value our courses and succeed beyond them. With just a year under our belts, we are happy to report that the vast majority of first-year students *elect* to take Writing 111 and that our assessment measures suggest that they find the course valuable. We see these as strong indicators that our courses are successful—and that a requirement may not be our only option for improving student writing.

8. For a list of our common texts and syllabus, see: http://compositionforum.com/issue/21/binghamton.php#appx2.

9. We wish to acknowledge one of Kinney's former colleagues at the University of Notre Dame, Connie Mick, whose writing center and writing program acumen helped inspire Writing 100.

10. As we expand our course offerings, we plan to diversify our instructor base to include graduate students from departments across campus *and* grow a critical mass of specialists committed to the teaching of writing; in our mind, a crucial step in this process will be the development of a graduate program in writing studies.

11. See http://compositionforum.com/issue/21/binghamton.php#appx3 to view the certificate requirements.

12. See http://compositionforum.com/issue/21/binghamton.php#appx4 for a description of the workshop and speaker series.

13. See http://compositionforum.com/issue/21/binghamton.php#appx5 for the "Introduction" to *Binghamton Writes*.

Works Cited

Bawarshi, Anis. *Genre and the Invention of the Writer: Reconsidering the Place of Invention in Composition.* Logan: Utah State, 2003. Print.

Beaufort, Anne. *College Writing and Beyond: A New Framework for University Writing Instruction.* Logan: U of Utah P, 2007. Print.

Crowley, Sharon. *Composition in the University: Historical and Polemical Essays.* Pittsburgh: U of Pittsburgh P, 1998. Print.

Devitt, Amy J. *Writing Genres.* Carbondale: Southern Illinois UP, 2004. Print.

Downs, Douglas, and Elizabeth Wardle. "Teaching about Writing, Righting Misconceptions: (Re)Envisioning 'First-Year Composition' as 'Introduction to Writing Studies.'" *CCC* 58.4 (2007): 552–584. Print.

Fishman, Jenn and Mary Jo Reiff. "Program Profile: Taking the High Road: Teaching for Transfer in an FYC Program." *Composition Forum* 18 (2008): <http://compositionforum.com/issue/18/tennessee.php>. Web.

Gay, Pamela and Elizabeth T. Tricomi. *The Binghamton University Writing Requirements Handbook.* Binghamton: State University of New York, 1996. Print.

Gradin, Sherrie. *Romancing Rhetorics: Social Expressivist Perspectives on the Teaching of Writing.* Portsmouth, NY: Heinemann Boynton/Cook, 1995. Print.

Hendricks, Bill. "Working Alone Together: Labor Agency and Professional Exchange in the Teaching of Composition." *Pedagogy: Critical Approaches to Teaching Literature, Language, Composition, and Culture* 9.2 (2009): 235–260. Print.

Herrington, Anne and Charles Moran. "The Idea of Genre in Theory and Practice: An Overview of the Work in Genre in the Fields of Composition and Rhetoric and New Genre Studies." *Genre Across the Curriculum.* Eds. Anne Herrington and Charles Moran. Logan: U of Utah P, 2005. 1–18. Print.

hooks, bell. *Feminist Theory: From Margin to Center.* Boston: South End Press, 1984. Print.

Kubota, Ryuko and Kimberly Abels. "Improving Institutional ESL/EAP Support for International Students: Seeking the Promised Land." *The Politics of Second Language Writing: In Search of the Promised Land.* Eds. Paul Kei Matsuda, Christina Ortmeier-Hooper, and Xiaoye You. West Lafayette, IN: Parlor, 2006. Print.

Marshall, Margaret J. "Teaching Circles: Supporting Shared Work and Professional Development." *Pedagogy: Critical Approaches to Teaching Literature, Language, Composition, and Culture* 8.3 (2008): 413–431. Print.

Pytlik, Betty P., and Sarah Liggett, eds. *Preparing College Teachers of Writing: Histories, Theories, Programs, and Practices*. NY: Oxford U P, 2001. Print.

Rose, Mike. *Lives on the Boundary: A Moving Account of the Struggles and Achievements of America's Educationally Underprepared*. New York: Penguin, 1989. Print.

Smit, David. *The End of Composition Studies*. Carbondale: Southern Illinois UP, 2004. Print.

Smith, Barbara Leigh, Jean MacGregor, Roberta S. Matthews, and Faith Gabelnick. *Learning Communities: Reforming Undergraduate Education*. San Francisco, CA: Jossey-Bass, 2004. Print.

Spellmeyer, Kurt and Richard E. Miller. "Teaching the Action Horizon." *The New Humanities Reader*. Boston: Houghton Mifflin/Cengage Wadsworth, 2008. Web.

7 Imagining a Writing and Rhetoric Program Based on Principles of Knowledge "Transfer": Dartmouth's Institute for Writing and Rhetoric

Stephanie Boone, Sara Biggs Chaney, Josh Compton, Christiane Donahue, and Karen Gocsik

Introduction

Dartmouth's Institute for Writing and Rhetoric (IWR) is a program in dynamic flux, with a long, strong history and an evolving vision for the future. We are collectively thinking about transforming our program design to intentionally foster writing knowledge transfer, operationalizing the available research about transfer while remaining open to the evolution of this research over time. This program profile is a collaborative attempt to identify the reasons for such a move and the shapes it might take, in hopes that other programs will find our process and thinking useful. The profile overviews our program, its current underpinnings, and the transfer research that might ground our program's future shape. We offer a sketch of an ideal program that might foster opportunities for students to transfer writing and speech knowledge across their studies and beyond. Finally, we discuss what we have learned so far about our students and our program design, the curricular and faculty development initiatives with which we are experimenting, and what remains to be considered.

We use the word "transfer" in this profile with full acknowledgement of its contested and complex nature as a term. Learning does not occur "in" an individual who moves it from context to context, but in the ongoing relationships between the individual and the activity systems she occupies and shapes. Writing knowledge and know-how do not simply move from one context to another; they adapt, transform, orient, are re-imagined and newly applied; they change the context in the process and are changed by the process. "Transfer," explored recently in a Fall 2012 special issue of *Composition Forum*, is the label we use to index this broader dynamic. We understand it as the very heart of learning—how it occurs and how it is sustained. For this reason, Dartmouth's Institute for Writing and Rhetoric is beginning to re-imagine a writing program that puts facilitating the possibility of transfer at its core.

Our Histories, Then and Now

Formally established in 2008, the IWR stands as the largest academic enterprise on Dartmouth's campus, offering 125 courses in writing and speech per year. While the Institute offers a healthy slate of speech courses and a growing list of upper-level writing courses,[1] the bulk of the Institute's course offerings are first-year writing courses, including Writing 5 and the writing-driven, discipline-inspired first-year seminar, which together comprise the two-term writing sequence required by the college. Underprepared writers are invited to enroll in Writing 2–3 in lieu of Writing 5, but before taking their first-year seminars.[2] Speech 20: Public Speaking is the core speech course—an optional course open to students from all disciplines and years of study. Optional upper-level speech courses (e.g., Intercultural Communication, Persuasive Public Speaking, Speechwriting, Rhetoric of Social Justice, Legal Rhetoric, Resistance to Influence) expand ways of thinking, theorizing, and doing speech. Upper-level writing courses currently include Writing with Media, Writing in the Workplace, Arguments in Context, Writing and Speaking Public Policy, The Art of Science Writing, and Composition Theory and Practice. Among Ivy League institutions, Dartmouth is unusual in requiring a sequence of first-year writing courses; our peer institutions typically require only one composition course (although they generally operate on semesters as opposed to Dartmouth's ten-week quarter system) or do not have a writing requirement at all.

The sequencing of our courses provides the IWR with the opportunity to examine two challenges: the challenge of understanding how students transfer their knowledge and know-how about writing and speaking from one course to another, and the challenge of determining how to create a coherent curriculum in which the possibility of transfer is fostered. It is interesting to note that although faculty have articulated them differently over time, these questions regarding transfer—why writing education hasn't seemed to "stick" after freshman English and what to do about it—have persisted at Dartmouth for more than a century. Even a cursory look at archival documents will reveal that Dartmouth faculty, like their counterparts in institutions of higher education nation-wide, were engaged in a decades-long preoccupation with whether writing, on the one hand, is a skill to be mastered *prior to learning*—and then easily transferred to other occasions for writing—or whether writing is an integral *part of learning itself*—in which case transfer becomes situational, more slippery to grasp and to apply. And now, we are beginning to ask similar questions about speaking and other modes of communication.

While early twentieth-century attempts at curriculum design fell firmly in the former camp, these questions found interesting articulation in the 1960s when two significant events occurred on the Dartmouth campus, both sponsored by the Carnegie Foundation. The first was Albert Kitzhaber's 1962 Report of the *Dartmouth Study of Student Writing;* the second was the 1966 Dartmouth Seminar, an Anglo-American conference organized around the question "What is English?"—a conference that many acknowledge as cementing the establishment of the field of rhetoric and composition as we know it today.

Kitzhaber's work at Dartmouth was driven by two questions: First, can composition at Dartmouth be taught more effectively? Second—and perhaps more pertinent to our discussion—is there a way to ensure that students will continue to write well after their first-year writing courses? Or, put in current terms, how might we design our courses to encourage transfer? In order to answer these questions, Kitzhaber counted errors in papers submitted across Dartmouth's previous two-course sequence, English 1 and English 2. He found that certain kinds of error occurred more frequently as students moved forward in their writing education—for instance, errors in focus and structure occurred more frequently at the end than at the beginning of English 1, and errors in punctuation and mechanics increased rather than decreased in frequency by the end of English 2 (Kitzhaber 58–59). Interestingly, while these findings of increased frequency of error suggested that writing is not a skill which, once learned, will inevitably be transferred (a finding which many studies would later confirm), Kitzhaber's advice for addressing the issue of failed transfer was most unsatisfying. Be strict in your grading, he instructed professors—and not only writing professors, but professors of subsequent classes (*The Writing of Dartmouth Students after the Freshman Year* 195).[3] Kitzhaber assumed that the presence of high and uniform standards would encourage (though not by themselves accomplish) the transfer of knowledge.

At the 1966 Dartmouth Seminar, Kitzhaber, though no longer a Dartmouth faculty member, was invited to deliver the conference's keynote speech, "What is English?" While the dynamics of the conference are too complicated to summarize tidily, it is perhaps fair to say that the Americans, represented by Kitzhaber, advocated in favor of defining English as a discipline that included language, literature, and composition. The cohort from the United Kingdom, represented by James Britton, countered that "the key question was not: 'What is the subject matter of English?'—but rather: 'What do we want students and teachers to be doing?' His answer was to define English as that space in the curriculum where students are encouraged to use language in more complex and expressive ways" (Harris 634). Britton's emphasis on *what students are doing* might be understood as promoting the second notion of writing and transfer mentioned above—i.e., that writing is part of learning itself, and that the transfer of knowledge and know-how from task to task is slippery, complex, and situational.

Although the stage at Dartmouth was set to explore how transfer works (or does not work) in the situated construction of knowledge, a lack of coherence in the administrative structure of the writing programs impeded our progress. In 1966, a few months before the Dartmouth Seminar, Dartmouth had adopted the aforementioned expository writing/first-year seminar writing sequence, adding a two-term option for underprepared students in the early 1980s. While individual course sections offered students effective and even excellent teaching, a different administrator directed each of the three writing programs (English 5, the two-term English 2–3, and the first-year seminar), and each of these administrators operated in isolation from the others. One faculty member described the administrative structure as "hydra-headed." Growth without communication created a structure that operated robustly in its parts, but was not set up to function effectively as a whole.

It wasn't until 2004 that, subsequent to an external review, the three programs—now Writing 5, Writing 2–3, and the first-year seminar—were brought under a single administrative entity in the freestanding Dartmouth Writing Program. When speech was added in 2008 after existing as a one-person "Office of Speech," suspended some years prior, the Writing Program became the current Institute for Writing and Rhetoric.[4] The aforementioned upper-level courses were added to the Institute's course offerings, along with Speech 20 and a number of upper-level rhetoric courses.

Overall, the establishment of the Institute as an autonomous entity proved to be an important moment in the evolution of Dartmouth's writing culture, in that the Institute provided an administrative structure upon which a cohesive writing and speech program might be built. While the Institute operates independently of other departments and programs and has its own core faculty who staff the first part of the two-term writing sequence, along with colleagues from the English department, the Institute also draws from faculty across the disciplines to staff the first-year seminars. The Institute's establishment as an autonomous entity, its strategic sequence of writing courses, its interdisciplinary composition—all affect the way we think about, theorize, and do writing and speech instruction, and all affect, in ways positive or not, our opportunities to facilitate writing transfer. This profile explores these issues.

Our Theories, Once and Future

Initial Foundations

Over time, the Institute for Writing and Rhetoric developed an understanding of our official theoretical underpinnings that became loosely assembled in the phrase "active learning"—a term that we defined (somewhat contrary to common use) as a student-centered pedagogy that engages students deeply and collaboratively in the various processes that comprise composition. In our writing and speech classrooms, active learning pedagogy manifested itself in a variety of ways, including collaborative learning (in the tradition of Kenneth Bruffee's "Conversation of Mankind" and further enlightened by the work of John Trimbur, Lisa Ede and Andrea Lunsford, and Rebecca Moore Howard); peer-to-peer instruction (as informed foundationally by Peter Elbow and Ann Ruggles Gere, and developed later through the work of Eric Mazur and many others); and learning via technology—which includes writing with wikis, or reading and writing multimodal composition (nourished by the various work of the members of The New London Group and validated by Kathleen Blake Yancey's keynote speech at the 2004 CCCCs).

Speech courses are similarly grounded in the IWR's particular "take" on active learning methodology. In Speech 20: Public Speaking, students enact a collaborative, active model of public speaking. A Speech 20 syllabus notes that students "test public speaking against a criteria of good dialogue" and "explore speechmaking processes, products, and connections between processes and products. Speech classes also feature applications of rhetorical, social scientific, and critical theories. As the current draft of a shared Speech learning outcome puts it: "Students will understand, and, when applicable, apply and extend, communication theory, concepts, or models, critically engaging theory into practice in societal contexts." Speech courses, like writing courses, reflect active processes of discovery and learning that should go well beyond the classroom walls of an individual class.

It is important to note here that this collection of methods has been more broadly inspired by scholarship on issues of power as they play out in the classroom. Specifically, the Institute has encouraged faculty to enact Paulo Freire's notion of the "teacher-learner" by designing student-driven learning environments that shift to students some portion of the authority that has traditionally been theirs. Students, in turn, have been expected to take increased responsibility for their writing educations by becoming "learner-teachers"—which typically requires students to lead discussions or engage in peer-to-peer instruction, and may go so far as to ask them to design their own writing or speaking assignments, or to collaborate with the instructor to determine course aims and assess course work. Best practices in active learning and results in the transfer scholarship clearly overlap: for example, responsibility for one's learning can increase the motivation that fosters transfer; student-designed assignments can foster the metacognition that facilitates transfer and offers students the opportunity to actively formulate transferable principles; and peer-to-peer instruction, well designed, might afford transfer across assignments or courses.

As we understand the term, active learning is also grounded in constructivist understandings of students' development. In the field of writing studies, we see these understandings in both the write-to-learn strand and the broader social constructivist strand. As George Newell suggests, "knowledge develops within particular instructional contexts when students are *actively engaged*" (our emphasis), and knowledge only "takes" when it becomes "knowledge in action," a step that writing and speaking can enable (236). Vygotskian models of zones of proximal development are particularly important theoretical frames that support our attention to the peer collaboration and dynamic classroom interactions we consider essential to active learning, although Lev Vygotsky's work has infrequently been an explicit topic of broad-based discussion for us. Our stated commitment to these various methods is proving invaluable to our program as we begin to navigate our way through the complicated terrain of knowledge transfer.

The brief overview of relevant transfer scholarship that follows hints at what might be, even as we recognize that the field and the IWR still have much to do in order to operationalize the research knowledge. The two sections following this overview offer a description of our current institutional structures and practices and their relationship to facilitating—or hampering—transfer, and then provide a sketch of an "ideal" program that could operationalize the research in ways that are compatible with our goals.

Evolving Directions

As we have already suggested, active learning principles sometimes intersect with the research to which the IWR is considering turning: research on fostering knowledge transfer. Active learning approaches in general have been justified, in terms of neuroscience and cognitive psychology, by the work of John Bransford and colleagues, whose research also figures centrally in transfer discussions. His team emphasizes the role of activity and engagement in student learning, as well as the importance of scaffolding, practice, analogy-driven learning, engagement with preconceptions, attention to the combination of factual knowledge, conceptual knowledge, knowledge organization, and development of metacognitive approaches. Writing studies have further developed many of these concepts that the transfer research has been developing in other fields for decades.

Several longitudinal studies of student writing in the past couple of decades have offered warnings to the field of composition about the challenges of transfer. Even as these studies were not necessarily focused on "transfer" by name, they have given sharp insight into successes, failures, and unintended consequences in the program structures and student experiences they have analyzed. The smaller-scale case studies (see, for example, Beaufort; Herrington and Curtis; Haas; McCarthy; Chiseri-Strater) identify individual students' experiences of disconnect between first-year writing or general education and their majors, between university writing and writing in their subsequent professional lives or graduate studies, and across disciplines in college. The larger scale studies (including Sommers and Saltz; Lunsford and Lunsford; Haswell; Carroll; Sternglass) report important nuances of students' development: a direct relationship between deepening expertise in their majors and writing competence; disconnects between in-school and out-of-school writing experiences; "regression" in writing when students are faced with tasks of increased complexity; and direct, unintended consequences in students' development because particular teaching approaches led students to back away, in new contexts, from strategies that posed problems for them in the past rather than mastering the strategies.

The rich evidence of longitudinal studies also indirectly supports what we have learned from studying the research on transfer, namely, that transfer depends heavily, although not exclusively, on our teaching: when we teach with analogies, encourage metacognition, scaffold student learning, motivate our students effectively, and provide sufficient time for the learning to "take," our teaching potentially enables transfer (Gray and Orasanu; Bransford et al; Dias et al; Haskell). Additional key elements from the broader discussions of transfer that matter to us here include understanding that vertical transfer and scaffolding (first explored by Vygotsky) are essential to effective transfer; understanding the crucial role of affordances; distinguishing among "near" and "far" transfer, as well as "forward" and "backward" transfer, and how each contributes to learning; studying the possibility that some automated learning is quite useful to overall writing development; understanding the relationship between "spontaneous" transfer and deep expertise as well as the types of expertise in question (Postman; Simon and Hayes; Perfetto, Bransford and Franks); exploring the emerging notion of "threshold" concepts in different disciplines; and recognizing the role of antecedent genres in both facilitating and hampering transfer (Bawarshi and Reiff; Reiff and Bawarshi).

When we help students to see a connection between classroom knowledge and the "real world," we are also better supporting their ability to transfer knowledge. In terms of the "real world" to which learning should be tied, we know that today's students will enter a world of unprecedented complexity that requires them to communicate with multiple audiences in multiple modes. Research suggests that students cannot transfer knowledge that is only associated with one mode or context (Eich; Bjork and Richardson-Klavhen). Based on this understanding, we have a working hypothesis that multimodal assignments can foster transfer in both of the ways mentioned above: these assignments connect to the "real world" that students will enter by requiring them to reuse abilities—such as the ability to make an argument and support it with evidence—in a new context.

Studies are also beginning to suggest that the competencies required to be digitally literate are in some ways similar to the competencies required by traditional literacies—in other words, students become digitally literate by drawing on familiar cognitive abilities and transferring these abilities to a new medium's particular set of features (Bruce; Coiro et al; Lunsford and Lunsford; Kress). To do *this* kind of transfer more intentionally, our students need to acquire the rhetorical flexibility—adaptability to context and need—that allows them to use digital media precisely as we have always helped them to use writing: to solve problems, to explore ideas, and to communicate both with themselves and with a variety of audiences around the globe.

Perhaps most important of all, we are learning in studies of transfer that how something transfers is not the same depending on what—what kind of knowledge or know-how—is being transferred (Donahue). This distinction leads naturally to the question: What kinds of knowledge are writing and speaking?—a rarely considered topic that will be central to our ongoing inquiry here at Dartmouth. In addressing this question, we have found research on the nature of expertise in relation to writing in the disciplines to be particularly helpful. While not always considered to be directly a part of the general "transfer" literature, this work offers additional ways to model a curriculum that fosters transfer opportunities.

In particular, the research on situated cognition and the model of "communities of practice" can inform a program design that seeks to intentionally foster knowledge transformation, helping us to re-imagine the nature of students' experiences moving into and out of multiple college communities. Specifically, Jean Lave and Etienne Wenger offer a model of writing development in which we do not conceptualize college as "an" academic community but rather as a "set of relations among persons, activity, and world, over time and in relation with other tangential and overlapping communities of practice" (135). Paul Prior suggests that this approach allows us to understand disciplines as slippery, open networks of human activity rather than unified social territories with stable epistemological and rhetorical moves (xi). This understanding changes our approach to students' status as novices as they move beyond the first year: "disciplinary enculturation thus refers not to novices being initiated, but to the continual processes whereby an ambiguous cast of relative newcomers and relative oldtimers (re)produce themselves, their practices, and their communities" (xi). In fact, in this process the "oldtimers" are defined by the fact that they *expect* to negotiate, to work with gray areas, to critically apply general knowledge to context—this defines, at least in part, their expertise. Rather than acquiring a set of conventions or approaches associated with a fixed discourse community, our students would need to develop that flexible expertise in order to successfully transfer their writing and speaking knowledge.

Our Structures and Practices

The IWR today has an institutional structure, course sequencing, diverse disciplinary inflections, and faculty practices that are all already priming us for a curriculum that intentionally fosters knowledge transfer. We will consider these here, noting the degree to which these conversations have begun to infuse our work and what has not yet developed, before describing, in the following section, our imagined program directions.

As noted earlier, the establishment of the Institute for Writing and Rhetoric as an autonomous structure was an important moment in the evolution of our program, in that our autonomy created new opportunities to define our outcomes, to shape our conversations, to determine our practices, and to assess our students' work. This autonomous structure also brought together Dartmouth's disparate writing courses under a single administrative umbrella, enabling us to begin building a coherent program—one that can permit, among other things, regular opportunities in the program's faculty development workshops to discuss and assess transfer of knowledge and know-how. For instance, faculty teaching in our first-year writing sequence have begun to discuss their teaching practices and how these practices might encourage (or discourage) transfer. Speech professors in the Institute are learning about possibilities of speech transfer from their writing faculty colleagues. Seminar faculty also teach courses at the upper levels, and so can bring to our conversation about transfer their perspectives on how students are adapting, transforming, orienting, and re-imagining their knowledge about writing both in the first-year seminar and beyond.

While the structure of the IWR as an autonomous program has helped to create an exciting environment in which to explore the transfer of knowledge and know-how, it is the unusual structure of our two-term sequence that deserves closer consideration. The Institute's course sequences pose both challenges and opportunities with transfer. We've learned from our own research that any program aiming for coherence across its course sequence must understand both the *hows* and *whys* of transfer: not only how learners do transfer knowledge from activity to activity, but also how and why they often do not transfer knowledge successfully to a new context. This understanding is essential if we ever hope to build the necessary conditions for learners to transfer knowledge successfully. What makes Dartmouth's two-term sequence so potentially ripe for enabling transfer (and the study of transfer) is that our students, moving from one classroom to another, discover that writing is a way to move among different communities of practice—the first introducing them to the practices of academic writing more generally; the second situating them within a discipline that employs practices that are relevant not only to the discipline at hand but that exist in relationship to other disciplines as well. Across the two writing courses, students, therefore, come to understand college not as "an" academic community but rather as the "set of relations among persons, activity, and world, over time and in relation with other tangential and overlapping communities of practice" mentioned earlier (Lave and Wenger 135).

Students are thus encouraged to work deliberately across situation and across domains. We know from the scholarship that bringing ideas, concepts, and practices from one domain into another demands a "significant cognitive retooling" (Tuomi-Grohn, Engeström, and Young 4) that appears to foster transfer. It also more easily invites students to undertake an exercise in metacognition.

In addition, in Writing 5 and Writing 2–3, students can be encouraged to look ahead to the seminar; in the seminar, they can be encouraged to look back to Writing 2–3 and 5. With this forward and backward looking should come a bigger, better understanding of how knowledge and practice work together to enable a student to write effectively in many different situations and contexts. The fact that first-year seminar

(FYS) faculty are from almost every discipline at Dartmouth contributes directly to this possibility. Writing 2–3 and 5 faculty are in conversation at least some of the time with FYS faculty, who are themselves grappling with the discipline (including the content), the writing, and the in-betweenness of a seminar (whose objective is not full disciplinary integration). In addition, FYS faculty can see anew both the writing they ask their students to do in their other contexts and the unique challenges of the first-year writing classroom. By working through the kind of knowledge and practice that each writing task requires, students and faculty should be able to collaborate to construct understandings of what does and does not transfer from discipline to discipline, from task to task, or from course to course.

Unlike writing courses, all speech courses are optional, and no speech courses have prerequisites. Consequently, the speech component of the IWR does not have a system that enables us to track evidence of transfer from speech course to speech course. And yet, we note that some students *do* take multiple speech courses, by choice. In these instances, a student will usually complete Speech 20: Public Speaking, and then later enroll in an upper-level speech course, such as Speech 30: Speechwriting or Speech 31: Rhetoric of Social Justice. Anecdotal evidence reported by Speech faculty suggests that students are successfully adapting their speech work to changing rhetorical situations and expectations. Informal conversations with students also suggest evidence of knowledge transfer from speech courses to their programs of study across disciplines and beyond the classroom, with student self-reported success in scientific presentations, for example, or in extra-curricular speaking engagements. That is, at least anecdotally, we see potential evidence of successful speech transfer. Also, student discussions in speech courses often turn to comparisons and contrasts with their writing experiences in FYS, Writing 2–3, or Writing 5, as students explore for themselves how their writing informs—or does not seem to inform—their speech, and vice versa.

Other issues of writing transfer have what appear to be clearer analogues to speech. The same autonomy of the Institute that encourages conversation and collaboration across disciplines fosters conversation and collaboration across the Institute—bringing together writing and speech faculty with similar intellectual pursuits and pedagogical aims. Scaffolding and active learning-informed pedagogy—fundamental to writing transfer across courses and disciplinary boundaries—are reflected in individual speech courses, as students build toward more complex speeches through carefully sequenced assignments and peer collaboration. The Institute's speech curriculum encourages metacognition, with reflective assignments (e.g., evaluation activities before and after speeches) that foster students' awareness of their decision-making. Faculty from across disciplines teach some writing courses; speech courses are taught only by speech faculty in the Institute. Nevertheless, through professional development sessions, interactive faculty workshops, and conversations with the disciplines (Compton), we are exploring issues of transfer that go well beyond speech courses.

In terms of the Institute's structure and implications for speech knowledge transfer, the main point is this: Writing and speech are inextricably linked in the Institute; writing and speech faculty are pursuing similar intellectual inquiries, grappling with similar challenges, and sharing similar pedagogical aims. And because we are working together, in conversation and collaboration, what we discover about transfer in writing helps to inform what we discover—or might discover next—about transfer in speech. It's quite possible that we'll learn that writing transfer does not always parallel speech transfer—but thanks to the structure and composition of the Institute, we're in a unique position to find out.

But our programmatic structure is not the only thing to consider as we look for opportunities for (and obstruction to) transfer in our writing and speech courses. We must also look to our practices. In terms of our current on-the-ground faculty practices in Writing 2–3 and 5, some methods and assignments are already designed to facilitate students' knowledge transfer and others are not. Nevertheless, we do have the advantage of collaboratively designed course outcomes that help us to think about how transfer might be better facilitated.[5] Broadly, the purpose of an outcomes document is to identify the capabilities students should be able to demonstrate while completing a particular course. However, while these documents offer insights into the desired outcomes, they do not directly define practices. A recent self-study for a regular external review helped us to identify the most common faculty practices in the Writing 2–3, 5, and Speech parts of our program. The Writing 2–3 and 5 courses and speech courses are using practices that could easily shape to fostering transfer—and perhaps already do: discussion, small group discussion, peer review, Blackboard-based and other web-based interactive tools, performance-based or role playing methods, writing workshops, debates, student-led discussion, oral presentations, collaborative learning, interdisciplinary approaches, regular group work activities, and frequent external visits to many college events and museum exhibitions, for example. These are all practices that lend themselves to scaffolded, metacognitive work and enable the "legitimate peripheral participation" that evolves into expertise—at least into contingent expertise that will itself need to transform across contexts. Particularly interesting is that 50 percent of the IWR writing and speech faculty report incorporating multimodal assignments into their classes, an activity that can connect students to "real world" motivations and can develop knowledge transfer across modes of expression.

While none of these practices is surprising for good writing and speech instruction, few have been intentionally designed for fostering the possibility of knowledge transfer. They are grounded in goals that may not be explicitly identified but are theoretically rich. Writing and speaking, our faculty report, are ways of coming to terms with complexity; they offer critical examination of epistemological and cultural assumptions; they develop intercultural awareness and foster active learning. Writing and speech *courses* enable the democratization of knowledge; they teach the methodological process of conscientious inquiry, logical analysis, and creative interpretation; they allow students to enter the ongoing conversation of academic scholarship, developing them as rhetorically flexible, active producers of knowledge; they help students to become deeply attentive citizens. It is not difficult to see how these practices and aims are concordant with a programmatic attention to transfer as a core underlying objective and a guiding force—especially in their emphasis on the "rooted relevance" of learning and their attention to the relationships among students, classroom contexts, and the real world.

We also see which transfer-fostering practices noted in the research might be less in evidence: notably, teaching by analogy, teaching for "boundary-crossing," and certainly ensuring ample time for initial knowledge development, given our ten-week terms. And finally, we recognize in these programmatic practices what might be hampering transfer: most importantly, the degree to which courses are insular, and the fact that learning "to write" or "to speak" is still sometimes seen as a generic ability that can be acquired in one context and reused everywhere. To put it another way, perhaps a key obstacle to facilitating the possibility of transfer in our courses is the general practice of teaching students in a given course without deep understanding of the diverse ways their knowledge will be called upon, will need to transform, will grow and evolve in new contexts. And the second key obstacle might be the lack, in each new context, of affordances that allow students to put their knowledge to use.

Our broader institutional goals and the faculty's stated practices indicate clearly that further investment in the study and practice of knowledge transfer makes sense for the Institute. This direction builds on the strengths of our program, but also on current needs. We acknowledge that the practices in use in the FYSs are less broadly known, a key area for development in the work that lies ahead. Our courses include diversity of practices, and faculty seem to recognize—most clearly when they teach in more than one of our courses—the possibility that students are not always transforming and reusing knowledge from one course to the next. These conditions are important to imagining our next steps. In this environment, explicit discussions about transfer can enable faculty to engage with the theories that underpin their practice. It is a logical progression from these frames to an intentional design based on evolving research about knowledge transfer and what we have begun to call "rhetorical flexibility" in speech and writing.

A Program Broadly Imagined

To design a program that intentionally fosters the possibility for transfer as we define it and as the transfer literature frames it, the IWR needs to collectively explore all facets of our program. The actual structural and curricular decisions that the IWR faculty and leaders make would, we believe, impact curriculum, faculty development, outcomes, and cross-curricular faculty communication. Faculty teaching the two-course requirement at Dartmouth would need to be intimately familiar with the criteria for intentionally fostering transfer. Institute faculty would require opportunities to further immerse themselves in this work—drawing on and discussing scholarship from various fields that will help them to frame an assessment project, generate and answer assessment questions, carry out assessment, interpret assessment results, and implement potential improvements. Many of these activities would be possible through ongoing professional development, through new initiatives, through growing attention to writing and speaking across the disciplines, and through progressive re-imagination of our curricular designs in writing and in speech. The project's central assessment questions could help faculty understand better how transfer functions in these new contexts.

Work with writing could inform future transfer work with other modes of communication, including speaking and multimodal communication.

If the knowledge about how to foster opportunities for transfer were to be the foundation for Dartmouth's program, we would imagine it to look something like this:

Guiding Principles, Core Values, and Outcomes:

- Writing and speaking are collectively understood as ways to acquire and construct meaning, to navigate the slippery networks of human activity.

- Students' growth as writers and speakers is not expected to be linear but instead is understood as an expansion of abilities that move back-and-forth between global strategies and specific expertise.

- Rhetorical flexibility becomes a core value in all contexts; students who are writing experts expect to negotiate new contexts, to create hybrids, to tolerate ambiguity, and to persist when tasks are challenging.

- Every course intentionally affords writing and speaking knowledge transfer.

- Discussions of program outcomes are ongoing, transparent, and publically negotiated.

Curriculum, Course Design, and Pedagogy:

- The overall curriculum offers frequent intentional opportunities for "boundary-crossing"—which research suggests as a way to afford transfer.

- First-year writing is required in a two- or three-course sequence; students are broadly encouraged to take speech courses; every first-year student can choose an additional term of direct writing instruction (Writing 2–3).

- Every teaching practice—from syllabus and assignment design, to course activity, to response to student work, to evaluation—draws on the acknowledged principles for enabling transfer, including teaching driven by analogy, scaffolded learning, at-

tention to inspiring motivation, and sufficient time given to cementing initial learning for new knowledge.
- Students are offered sustained opportunities to work on multimodal meaning-construction and communication.
- A portfolio (or similar tool) is designed to move students to be conscious of their growth across courses and to account for their knowledge as it evolves. More than just a token moment of self-reflection, this portfolio is carefully integrated to emphasize intentional learning as a central goal of the program.

Faculty Support, Development, and Research:

- In order to accomplish all of this, faculty development builds upon the foundations of knowledge transfer, engaging faculty from all disciplines and courses in ongoing exchange. Ongoing research—about the ways students are transferring their writing knowledge across modes and disciplines and from the classroom into "real-world" settings—is actively supported and reported. We thus envision a broad research agenda, one that tackles questions of transfer among modes, disciplinary settings, and even workplace or other non-academic contexts.
- We also make better use of the programmatic structures we have. Our unique sequence of Writing 2–3 or 5 and then the discipline-infused FYS is the perfect sequence for identifying the knowledge and know-how that thread through both courses, that are specific to each, or that thread through but also transform from one course to the next in diverse ways. This understanding serves as a foundation for thinking and writing across the rest of the curriculum.

Student Support and Institutional Participation:

- Student writing support staff participate in the conversation about transfer and are both educated in the principles of transfer and schooled in the practices that encourage transfer.
- Writing and speech in courses beyond the first year receive sustained, embedded attention; designated writing courses are available across disciplines; deep attention to the writing and speech involved in culminating experiences is offered to all, with that

experience simultaneously pointing forward to workplace, graduate, and other next-stage experiences.

- Writing and speaking in other student experiences—off-campus programs, co-curricular activities, etc.—are attended to and celebrated.

To be clear, Dartmouth's IWR has not yet determined that this is the path to follow. We must collectively explore the values and benefits, as well as the predicted effects, of this global approach. We are, however, already exploring, testing, and implementing parts of this approach in order to determine our best choices; the rest of the profile will discuss our current activities.

Our Processes, Now and Again

The program just outlined is in the initial phases of implementation. In this section of the profile, we focus on first-year initiatives and developments that we hope will become the foundation for a future program informed by our understanding of transfer. Our work over the past few years has focused particularly on stimulating discussion of course outcomes, promoting a research-based understanding of student writing development, and increasing opportunities for multimodal composition in the first-year writing classroom. These efforts have been important in helping us build toward our future programmatic goals.

The Davis Study of Student Writing

In 2009, Dartmouth's Institute for Writing and Rhetoric was competitively awarded a $200,000 grant from the Davis Educational Foundation to address questions regarding course coherence and transfer in order to improve the effectiveness of our first-year writing program. Supported by Davis funds, the IWR launched a three-year study of first-year writing at Dartmouth, motivated by two important and related questions—*How do students transfer knowledge about writing from course to course and task to task? How does composing with new technologies improve the transfer of more traditional writing abilities?* To answer these questions, we have undertaken a large-scale study of first-year student essays. This study has given our leadership team, and our Institute as a whole, an opportunity to map the present and cast an eye to the future. Fundamentally, we understand the study as a flexible and comprehensive approach to program and faculty development that relates closely to the goal of "building upon the foundations of knowledge transfer, engaging faculty from all disciplines and courses in ongoing exchange" (from the Davis Foundation grant proposal). By giving our faculty an opportunity to explore research-based perspectives on student writing, to design and test new approaches, and to engage in ongoing conversation about the results of their efforts, we hope to promote a flexible and informed teaching culture that will facilitate the possibility of transfer. (This, incidentally, models active learning itself; that is, through the work of this research, we are practicing what and how we teach.)

Preliminary Work. During the summers of 2010–2012, volunteer groups of faculty gathered to identify outcomes shared across our first-year courses and to read a stratified random sample of first-year student papers for evidence of those outcomes.[6] Knowing that students are better able to transfer learning when first guided to understand the relation between previous and current writing tasks, we feel confident that the initial step of identifying the outcomes that span our courses is invaluable. In identifying a shared outcome across first-year courses—for example, the ability to integrate and use sources effectively—we are earmarking a general similarity among course contexts and leaving room for future discussion of specific, contextual differences. In doing this work together as a faculty, we are also developing our ability to see the question of transfer in shades of gray. Students may be using a similar strategy, for example, but are using it in a different way, to a different purpose, and with a different set of teacher expectations. The work also provided a useful model for creating speech program learning outcomes. Early and ongoing work already suggests interesting areas of overlap, but also divergence, between outcomes in writing and speech.

Methods and Results. With its central interest in describing college students' writing development across course contexts, our study shares common ground with qualitative studies like Elizabeth Wardle's work on transfer in FYC, Lee Ann Carroll's *Rehearsing New Roles,* and the host of other excellent longitudinal studies mentioned in the section above. Ours, however, is a larger-scale study with a quantitative dimension. We are also analyzing our sample of student writing with less contextual information about the pieces of writing than is typical in longitudinal studies.[7] We also differ from previous studies in the degree of our emphasis on student self-report. While many related studies make extensive use of interviews, we are working primarily with student writing without student commentary. This approach fits our goals, as our study is as much about our own program and our ways of perceiving and evaluating student work as it is about the students' perspectives.

Since 2009, we have been collecting the first and last source-based assignment from all of our first-year writing courses. We define *source-based* broadly, including any writing in which students engage with a source, including summarizing sources, analyzing sources, and writing their own full-blown research papers. Out of those papers we generate a stratified random sample from across the four possible course sequences, which includes the first and final papers from approximately fifty students in each cohort across their first year. We then norm (achieving from 77 percent to 100 percent agreement in our coding, depending on the feature) and code this sample of full papers with a descriptive coding rubric derived from our course outcomes. The rubric captures a range of information about each paper's structure, the nature and location of its thesis statement, introductory and concluding strategies, and type of evidence used. We also code smaller excerpts from the same sample for information about paragraph structure, source integration, and some grammatical errors. The two approaches to coding work together to give us a range of information about student writing in first-year courses. Meanwhile, we gather information about how students read their sources from an even smaller sub-set of papers (five students from each course sequence), which we study using a modified version of Citation Project methods.[8] We plan to layer these results with other sources of information, including three years of responses to the writing-specific questions of the National Survey of Student Engagement survey and relevant institutional data.

The Davis study has already produced results that offer insight into student writing. Though quite tentative, the results to date are intriguing in terms of understanding what students appear to use and adapt from the first writing course to the second. They suggest that the students are using the general strategies identified by the outcomes—like the ability to craft a thesis, use evidence to support that thesis, and integrate source material—across their course sequences and across course groups. Indeed, we've discovered some clear trends in how they use these strategies:

Regarding argument. In overwhelming numbers, students use a few key structures for argument—particularly, the traditional thesis-driven essay structure with an explicit thesis at the beginning of the paper and body paragraphs supporting the thesis.

Regarding organization. More specifically, most students are writing using one of two major organizational strategies, which we identified as "linear argument-driven" or "critical thesis-driven." Both of these are thesis-driven arguments, but the first primarily reports source content while the second critically engages with sources. These two organizational strategies account for anywhere from 60 percent to 90 percent of the papers submitted, depending on the course sequence (60 percent for the first paper in Fall Writing 5; 90 percent for the final paper for first-year seminar students exempted in the fall).

Regarding evidence. The two most commonly used types of evidence are "interpretation of a primary object of analysis" (this includes a range of possible objects—literary texts, architecture, works of art, non-fiction documents . . .) and "report from an authoritative source." The least used is quantitative data. This is generally true across course types and course sequences.

Regarding content. Case study results of individual students suggest that students appear to seek out opportunities to write about similar topics from similar angles across the two courses, which might account for some of their re-use strategies.

At this point, we can say less about the level of sophistication or fluency with which students are using these general strategies, since we were not measuring these degrees in our initial reading. The fact that the majority of students draw on similar strategies for argument structure and evidence across the sequence of writing courses could also suggest a few possible hypotheses and areas for further inquiry:

- At the level of general, descriptive categorization, the first-year writing sample is somewhat uniform. For better or for worse, students are using similar organizational strategies from task to task and from course to course. This finding supports David Smit's hypothesis in *The End of Composition Studies* that students may be more likely to transfer general rather than specific strategies from first-year composition, which in turn may have less future potency for them in new situations than more context-specific strategies might.

- A preliminary look at the data also suggests that the epistemological and conventional differences in student writing across the sequence of writing courses may be more subtle than we might have expected. The fact that most students are neither using data as evidence nor using discipline-specific structures for organization, for instance, suggests that within their seminars they are not entering fully into the disciplinary conventions for writing that they may confront in the future.

- What our current coding system does not capture as clearly are the subtle modifications and revisions that students may be making within the broader strategies we've studied as they move further into college life. We have earmarked general similarities in, for example, the organizational strategies used in different courses, but we hypothesize that these strategies present differently when examined more specifically.

In our ongoing analysis of results, we are careful to keep in mind that the apparent resurfacing of common strategies in different contexts does not necessarily point to any evidence of transfer (defined as a transformative activity, closely tied-in to the student's learning process) in part because, as Elizabeth Wardle points out, "If we do not know how students understand and respond to tasks and contexts, we have no basis for identifying and interpreting generalizing behaviors that might be considered forms of 'transfer'" (72). It does, however, point to the power of these strategies, perhaps by virtue of their very generality, to re-emerge in very disparate contexts and different assignment types. As we mentioned when describing the study methods, the collected papers represented a huge range of assignment types, yet nevertheless, students favored a "preview style" introduction over all other types. This tendency prompts us to wonder: Are we really seeing uniformity, or subtle, not-yet-fully captured transformation, or some combination of the two? Our initial results hint at possible influences on students' composing choices, influences that might even work against the likelihood of students trying new approaches in new situations. In other words, they are in large numbers sticking to the same strategies, even as their writing circumstances transform.

Additional Davis Study Initiatives

In addition to the study of student writing as a way to engage faculty in curricular thinking about transfer, two curricular pilots supported by the Davis study have begun to support teaching for transfer by fostering faculty exchange and encouraging attention to rhetorical flexibility. The first, a multimodal pilot, supports faculty already committed to teaching digital composing as they work to create faculty development opportunities for fifteen of their colleagues who are new to teaching multimodal composition. These faculty leaders offer a workshop for their colleagues that provides them examples of tested multimodal assignments. The leaders then mentor their colleagues in designing and piloting their own assignment. In this process, the outcomes documents provide a crucial bridge between multimodal and more traditional writing assignments, encouraging faculty to recognize analogies between these different kinds of composing. They also provide a basis for later assessment and comparison between multimodal and traditional composing tasks. Recalling that students may be better able to transfer writing abilities when they are applied in multiple contexts and connected to real world experience, we sense that multimodal work might serve to help students increase their rhetorical flexibility, which will ultimately help them transfer learning to new contexts. To date, the assessments of pilot activities have borne this out, as faculty who teach these assignments report in post-pilot assessment reports that students do both reuse and adapt the multimodal strategies to other compositions, and note that they have better understood traditional rhetorical choices.

The second pilot builds explicitly on an advantage that we have underexploited: the two-term sequence. It supports pairs of faculty interested in linking their first-year writing and FYS courses into one cohesive learning experience for our students. Faculty are not co-teaching, in the sense that they are not in the classroom together. Rather, they are co-constructing learning environments that may improve students' ability to transfer writing competencies from one course context to the next. The research suggests that students need explicit scaffolding and extended time for learning to "take." It also suggests that students will be better able to transfer (or transform) old knowledge in new contexts when faculty create affordances for them to do so. Collaborative course design makes these criteria far easier to meet.

Professional development has also been redesigned to more actively work through questions of knowledge transfer, writing and speaking instruction, research instruction, and classroom work. Faculty from different types of courses attend together, to develop collective understandings of writing and speaking instruction, and to share across course types. In the process, they learn from each other—a key exchange, given the importance of faculty understanding what other faculty are doing so that they can explicitly draw on that knowledge in their work with students. The Davis Foundation-supported work served this same purpose, results aside. In the literature review groups, the norming and coding sessions, and the subsequent workshops with faculty who provided papers for the study, faculty from Writing 2–3, Writing 5, Speech, and FYS have come together to read scholarship and discuss it, to hash through criteria for coding, to explore and scrutinize student work with an eye to simply describing what we see, and to learn about the research methods used (We note again here that the activities we're using to study transfer and active learning model what we're trying to do in our classrooms: dialogue, action, critical analysis, theory-guided discoveries.)

One faculty member attests to the valuable effect of Davis study participation on her ability to teach for transfer:

> I've moved away from the very comfortable "sage on the stage" model to the much more anxious-making effort to transfer much of what happens in the class to my students. . . . As part of a Davis-funded research review project, I chose to read about "knowledge transfer" and summarize a series of readings for my colleagues. I was *astonished* to learn how I had been undermining knowledge transfer, not only in Writing 5, but in every class I offer. In the last year, I've begun to incorporate what I learned into each course.

Additional initiatives are well underway. For example:

- We have begun to work with an existing linked program, Humanities 1–2, to develop ways to ensure long-term investment in the linked course model, which is so clearly likely to foster students' reuse and transformation across two courses.

- We are considering developing a first-year portfolio that will encourage faculty across courses to enhance their knowledge of students' work, and will encourage students to see the growth in depth and diversity of their abilities.

- We see potential in communication between writing and speech faculty to foster understanding of how spoken and written competencies might inform each other, both in what they share and in each one's unique nature.

Finally, to ensure the long-term value of what we teach, we know we must design course materials and teaching approaches that facilitate knowledge transfer *beyond* the first-year sequence and the classroom. This phase of program design is in its earliest stages. In 2012, Dartmouth will focus on building faculty attention and awareness to writing beyond the first year. The college is sponsoring a group of faculty from across disciplines, in collaboration with consultants, who will work together to share the methods they use for teaching writing and to develop useful curricular models for their colleagues. The knowledge built about first-year students' work—and in particular the knowledge we gain about transfer—will be foundational to this project.

These initiatives do face real material constraints, likely to play into the program's growth and development.

- Faculty perception outside of the IWR of Writing and Speech as "service" and not disciplines in their own right can contribute to resistance to initiatives that build on research and demand investment; similarly, the IWR faculty as a body are not convinced that building from research to develop a coherent curriculum is the best avenue.

- The faculty who teach in the different segments of the program are a mix of tenure-track, full-time lecturer, and adjunct faculty, a mix that poses interesting challenges in terms of faculty exchange and development, in part due to differences in time investment available to those who teach writing occasionally and those whose entire focus is on Writing or Speech. The challenge posed by different levels of investment is already being addressed through projects like the Davis Foundation work and through ongoing professional development, as we've mentioned.

- On a more basic level, challenges are logistical. For example, the introductory writing courses' class size did not match the FYS size until 2012, making linked courses harder to arrange.

We suspect these are challenges far from unique to Dartmouth, sure to be shared by any program considering building a curriculum that intentionally fosters opportunities for transfer. And if these challenges cross campuses and disciplines, programs and departments, program philosophies and research agendas—then that's all the more reason to take them on.

Conclusions

The stages of development described here have shown us some key insights into the kinds of planning that might be useful to other programs thinking about structuring a program so as to intentionally foster transfer. What is perhaps most interesting is the degree to which these kinds of planning apply to almost any type of program development. First, planners must enable broad faculty investment in the questions to be addressed and the knowledge base to be developed. Second, a smaller team must coordinate efforts and activities. That team must include members with strong relationships with faculty more broadly and must be open to learning. Third, every part of the process must be imagined as a form of faculty development. Fourth, there must be space (time, resources) for initiating on-the-ground research or at least solid research review.

We can imagine many programs within Dartmouth that might engage in a curricular project using the principles of transfer to support stronger programs: language departments looking to create the possibility of fostering transfer across language levels; specific disciplines like linguistics, English, history, and others that require writing across almost every required course for majors; intra-disciplinary initiatives, such as a recent pilot biology-chemistry course sequence using writing as a key tool for learning; and inter-disciplinary programs and departments that have cross-listed courses, such as comparative literature. Other writing programs across the country could find our process useful as they work through shared issues of program design. Although every program is unique, offering different ways to teach and learn writing and speaking, we are all looking for the golden key of coherent, sustained student development over time.

Perhaps equally important is our acknowledgement of what we don't know here in the IWR, and what the field more broadly does not know. The dangers of developing any program based on research knowledge include a too-quick "application" of scholarly ideas and research conclusions to classroom activities. While the subject of knowledge transfer now has a long history, its specific exploration in terms of students' writing and speaking in college is more recent. Longitudinal studies that should provide strong insights have not been widespread in the past, and the scholarship about knowledge transfer more generally is dynamic, in flux, acquiring new layers all the time (see other articles in this issue for examples). Writing programs, on the other hand, are not known to be nimble adapters; the institutions into which they are integrated are more often cargo ships than schooners. A new program design might not be able to flex with the new knowledge that develops in the field; simultaneously, new knowledge might be too quickly accepted. Our overall path has thus been, and will continue to be, one of cautious building from the strong foundation we have, persistent exploration of new possibilities, deep attention to evolving knowledge, and creative testing and implementation of the pieces that can lead to a coherent overall institution-wide investment in the programmatic design that offers our students the best opportunity to "transfer" their writing and speaking knowledge over time and context. Stay tuned!

Coda: Where We Are Now

In the time since our profile was published in 2012, we have evolved in some ways, not yet in others. We are as excited by the questions and efforts of the program as we were initially, but we continue to imagine the work of the program to be a long-term and productive project. In this way, we exemplify the ecological features of both fluctuation and emergence described in the Introduction.

One key change has been the college- and endowment-funded creation of the possibility for every student to take our Writing 5 course. We no longer exempt any students and can now develop writing abilities over two terms, or three for our developmental students. The embeddedness of this sequence in the broader college culture suggests an interconnectedness that hasn't always been recognized.

We have introduced a pilot portfolio project. Our Davis study of student writing raised questions for us that we feel a two- or three-term writing portfolio might answer. Sixty students from our first-year program collect all work, select best pieces, reflect on the best pieces, meet with the second-term faculty member to discuss the pieces, and then reflect again at the end of the second course. Students are helping us to imagine and create new thinking and approaches from their reflections; faculty reading and analyzing the portfolios are creating new knowledge about learning and curricular impact.

We are taking a much closer look at the first-year seminar program staffing patterns, syllabi, and assignments; drawing on strong first-year seminar faculty to offer faculty development; exploring extended development sessions for faculty in the summer; and supporting a full proposal from a group of faculty across the disciplines to integrate writing instruction intentionally in the curriculum after year one.

Finally, we tailored our 2012 to 2013 faculty development series to draw a broad range of faculty into the "transfer" discussion, hosting sessions focused on the theme of re-thinking what we do and thus emphasizing the relationship between the program as an entity, the college overall, and the individual faculty members' practices. Sessions included: "Where Does the Learning Go? How Students Implement Knowledge about Writing in New Environments" or "Carrying (and Transforming) Research Capabilities Across Courses." Faculty development sessions also explored theoretical and conceptual models of speech, with special attention to how speech can transform knowledge and practices across disciplinary contexts. For example, Claudia Anguiano offered a session to explore how speech theory can inform classroom practices and content, with special attention to opportunities for reflection—core process of knowledge transfer.

The speech faculty has evolved in its focus on the kinds of questions raised in our profile. In our original piece, which considered how "transfer" informs the Institute's work, we referenced four speech courses taught in 2011 to 2012. Since 2011 to 2012, we have added six new speech courses to the Institute's curriculum. In creating each course, speech professors were mindful of shared learning objectives, drawing out themes that cross such diverse topics as intercultural communication and political humor, image repair work, and social protest. Although we do not yet have a system for tracking evidence of knowledge transformation across speech courses, a mindfulness of shared learning outcomes creates a foundation, a set of benchmarks, for eventually exploring transfer across them.

Speech has also begun the work of proposing a minor. We find ourselves returning to the literature of transfer as we explore issues of cohesion, scaffolding, and programmatic development of student competencies, awareness, practices, and theorizing in sequenced courses of study. We also see the minor as a chance to begin exploring, with empirical evidence, knowledge transformation among courses for students minoring in speech.

Thus, both our speech and our writing programs are in dynamic, self-reflective flux, adapting to new knowledge as well as to new institutional demands, positioning ourselves as always seeking to evolve.

Finally, individual faculty in our program have developed their research agendas about "transfer." Contributing author Sara Chaney has begun to study patterns of transfer in the writing processes and products of developmental writing students. Contributing author Christiane Donahue has turned her attention to showing the ways in which transfer is, in fact, more aptly understood as transformation or orientation and is studying the ways in which linguistics methods for analysis of this transformation can be a powerful tool. In particular she is tracing the parallels between language acquisition and writing acquisition. This evolution in our understandings of knowledge transformation parallel the evolution in our program design.

Notes

1. For course descriptions, see: http://compositionforum.com/issue/26/dartmouth-course-descriptions.pdf.

2. For sample syllabi, see: http://compositionforum.com/issue/26/dartmouth-sample-syllabi.pdf.

3. Kitzhaber's report contained a sixty-page addendum, *The Writing of Dartmouth Students After the Freshman Year.* Interestingly, while Kitzhaber carefully details the worsening of student writing after their first-year writing courses, he nowhere argues that professors outside of English be responsible for teaching writing. Rather, he advises professors in other departments to assign more papers and to give punishing grades as a way of ensuring that high standards/good writing are maintained at Dartmouth.

4. The general descriptions of the Institute for Writing and Rhetoric (IWR) and its three writing programs are as follows:

- IWR http://www.dartmouth.edu/~writing/courses/

- Writing 5 http://www.dartmouth.edu/~writing/courses/writing5/about.html

- Writing 2–3 http://www.dartmouth.edu/~writing/courses/writing2-3/about.html

- FY Seminar http://www.dartmouth.edu/~writing/courses/firstyearseminars/about.html

5. The outcomes for each writing course are as follows:

- Writing 5: http://www.dartmouth.edu/~writing/writingfiveoutcomes.html

- Writing 2–3: http://www.dartmouth.edu/~writing/writingtwothreeoutcomes.html

- FY Seminar: http://www.dartmouth.edu/~writing/first-yearseminaroutcomes.html

6. As a preliminary step to this activity, we collected the first and last paper that responded to a text from every student in every first-year course (including the Writing 2–3, Writing 5, first-year seminar and Humanities 1–2). We have been fortunate to have a very high rate of collection in both years.

7. IRB approval for this study requires complete anonymity of student work, a fact that is important for several reasons. Primarily, since we see this work as an avenue for faculty development and precursor to future assessment, it is vital that all concerned understand this work to be a study of student writers, and not courses or instructors. This understanding is important to encourage a high rate of paper collection and ensure good faith within the Institute as a whole.

8. In the first year of the study, we were fortunate to be a participating institution in the Citation Project, led by Rebecca Moore Howard and Sandra Jamieson. In the second year of our study, we have undertaken follow-up research that looks at patterns of source use across first-year course contexts.

Works Cited

Bawarshi, Anis, and Mary Jo Reiff. *Genre: an Introduction to History, Theory, Research and Pedagogy*. West Lafayette: WAC Clearinghouse/Parlor Press, 2010. Print.

Bjork, Robert, and Alan Richardson-Klavhen. "On the Puzzling Relationship between Environment Context and Human Memory." *Current Issues in Cognitive Processes*. Ed. C. Izawa. Hillsdale NJ: Erlbaum, 1989. Print.

Beaufort, Anne. *College Writing and Beyond: A New Framework for University Writing Instruction*. Logan: Utah State UP, 2007. Print.

Bransford, John, Anne Brown, and Rodney Cocking. *How People Learn: Brain, Mind, Experience, and School*. Washington DC: National Academy Press, 2000. Print.

Bruce, Bertram. *Literacy in the Information Age: Inquiries into Meaning Making with New Technologies*. Newark, Delaware: International Reading Association, 2003. Print.

Bruffee, Kenneth. "Collaborative Learning and the 'Conversation of Mankind.'" *College English* 46.7 (1984): 635–52. Print.

Carroll, Lee Ann. *Rehearsing New Roles: How College Students Develop as Writers*. Carbondale IL: Southern Illinois UP, 2002. Print.

Chiseri-Strater, Elizabeth. *Academic Literacies: The Public and Private Discourse of University Students*. Portsmouth, NH: Heinemann, 1991. Print.

Coiro, Julie, et al, eds. *Handbook of Research on New Literacies*. Mahwah, NJ: Erlbaum, 2008. Print.

Compton, Josh. Speaking of Speech with the Disciplines. *Arts and Humanities in Higher Education* 9.2 (2010): 243–55. Print.

Curtis, Marcia, and Anne Herrington. "Writing Development in the College Years: By Whose Definition?" *College Composition and Communication* 55.1 (2003): 69–90. Print.

Dias, Patrick, Aviva Freedman, Anthony Paré, and Peter Medway. *Worlds Apart: Acting and Writing in Academic and Workplace Contexts.* Mahwah, NJ: Erlbaum, 1999. Print.

Donahue, Christiane. "Transfer, Portability, Generalization: (How) Does Composition Expertise 'Carry'?" *Exploring Composition Studies.* Eds. Kelly Ritter and Paul Kei Matsuda. Logan, UT: Utah State UP, 2012. Print.

Ede, Lisa, and Andrea Lunsford. *Singular Texts/Plural Authors: Perspectives on Collaborative Writing.* Carbondale IL: Southern Illinois UP, 1990. Print.

Eich, Eric. "Context, Memory, and Integrated Item/Context Imagery." *Journal of Experimental Psychology* 11 (1985): 764–70. Print.

Elbow, Peter. *Writing With Power.* New York: Oxford UP, 1981. Print.

Freire, Paulo. *Pedagogy of the Oppressed.* Trans. Myra Bergman Ramos. New York: Continuum, 2000.

Gere, Anne Ruggles. *Writing Groups: History, Theory, and Implications.* Carbondale: Southern Illinois UP, 1987.

Gray, Wayne D., and Judith Orasanu. "Transfer of Learning: Contemporary Research and Applications." *Transfer of Training.* Eds. Stephen M. Cormier and Joseph D. Hagman. San Diego, CA: Academic Press, 1987. Print.

Haas, Christina. "Learning to Read Biology: One Student's Rhetorical Development in College." *Written Communication* 11.1 (1994): 43–84. Print.

Harris, Joseph. "After Dartmouth: Growth and Conflict in English." *College English* 53.6 (1991): 631–46. Print.

Haskell, Robert. *Transfer of Learning: Cognition, Instruction, and Reasoning.* San Diego CA: Academic Press, 2001. Print.

Haswell, Rich. *Gaining Ground in College Writing: Tales of Development and Interpretation.* Dallas, TX: Southern Methodist UP, 1991. Print.

Howard, Rebecca Moore. "Collaborative Pedagogy." *A Guide to Composition Pedagogies.* Eds. Gary Tate, et al. New York: Oxford UP, 2001. Print.

Kitzhaber, Albert. *The Dartmouth Study of Student of Student Writing,* 1962. Report to the Carnegie Foundation. Print.

—. *The Dartmouth Study of Student Writing: The Writing of Dartmouth Students After the Freshman Year,* 1962. Report to the Carnegie Foundation. Print.

Kress, Gunther. *Literacy in the New Media Age.* New York: Routledge, 2003. Print.

Lave, Jean, and Etienne Wenger. *Situated Learning: Legitimate Peripheral Participation.* Cambridge, UK: Cambridge UP, 1991. Print.

Lunsford, Andrea A., and Karen J. Lunsford. "'Mistakes are a Fact of Life': A National Comparative Study." *College Composition and Communication* 59.4 (2008): 781–806. Print.

Mazur, Eric. *Peer Instruction: A User's Manual.* San Francisco, CA: Benjamin/ Cummings Publishing Company, Inc., 1996. Print.

McCarthy, Lucille. "A Stranger in Strange Lands: A College Student Writing Across the Curriculum." *Research in the Teaching of English* 21.3 (1987): 233–65. Print.

New London Group. "A Pedagogy of Multiliteracies: Designing Social Futures." *Harvard Educational Review* 66.1 (1996): 60–92. Print.

Newell, George. "Writing to Learn." *Handbook of Writing Research.* Eds. Charles MacArthur, Steve Graham, and Jill Fitzgerald. New York: Guilford Press, 2006. Print.

Perfetto, Greg A., John D. Bransford, and Jeffrey J. Franks. "Constraints on Access in a Problem Solving Context." *Memory and Cognition* 11.1 (1983): 24–31. Print.

Postman, Leo. "Organization and Interference." *Psychological Review* 78.4 (1971): 290–302. Print.

Prior, Paul. *Writing/Disciplinarity: A Sociohistoric Account of Literate Activity in the Academy.* Hillsdale, NJ: Erlbaum, 1998. Print.

Reiff, Mary Jo, and Anis Bawarshi. "Tracing Discursive Resources: How Students Use Prior Genre Knowledge to Negotiate New Writing Contexts in First-Year Composition." *Written Communication* 28.3 (2011): 312–337.

Simon, Herbert A., and John R. Hayes. "The Understanding Process: Problem Isomorphs." *Cognitive Psychology* 8.2 (1976): 165–90. Print.

Smit, David W. *The End of Composition Studies.* Carbondale, IL: Southern Illinois UP, 2004. Print.

Sommers, Nancy, and Laura Saltz. "The Novice as Expert: Writing the Freshman Year." *College Composition and Communication* 56.1 (2004): 124–149. Print.

Sternglass, Marilyn. *Time to Know Them: A Longitudinal Study of Writing and Learning at the College Level.* Hillsdale NJ: Erlbaum, 1998. Print.

Trimbur, John. "Consensus and Difference in Collaborative Learning." *College English* 51.6 (1989): 602–16. Print.

Tuomi-Groehn, Terttu, Yrgo Engestroem, and Michael Young. "From Transfer to Boundary-Crossing between School and Work as a Tool for Developing Vocational Education: An Introduction." *Between School and Work: New Perspectives on Transfer and Boundary-Crossing.* Eds. Terttu Tuomi-Groehn and Yrgo Engestroem. New York: Pergamon, 2003. 1–15. Print.

Vygotsky, Lev. *Thought and Language.* Cambridge MA: MIT, 1969. Print.

Wardle, Elizabeth. "Understanding Transfer from FYC: Preliminary Results of a Longitudinal Study." *Writing Program Administration* 31.1/2 (2007): 65–85. Print.

Yancey, Kathleen Blake. "Made Not Only in Words: Composition in a New Key." *College Composition and Communication* 56.2 (2004): 297–328. Print.

Part III. Claiming Disciplinary Locations: The Undergraduate Major in Rhetoric and Composition

The profiles in this section focus on the undergraduate writing major (UWM) and its development at universities across the nation. The UWMs described here highlight three of the hallmarks of a discursive ecology—emergence, interconnectedness, and fluctuation—and it is useful to examine them in that context.

In many ways, emergence and growth are the predominant themes in recent scholarship about UWMs. The CCC Committee on the Major in Writing notes that "the number of writing majors is increasing rapidly," and that the number of such majors nationwide has increased from a handful to more than seventy within the past decade.[1] Emergence is a key ecological attribute, and in this respect, the undergraduate writing major is coming into its own through continued expansion and development. Similarly, as the profiles in this section indicate, interconnectedness is also vital in the continued development of the writing major. Many undergraduate writing majors draw upon, intersect with, and sometimes compete against related majors and programs within their universities; some have developed through cooperative relationships and shared resources with similar programs at other schools and locations. The three profiles presented here demonstrate just a few of the many ways in which UWMs are enmeshed and interconnected with other academic programs and institutions. And finally, these profiles reveal the importance of fluidity in the success of the writing major. Like most undergraduate writing majors, the programs profiled here demonstrate fluidity, adaptation, and transformation in the face of a range of material, cultural, and economic challenges. In many ways, the undergraduate writing major is unique in its malleability; nearly every profile of an undergraduate writing major underscores constant change and the revision of goals, curricula, assessment practices, and other aspects of the major. Ultimately, the undergraduate writing major is a revealing lens through which to examine the ecology of writing, and these profiles bring that into focus.

As one of the first published profiles of an undergraduate writing major, Stephanie L. Kerschbaum and M. Jimmie Killingsworth's "Diverse Lessons" provides a distinct vantage point from the perspective of a pioneering undergraduate program. Published in 2007, the profile describes the history and development of the undergraduate rhetoric concentration within the English department at Texas A&M University—a groundbreaking UWM that was developed prior to the widespread emergence of the writing major nationwide. In their introduction to the profile, Kerschbaum and Killingsworth describe the fluid growth and evolution of FYW and WAC programs across the United States while noting that "a similar rise has not been seen in the growth of concentrations in rhetoric and writing as an undergraduate major or minor." Little did they know of the diverse and widespread undergraduate programs that would appear just a few years later. This profile emphasizes the important role that rhetorical study can play for undergraduates, and it is likely that the publication helped to shape the curricular grounding of rhetoric within the majors that would soon follow. Likewise, the "Where We Are Now" coda describes the marginalization and material adversities that undergraduate writing majors often face, highlighting the ways in which faculty in such programs must adapt and transform their work and their programs—a challenge that may be all too familiar to others working in undergraduate writing majors.

As a corollary to the Texas A&M profile, Lori Ostergaard, Greg Giberson, and Jim Nugent's "Reflections on the Major in Writing and Rhetoric at Oakland University" (published in 2010) describes the newer and more recent development and implementation of a writing and rhetoric major. The ecological themes of emergence, interconnectedness, and fluctuation are evident in this profile as well. Ostergaard, Giberson, and Nugent's work reflects on the separate, yet interconnected issues involved in creating an undergraduate writing program, including the development and naming of a department of writing and rhetoric, the impact the major had on Oakland University's first-year writing program, the theoretical and practical structure of a three-track major, as well as the institutional impact of the program. As the authors write in their "Where We Are Now" coda, the Oakland program was marked by a period of initial growth, followed by a series of logistical challenges in course scheduling, advising, assessment, governance, and other matters. The authors write that "there have been times of growth, happiness, and discord for our major and our program" and that "change has also been a constant in our first five years."

Published in the same year (2010), David Beard's "The Case for a Major in Writing Studies" describes the shifting landscapes and ecological debates surrounding the writing studies major at the University of Minnesota Duluth. Those arguments included the desire for differentiation from the contested spaces of other disciplines (literary studies and communication studies) as well as from the prior disciplinary identity of the department (composition studies) while at the same time incorporating a positive identification of writing studies as one of the disciplines defined by its object. In this way, the profile provides examples of emergence, interconnectedness, and fluctuation in the creation of a writing major, since so much of the debate described within it addresses the growth of a new program amidst pre-existing ones, the connections and competition for resources among those programs, and the transformative powers of discourse. In the "Where We Are Now" coda, Beard writes of the paradoxes and challenges of bringing together faculty and students with different goals and motivations into the same discursive ecosystem, noting that "the collaborative project of the major remains the subject of persistent discussion." Like the other profiles in this book, we can learn much about the ecology of writing by studying the discursive ecosystems of undergraduate writing majors.

Note

1. See http://www.ncte.org/cccc/committees/majorrhetcomp

8 Diverse Lessons: Developing an Undergraduate Program in Rhetoric, Writing, and Culture at Texas A&M

Stephanie L. Kerschbaum and M. Jimmie Killingsworth

The number of first-year writing and writing across the curriculum programs has been increasing at institutions across the United States, but a similar rise has not been seen in the growth of concentrations in rhetoric and writing as an undergraduate major or minor. As David Fleming noted nearly ten years ago in "Rhetoric as a Course of Study," despite an upturn in scholarly attention to rhetoric generally, the effects of this increased attention have not yet been fully felt at the undergraduate level, where rhetoric has not re-established a presence as a "coherent and attractive course of study" (169). Since the publication of Fleming's article, at many institutions, rhetorical study at the undergraduate level remains concentrated in first-year composition courses, and English department faculty members are still asking what an undergraduate program of rhetorical study might look like.

In this program profile, we describe how the discourse studies faculty at Texas A&M University has worked to develop a concentration in rhetoric that stretches across the undergraduate curriculum, offering the opportunity for undergraduate English majors and minors to select rhetoric as one of their areas of specialty. Students specialize in rhetoric by choosing to follow what is called a concentration in "Rhetoric, Writing, and Culture." This profile addresses the emergence of rhetoric in the English department's undergraduate curriculum, discusses the faculty's efforts to expand and revise course offerings for a rhetoric program that better draws on the teaching and research strengths in our department as well as the unique needs of our students, and finally, points to some of the challenges in assessing the program's success.

History of Undergraduate Rhetoric in the English Department

As late as the early 1960s, Texas A&M was still a small military institute with an all-male student body, as well as the state land-grant college. Its long and distinguished tradition in agriculture and engineering formed the basis for what it would become when it opened its doors to nonmilitary students and women in the sixties: one of the state's two "flagship universities," a full-scale Research I school with a varied and diverse student body of 45,000 students (including a female majority) and a full complement of doctoral programs in various disciplines, including the liberal arts. Since the early 1970s, the English department has grown from a unit that mainly provided service courses in composition and technical writing for engineering, agriculture, science, and business majors to a world-class research department with 750 majors and more than one hundred MA and doctoral students.

As happens at many universities, rhetoric and composition grew up at Texas A&M primarily as a service program at the undergrad level and a research specialty at the grad level. The opportunities to teach comp and tech writing on a regular basis and to help with the heavy administrative duties associated with the large writing program directed mainly to undergrad non-majors contributed to the growth of interest among graduate students, who often moved from a specialty in literature into the study of rhetoric after good teaching experiences in the comp program. Literature dominated the undergraduate offerings and also predominated in the graduate program though demand was high enough among grad students to support a three-course sequence that provided the necessary background for thesis work in rhetoric. The courses—"History of Rhetoric to 1900," "Modern Rhetorical Theory," and "Contemporary Composition Theory"—were taught once a year on a regular rotation.

In 1989, an important shift occurred at the undergraduate level. To accommodate an increasing interest in writing among the growing population of undergraduate students, the faculty proposed to offer alternatives to a concentration in literary study. What emerged was a three-track undergraduate program. In addition to the literature track, which continued to be the main choice, students could opt to focus in creative writing or in rhetoric.

Interestingly, the development of the rhetoric track was partly the result of some resistance among the majority group of literature faculty to the idea of offering an undergraduate specialization in expository or professional writing—the kind of program developing at that time in other universities. The faculty wanted to keep the growth of the program in check and keep the emphasis fairly traditional. In other words, the small group of rhetoric and comp professors found it easier to "sell" the idea of a rhetoric program. It seemed to fill a gap left by the departure of speech communication, which had left English to form its own department only a few years earlier, a good deal later than the national norm for such divisions. As in the graduate program, three core courses were approved in the new concentration—"History of Rhetoric," "Modern Rhetoric," and "Rhetoric of Style." Students in the rhetoric track were still required to take a significant number of hours in literature, so the program seemed something of a compromise at the time.

As students came into the rhetoric track, however, many of us teaching in the program came to see its value. It proved a particularly strong option for pre-law students. If, as James Boyd White suggests, every lawyer is a "professional rhetorician" (4), we should not be surprised to find that a concentration in rhetoric backed by the study of literature and critical theory provides a preparation for law school every bit as good as the kind of preparation students might receive in a major in political science or some other social science. The law schools to which we recommended our students seemed to be especially impressed by the component of research and written communication that distinguished our program from rhetoric programs in speech departments. It was a good sign when one of our first graduates in the rhetoric track made law review at the University of Texas School of Law. Others found the rhetoric track good preparation for graduate school, teaching, and professional writing.

In the mid-1990s, we were able to convince our colleagues that our students would benefit by a more practical program in writing, that they were in fact already landing jobs in business and industry, and, as a result, we added a certificate and minor program in professional writing. The core rhetoric courses, which every professional writing student was required to sample, provided a strong theoretical and historical context for the student of technical writing and editing. Many students already in the rhetoric track also took the certificate in professional writing and went on to graduate programs in technical writing or to jobs in a wide range of fields.

THE RHETORIC TRACK: THEN AND NOW

In the 2005–2006 undergraduate bulletin, the rhetoric track is described as meeting the needs of "English majors who wish to concentrate their studies on the theory and practice of written communication." The track requires fifteen hours of course work in the theory and practice of rhetoric and writing alongside twelve hours of literature courses and a senior seminar.[1]

According to numbers provided by the undergraduate office, 18.7 percent of the English department's majors are enrolled in the rhetoric track, with 65.7 percent in literature and 13.7 percent in creative writing, and about 2 percent in a certificate program for teaching middle school. In addition, approximately 300 minors are pursuing a professional writing certificate. These numbers reflect nearly 50 percent growth in the rhetoric track over the last three years, as the number of students has exploded from ninety-one majors in fall 2004 to 135 during fall 2006.

All students in the rhetoric track must take English 353 ("History of Rhetoric") and 354 ("Modern Rhetorical Theory"). Along with these two required courses, students can select from a variety of options in rhetoric, professional writing, creative writing, and literary studies. Regular rhetoric and writing course offerings include English 201 ("Introduction to Literacy"); English 210 ("Scientific and Technical Writing"); English 235 and 236 ("Creative Writing, Prose"; "Creative Writing, Poetry"); English 241 ("Advanced Composition"); English 301 ("Technical Writing"); English 320 ("Technical Editing and Writing"); English 355 ("Rhetoric of Style"); and English 461 ("Advanced Syntax").

To complete the professional writing minor, students take eighteen hours, comprised of the following courses: 210 or 301 ("Scientific and Technical Writing" or "Technical Writing"); 320 ("Technical Editing and Writing"); 353 ("History of Rhetoric"), 354 ("Modern Rhetorical Theory), and 355 ("Rhetoric of Style"). Students who earn the Professional Writing Certificate must take 210 or 301 ("Scientific and Technical Writing" or "Technical Writing"); 241 ("Advanced Composition"); 320 ("'Technical Editing and Writing:"); and 355 ("Rhetoric of Style"), along with six hours of elective courses chosen from a list of courses offered in English and communications. While the rhetoric track and professional writing minor and certificate program overlap in many ways, they remain separate from one another, and the changes we describe below do not affect the Professional Writing Minor or Certificate Program.

New Directions

Beginning in the 2004–2005 academic year, as part of a response to the continuing growth of the undergraduate rhetoric program, as well as the need to prepare students for a diverse array of careers, the faculty responsible for teaching these rhetoric courses opted to change the name of the rhetoric track to the "rhetoric, writing, and culture track." Along with the name change, the faculty also began planning significant revisions to the major course offerings. In some ways, the revised program title offers a better portrait of the work that has *already* been going on in our teaching and research, and in other ways, it reflects new directions in which we see the program moving in upcoming years. We should note here that these revisions, while accepted by the RWC program faculty, have yet to be revised in the university's course catalog and officially implemented in the English department's major program. In particular, the process of adding a brand-new course to the university catalog is a lengthy process, which we are beginning by offering the course (described below) for the first time in fall 2007.

To reflect the current emphasis on writing and culture, classes traditionally listed as "surveys" in the course catalog (e.g., "History of Rhetoric"; "Modern Rhetorical Theory") are now treated as more focused topical studies of significant issues in rhetorical study. This change calls attention to the ways that our courses can foreground the use and application of rhetoric to understand writing and culture, rather than surveying major figures or texts. Other courses in the program, such as the "Rhetoric of Style" course, will more explicitly address comparative and cross-cultural approaches to rhetoric and writing. A new 400-level capstone course, currently titled "Writing Culture," which will ask students to treat writing in its fullest possible context, is also in development.

Writing is at the heart of the rhetoric, writing, and culture track, and all of the RWC courses are university "W" courses, or writing-intensive courses. We are also in the process of starting an annual essay competition for the best essay written in a rhetoric or linguistics course. These changes, we believe, promise a more cohesive program of study, one that will immerse students not only in the practices and processes of writing, but which will also prepare students to conduct more advanced study in rhetoric, to cultivate self-consciousness about their own uses of rhetoric and writing, and to explore the interrelationships between language and culture.

Theory Underlying Development of Rhetoric, Writing, and Culture Program

One of the key assumptions underlying this program is that first-year composition does not have to be the only source for rhetorical theory and attention to writing. While Texas A&M's first-year writing course is titled "Rhetoric and Composition," students find it valuable to have these principles reinforced—and more deeply grounded—through further coursework in rhetoric and rhetorical theory. Offering a wide range of classes that facilitate students' rhetorical development takes some of the burden off of the first-year writing course. Longitudinal research on students' writing development (e.g., Herrington & Curtis; Carroll) has shown that writing skills develop over time, through continued reinforcement and involvement, precisely what our RWC program aspires to do.

We have also worked to draw on the strengths of our faculty in our program development. Taking into account the strengths that our faculty bring to the table means acknowledging a wide range of methodological and scholarly backgrounds, including linguistics, literacy studies, ethnography, rhetorical history, modern rhetorical analysis, and classroom discourse analysis, while at the same time recognizing that we all share a concern with how language and culture are intertwined, whether such study takes shape in historical, literary, rhetorical, or ethnographic ways.

As we work with students in this program, we have become convinced of the important role that rhetorical study can play for undergraduate students who go on to pursue a wide variety of professional and career options. The traditional "U-shape" of rhetoric concentrations—where it is highly studied at the freshman level and in graduate programs of study, but scantily represented at the mid to upper-undergraduate level—has resulted in a great deal of scholarship that looks at the influence that first-year writing can have on students' subsequent writing, as well as a great deal of attention to graduate seminars in rhetoric. However, less is known about what an ongoing emphasis on rhetoric at the undergraduate level can contribute to students' linguistic, cultural, and rhetorical self-awareness. We hope to see more and more programs reinvigorate their course offerings by providing greater attention to rhetoric throughout the undergraduate curriculum. We also believe that the interrelationships between cultural studies, literacy, and rhetorical study make for a powerful course sequence that can have an impact on students' attentiveness to issues of communication and difference more generally at Texas A&M.

Assessment

One of the most significant elements of our program design has been the building-in of regular opportunities for ongoing assessment on multiple levels. In this section, we talk briefly about our efforts at designing an assessment program for our undergraduate rhetoric program.

In developing the rhetoric, writing, and culture track, the RWC program faculty created a set of ten outcomes that detail what we believe students should take away from our course sequence. These ten outcomes were developed collaboratively by the RWC program faculty, along with input from various colleagues throughout the English department, with an eye towards university and state educational objectives. These outcomes anchor a two-pronged assessment procedure: looking to understand how well students feel they are meeting these outcomes through their coursework as well as how well the faculty feel students' work meets the program's expectations. From these outcomes, we generated two rubrics: one for a student self-assessment (i.e. did students think their courses were helping them meet the learning objectives we set for the program?) and a second in which faculty would assess the work students produced (i.e. did students' work demonstrate their having met the program's learning objectives?). (See Appendix 1 for the learning outcomes and Appendix 2 for the two rubrics used in this assessment.)

While many of us were comfortable with using rubrics in our own grading, it was something of an experiment to see how a rubric would work in assessing papers from one another's classes—but a successful one. At this point, we have completed a pilot assessment in each of the four core rhetoric courses currently being offered (201: Approaches to Literacy; 353: History of Rhetoric; 354: Modern Rhetorical Theory; and 355: Rhetoric of Style). For each course, all the students completed the self-assessment, and a team of three professors who were not teaching the courses read a random sample of ten papers and evaluated it using the collaboratively designed rubric.

Tentative results from the pilot assessment showed that students generally agreed or strongly agreed that the course fulfilled the learning objectives, with a slight downturn—a shift towards somewhat agree—for the learning objectives focused on methods, cultural awareness, and writing practice. The faculty review of student papers found a similar trend, with a shift away from strongly agree and agree for questions related to method, cultural awareness, applying principles to writing, and ability to argue, particularly in the "History of Rhetoric" and "Modern Rhetorical Theory" courses. One hypothesis for this shift has to do with the different kind of work that students do when the course is heavily slanted towards absorbing theoretical material.

These findings are promising with regard to our sense of the program's goals and how our courses meet those goals. However, they also point to some important areas for improvement. In particular, they suggest that, especially in our theory-heavy courses, we have more work to do in order to improve students' engagement with issues of cultural difference in their writing. The development of our capstone course ("Writing Culture") may provide one way of reinforcing for our students the role that culture plays in rhetorical performances.

Institutional Detail of Restraints Guiding the Program's Development

The restraints that guide the development of the rhetoric, writing, and culture program are primarily institutional. The first constraint is the understaffing of the RWC program, in no small part due to the administrative and other programmatic responsibilities the faculty have across the university, including work in the English department's writing programs office (which administers first-year composition as well as a required "writing about literature" course) and the university writing center.

A second, and closely related, constraint as the program develops is the continual need for the faculty involved with the program to meet a set of diverse needs within the English department: RWC undergraduate course staffing, service in the writing programs office and the university writing center, and scholarship and mentoring in the graduate Discourse

Studies program. While many of these needs overlap, at times they pull in different directions.

If We Knew Then What We Know Now

What have we learned through participating in this ongoing process? Perhaps the most important lesson addresses the way we talk about and integrate issues of culture, difference, and diversity into our courses. We recognize that this work needs to go beyond simply incorporating content that deals with cultural sensitivity and awareness, but our initial assessment findings have also suggested that this is an area where we can continue to improve. We need to incorporate into our courses experiences that engage students with one another as they address highly sensitive issues of difference and diversity. We have also learned the difficulty of talking about how to assess students' receptiveness to cultural difference. This was perhaps the learning outcome that we spent the most time discussing how to describe in a measurable fashion. Indeed, during student paper evaluation procedures, determining whether a student "demonstrated a sensitivity to cultural differences" proved to be one of the more difficult rubric items to measure. This experience re-emphasizes for us the importance of continuing to talk about ways to make such concerns an ever-more central part of what we do—despite the fact that studying diverse rhetorical practices and cultural orientations to communication are at the heart of our scholarship and teaching. How well we engage our students in these pursuits is the real test of the program's success.

Coda: Where We Are Now

In this coda, we highlight two key aspects of the ecological nature of this writing program. First, Stephanie's experiences at Texas A&M and the University of Delaware have solidified her sense that undergraduate programs focused on rhetoric and culture are of immense value for students and that such emphases emerge out of faculty interests and commitments as well as student involvement. Second, Jimmie's perspective reveals that the rich rewards of such programs cannot altogether offset the challenges that their complexity offers for implementing and maintaining them.

Stephanie: It remains relatively rare to find programs that focus on rhetoric and culture at the undergraduate level in English departments. Teaching courses in modern rhetorical theory and the history of rhetoric at Texas A&M not only enriched my scholarly attention to these fields but also gave me an important vantage point on the ways that undergraduate students benefit from such coursework. Rhetorical study prepares students for a variety of career approaches, complements the work they do in other undergraduate programs, and can favorably position them for post-graduate study. Interest in rhetorical study for undergraduates has been growing in my current department at the University of Delaware as colleagues in writing studies design and offer elective courses focused on rhetoric and writing in new media and literacy studies. But the growth of these courses within the curriculum is still dependent on institutional structures as well as faculty ability to teach them and regularly build them into course offerings.

Jimmie: When we published our program profile, there was a strong foundation for the undergraduate study of rhetoric at Texas A&M, and faculty built on programmatic strengths to integrate culture and literacy within our course offerings. In this way, our program was grounded in and emerged out of specific institutional and departmental contexts. Despite these strengths, the program has faced significant challenges. Above all, a rhetoric and composition faculty cannot choose between standard offerings in composition and technical writing on the one hand and an inventive rhetoric program for majors and graduate students on the other. We must take a *both/and* rather than an *either/or* approach. We are responsible for writing program administration, the training of graduate student teachers, and the support of non-tenure-track instructors, as well as staffing a variety of courses for majors, graduate students, and non-majors. We've lost two other tenure-track faculty members in rhetoric and composition besides Stephanie and have been unable to replace them. We are down to three faculty, all full professors, but one is devoted full-time to the university writing center. Since the budget crash in 2009, we have increasingly staffed even advanced courses with graduate students and temporary non-tenure-track faculty. Cuts in all kinds of teaching positions have led us to cancel courses on many occasions. A recent undergraduate curriculum review asked us to defend our ability to staff specialized courses while meeting other program needs. In short, the ambitious version of our program described in the original profile is in peril—not because of intellectual merit but for primarily economic reasons, highlighting the ways that material conditions and the relations between them reverberate throughout and beyond the system itself. Here, the old politics of existing as a minority faculty (along with linguistics and creative writing, also under fire) in a department still mainly populated by literature faculty—a conflict that goes underground during good times—again becomes an issue.

Appendices

Appendix 1: Program Outcomes for Rhetoric, Writing, and Culture

By the end of the course, you (the student) should be able to apply your studies in the history and theory of literacy, rhetoric, and/or culture in the following ways:

- Identify and demonstrate an understanding of key ideas, authors, and texts in the field.
- Demonstrate an understanding of methods for studying texts and ideas.
- Demonstrate an awareness of how cultural differences affect reading, writing, interpretation, and other forms of communication.
- Apply methods and key ideas to the analysis or criticism of written texts, images, films, cultural practices, or other forms of communication and art from a variety of cultures.
- Apply methods and key ideas to the improvement of your own writing and other communication practices.
- Demonstrate an understanding of basic elements of communication; an appropriate competence in grammar, diction, and standard usage; and a willingness to revise and edit your papers as needed.
- Demonstrate a general ability to interpret texts and construct explanations and arguments in writing (composing papers with a thesis, supporting evidence, appropriate documentation, and other elements of good academic writing).
- Demonstrate creativity and critical insight in writing.

Appendix 2: Rubrics

Student Self-Assessment Survey

Strongly Agree / Agree / Somewhat Agree / Disagree / Strongly Disagree / Question Does Not Apply

- In the course, I had the opportunity to study texts that improved my understanding of one or more of the following areas: literacy, rhetoric, and/or culture.
- In the course, I had the opportunity to learn or practice sound methods for the study of ideas, authors, and texts.
- The course increased or reinforced my awareness of the cultural differences in reading, writing, and communication in general.
- In the course, I was able to apply what I learned to the analysis or interpretation of written texts, images, films, cultural practices, or other forms of communication and art.
- In the course, I was able to apply what I learned to my own writing or communication practices.
- In the course, I was able to demonstrate my effective use of basic elements of communication and to show my competence in grammar, diction, and standard usage.
- In the course, I had the opportunity to revise and edit my papers as needed.
- In the course, I have had the opportunity to show that I can interpret and construct explanations and arguments (composing papers with a thesis, supporting evidence, appropriate documentation, and other elements of good academic writing).

Faculty Rubric for Evaluating Course Papers

Strongly Agree / Agree / Somewhat Agree / Disagree / Strongly Disagree / Question Does Not Apply

From the evidence of this paper alone, the student seems able to:

- Identify key ideas, authors, and texts in the field.
- Demonstrate an understanding of these key ideas, authors, and texts.

- Demonstrate an understanding of the methodology used in the formal study of these ideas, authors, and texts.
- Articulate a personal response to these ideas, authors, and texts.
- Demonstrate sensitivity to cultural differences.
- Identify key issues or principles of rhetoric, writing, and/or cultural studies.
- Apply key principles to the critical analysis of written texts, images, films, cultural practices, or other forms of communication and art.
- Apply successfully these key issues/principles to his or her own communication practices.
- Demonstrate an understanding of basic elements of communication and appropriate competence in grammar, diction, and standard usage.
- Interpret and construct arguments.

NOTE

1. For a chart of course offerings and brief descriptions of the courses, see: http://compositionforum.com/issue/17/undergrad-texasam.php#appx1.

WORKS CITED

Carroll, Lee Ann. *Rehearsing New Roles: How College Students Develop as Writers.* Carbondale: Southern Illinois UP, 2002. Print.

Fleming, David. "Rhetoric as a Course of Study." *College English* 61.2 (1998): 169–191. Print.

Herrington, Anne & Marcia Curtis. *Persons in Process: Four Stories of Writing and Personal Development in College.* Urbana, IL: NCTE, 2000. Print.

White, James Boyd. *Heracles' Bow: Essays in the Rhetoric and Poetics of the Law.* Madison: U of Wisconsin P, 1985. Print.

9 Reflections on the Major in Writing and Rhetoric at Oakland University

Lori Ostergaard, Greg A. Giberson, and Jim Nugent

In December 2007, three of the rhetoric faculty at Oakland University (OU)—Greg, Lori, and Marshall Kitchens—began revising our program's proposal for a major in writing and rhetoric, which had been initiated by our senior faculty some ten years earlier. As we reconceptualized the proposal and consulted the available scholarly materials on the topic of undergraduate writing majors, we recognized that while there was a good deal of work on the major, there was still much work to be done in terms of breadth, depth, and consistency in the scholarly conversation. We were lucky to have access to early drafts of the now published collection *What We Are Becoming: Developments in Undergraduate Writing Majors*, as Greg was lead editor on that project, and we found that *Coming of Age: The Advanced Writing Curriculum* provided valuable insights into developing our upper-level writing courses and core curriculum. But for the purposes of proposing and designing our own major, the three of us felt there was much more we needed to know about designing, proposing, and implementing BA programs.

As we rewrote the proposal for our own major, we also decided to begin contributing to the small, but growing, corpus of material about the writing major, to offer our own advice to composition-rhetoric faculty who are interested in developing these degree programs. This program profile grew out of that initial desire to share what we have learned, as we believe program profiles like this one are useful for three reasons. First, local descriptions can be helpful to others when those descriptions focus not only on praxis, but also on gnosis, which is what we have tried to do in what follows. Second, we believe undergraduate writing degrees have the potential to impact the field in significant ways, so we want to encourage the purposeful, intellectual, and scholarly development of those degrees. And last, we wish to share our model for what undergraduate degrees in writing studies can be by describing the choices we made during the proposal process and by discussing why we made those choices. Ultimately, we are offering this program profile because we agree with Deb Balzhiser and Sue McLeod, members of the CCCC's Committee on the Major in Writing and Rhetoric, that "as a field we could and should discuss the general outlines of what our major should be, and . . . a national conversation on this topic is in order" (416).

As we describe in the following section, Oakland University's writing program has a unique history that has made it possible for us to develop a writing degree that has very few historical, institutional, or political strings attached. Our new degree program was developed within a department of rhetoric, communication, and journalism, rather than a department of English; was approved in May 2008; and is currently housed in a new department of writing and rhetoric.[1] But while the conditions surrounding the development of OU's undergraduate writing and rhetoric major were unique, we believe our experiences may help to fill in some of the gaps in the existing scholarship about writing majors. In this profile we briefly address the history of our writing and rhetoric department, describe the major we developed just two short years ago, examine the connections we have created between our first-year writing program and the major, and offer some advice for faculty who are considering developing their own writing major. We do not devote much space here to providing an argument supporting undergraduate degrees in writing, as several already exist (see Beard; Delli Carpini; Moore Howard; Baker and Henning; Yancey).

Departmental Values and Mission

The mission of the department of writing and rhetoric is to develop students' abilities to write independently and collaboratively, to become engaged participants in a democracy, and to be critical readers and thinkers in academic, community, national, and global environments. Our faculty, predominantly trained in writing and rhetoric, view rhetoric and literacy as subjects that must be studied in the context of broader cultural and public interests, and we are committed to offering students opportunities to write and read diverse kinds of texts. Therefore, our courses integrate principles of academic inquiry and encourage students to become critical consumers and producers of texts. Because we view written language as a form of action, worthy of careful consideration by students, teachers, and citizens, we affirm its ability to create common interests and foster the understanding of differences. Our curriculum is ethically and intellectually grounded, requiring that students reflect on the forms and purposes of writing and on the ways written communication is shaped to suit particular rhetorical contexts inside and outside the university. In short, our department seeks to create "thoughtful, informed, technologically adept writing publics," as Kathleen Blake Yancey put it in her call for a *"new [writing] curriculum for the 21st century"* (308).

THE HISTORY OF THE WRITING PROGRAM AT OAKLAND UNIVERSITY

The history of writing instruction at OU begins in the early 1970s with the creation of a department of learning skills. Prior to this time, our university was an honors college for Michigan State, and OU students were assumed to be adept writers with no need for early intervention in their writing practices. With the designation of OU as a separate state institution with its own, less strict admissions requirements, writing studies entered the university as the perceived need for it increased. Like in many writing programs around the country, writing and rhetoric faculty at OU have struggled to work against the assumption that our discipline deals with remedial, skills-based knowledge. For example, to counter the seemingly inevitable marginalization of writing in its "learning skills" department, rhetoric faculty were formed into their own department of rhetoric, which was later combined with programs in communication and journalism to form a third, more powerful, departmental configuration (Rhetoric, Communication, and Journalism or RCJ). To enhance the program's legitimacy and provide a stable and equal platform for the growth of writing studies at OU, the dean of the college of arts and sciences removed the rhetoric program from RCJ and created an independent department of writing and rhetoric in 2008.

The formation of a separate department of writing and rhetoric accompanied a number of other significant changes within the last five years that support a culture of writing at OU, including the creation of our university's first writing center, the transfer of OU's business and technical writing classes from the English (literature and creative writing) department to our new department, the implementation of writing-intensive requirements for general education, and the opening up of five new tenure-line positions in writing. And these changes were complemented by the university board of directors' approval of our proposal for a new major in writing and rhetoric at the same time as the new department was formed.

While brief, this history is detailed enough to show how unique the circumstances were for the creation of our department and our major. Wallace May Andersen provides a much more detailed history of the program in her chapter "Outside the English Department: Oakland University's Writing Program and the Writing and Rhetoric Major" in *What We Are Becoming*.

On Naming the New Department and the New Major

On the national level, departments similar to ours tend to call themselves departments of writing, composition, professional/technical writing, or rhetoric. In developing both a new department and a new major simultaneously, our faculty were acutely aware of the importance of choosing a name that would serve our purposes as both a department providing liberal arts and general education in writing as well as a rigorous disciplinary program leading to a bachelor of arts degree. After much discussion, we chose "writing and rhetoric" as the key terms for our department and degree program. While most of us agreed that "rhetoric" was a more appropriate term to describe the work of the department, we also felt that the term "writing" would be more student and employer friendly than composition or rhetoric, as it is commonly identified by both groups as a valuable and desirable skill. While we spent a good deal of time debating the historical significance of the term writing and its institutional connection with remediation, we concluded that the combination of writing and rhetoric provided us several opportunities and benefits. First, while writing might have remedial connotations within academic institutions, it is recognized in the professional world as a very desirable asset. Second, we hoped that by embracing and emphasizing "writing," we might begin the process of reviving the term institutionally. Third, given that OU's general education program requires all students to take one "writing intensive" course outside of their major, we believed that by having our courses listed under "writing and rhetoric" in the course catalog, we might attract more students to our upper-level courses as they looked to the catalog for "writing" courses. Based on anecdotal evidence, we feel that this choice has worked out quite well.

As is the case for most writing programs, rhetoric provides the theoretical base by emphasizing communication and argumentation within the context of specific discourse communities. Thus, by joining writing and rhetoric, we hoped to portray both an easily understandable and theoretically sophisticated branding of our department and our degree program. Our decision to place "writing" before "rhetoric" in both our department name and in the title of our major was purely strategic: our major now occupies the very last place in the college catalog, making it easier for students to find us and our classes.

The Major in Writing and Rhetoric

We believe that the major in writing and rhetoric prepares our graduates to perform the kinds of collaborative work in written communication that will be required of them for full participation in an increasingly global and high-tech society, whether they go on to professional writing in business, industry, and non-profits; production work in new media; or continue on to graduate studies in composition-rhetoric. All writing and rhetoric majors at OU study rhetorical theories and gain experience composing a variety of texts for multiple audiences, media, and contexts. Individual students also pursue one of three tracks that permit them to choose courses that fit with their unique academic and professional goals. The three tracks consist of writing for the professions, writing for new media, and writing as a discipline. Students who pursue the professional track take classes in business and technical writing and in writing for diverse disciplines; those who choose to follow the new media track take classes that require their critical engagement with and production of digital texts; and those who pursue writing as an academic discipline take classes ranging from classical rhetoric to peer tutoring in preparation for graduate study in composition-rhetoric.

Students majoring in writing and rhetoric learn to analyze the processes by which print and digital texts are produced in diverse contexts and communities. Through a group of four core courses, majors gain an understanding of the practices, conventions, theories, and ethics of written and visual communication and use that understanding to produce their own works for multiple audiences and contexts. Central to this understanding is an ability to think critically about emerging forms of literacy and to adapt to the rhetorical demands of new media. Consequently, coursework in the major involves immersion in online and digital forms of communication and consensus building. Required coursework in the major is comprised of three common core courses that emphasize the practical, theoretical, historical, and disciplinary place of writing and rhetoric studies in the university, in business, and in society: WRT 160 Composition II, WRT 340 Issues in Writing and Rhetoric, and WRT 394 Literacy, Technology, and Civic Engagement.[2] Students also complete a single core course introducing them to one of the three tracks in the major: WRT 331 Introduction to Professional Writing, WRT 330 Digital Cultures, or WRT 320 Peer Tutoring in Composition.

During the proposal revision process, we thought as much about the common experiences we wished our majors to have as we did about the unique experiences they should be offered in each of the three tracks in the major. Thus, our common core courses were developed to provide students with a solid foundation in disciplinary history, theory, and practice. WRT 340 Issues in Writing and Rhetoric, for example, is described in the course catalog as "[a]n introduction to important past and current issues in the field of Writing and Rhetoric." Designed as "an introduction to the discipline, the course will provide a theoretical and historical foundation for understanding current issues and challenges for the discipline." Students in this course engage with current issues in the field by considering the underlying historical, ideological, and disciplinary implications of those issues. For example, in his Issues in Writing and Rhetoric course, which functions as a gateway course, Greg's students are introduced to the discipline by reading several different foundational works by past and current scholars in the field to provide a general academic and scholarly base upon which the rest of the courses in the major can build. Because all of our majors will need to deal effectively and critically with evolving technologies, WRT 394 Literacy, Technology, and Civic Engagement engages students in the critical "exploration and application of technology in the discipline of writing and rhetoric." The course catalog further describes this course as one in which students examine "the uneven shifts from oral to print to digital literacy and how those shifts affect the production of knowledge, social relationships, and opportunities for civic engagement."[3] This course unites our writing and rhetoric department's dual focus on new media technologies and civic engagement, helping students to recognize, theorize, and make sense of these two strands in all of their other course work for the degree.

Each of the three tracks also requires a track-specific core course to introduce students to that track and guarantee some consistency in students' understanding of and experiences with that track. The required course for the professional writing track is WRT 331 Introduction to Professional Writing. Grounded in rhetorical theory, this required course prepares students to write effectively in a variety of contexts, examines the professional identity of professional and technical writers, and prepares students to consider the social and ethical responsibilities of professional writing in practice. Students pursuing the new media track are introduced to the theories, technologies and practices of writing in, with, and for new media through WRT 330 Digital Culture. This course focuses on the rhetoric and ethics of Internet technology and culture by introducing students to theories of digital culture and its effects on both online and actual identities and communities. Students in this class begin their work composing for new media by completing audio and video projects, composing web sites, developing wikis, and participating in a number of web-development projects. Finally, students pursuing our graduate-school track in writing and rhetoric as a discipline begin their study with WRT 320 Peer Tutoring in Composition. The Peer Tutoring in Composition course was chosen as the introductory course for the disciplinary track because it provides students with an opportunity to apply the theories they encounter in WRT 340 Issues in Writing. While they study current theory and best practices in writing center studies, students in Peer Tutoring in Composition also engage in a number of practical experiences (observations, interviews, co-tutoring, and tutoring sessions), working with developing writers in both our department's WRT 160 Composition II classes and in the writing center. The electives offered for each track occasionally overlap, but students are encouraged to focus their coursework on those classes that relate the most to their chosen field of study. A breakdown of the core and elective offerings for each track is provided in the Appendix.

After completing their core and elective coursework, our students take part in either a semester-long internship that synthesizes the knowledge and skills they have gained from their specific tracks in the major or a senior thesis project. The internship requires that students put what they have learned into practice by working in some capacity in the local community, in web development projects on campus, or in research, tutoring, or co-teaching positions. The department was in agreement over the importance of providing our majors with the practical, yet guided, experience of an internship; nevertheless, OU's student body, comprised of students who work an average of seventeen hours a week, meant that we needed to consider alternative experiences for our majors who are already working in their chosen fields while pursuing their degrees. Thus, our students who elect not to complete an internship for their capstone experience compose a senior thesis that synthesizes the work they have done in the major or addresses some issue related to their elective work in one of the three academic tracks in the major. The core (including the first-year writing course), elective tracks, and internship/thesis all reflect this program's dual commitment to theory and practice—to helping students use theory to be critical consumers and producers of text.

THE MAJOR AND THE FIRST-YEAR WRITING COURSE

Our new degree program has simultaneously necessitated an increase in our upper-level offerings and a reconsideration of the ways our required first-year course, WRT 160 Composition II, operates in conjunction with those upper-level courses. In many ways, including the introductory writing course in the collection of required courses for our major has helped us to view first-year-composition as wholly "our own course," rather than as a service course we provide for the best interest of faculty around the university. As writing has gained a legitimate place in the university curriculum through the implementation of our new major, so too has our first-year course improved its status. In this section we address the changes to our first-year writing course that we were able to make largely as a result of including this introductory writing course in the core curriculum for the major.

While the opportunity to propose and implement a writing major had a great appeal for all of our faculty, we were also interested in enacting a redesign of our first-year writing program that we hoped would bring the required composition course more in line with social constructivist teaching philosophies and rhetorical principles. We looked forward to working with our own majors, in other words, but the department also remained committed to strengthening our first-year writing program. In fact, at OU, all full-time faculty teach a first-year course every year, and this is a practice we are determined to continue. Thus, it made a lot of sense for us to make our required introductory writing class, WRT 160, one of the core courses for our new writing major.

Tying this first-year course to the major has also helped us to achieve a couple of important goals that might have been otherwise out of reach. First, by making this course a part of our own major, we illustrated to faculty across campus that first-year writing is only the introductory part of a much longer sequence of writing instruction that students may pursue throughout their college careers. Second, making WRT 160 a part of our core curriculum for the major has helped us to emphasize the importance of this course to the other departments and the administration by elevating its status on campus, much like we did by embracing the term "writing" and its institutional history when naming the department.

As evidence of the changing status of "writing" on campus and FYC in particular, the administration has provided significant funds over the last two years to revamp our required first-year writing course to bring it in line with the values of the first-year experience and to offer a variety of professional development opportunities for our faculty. For example, in summer 2009, we offered $500 stipends to faculty to attend a series of two-day syllabus revision workshops. During this successful series of workshops, more than half (close to thirty) of our faculty revised their syllabi to fit with the new program goals. We have also begun offering our graduate course, WRT 615 Teaching of Writing, and a newly developed graduate course, WRT 525 Teaching Writing with New Media, to our own part-time faculty. Part-time (special lecturer) faculty may take these semester-long, fully-online courses for free as a part of the university's tuition waiver benefit, and they receive graduate credit for the courses. In the years to come, we hope to have all of our first-year classes staffed by only full-time faculty or special lecturers who have completed this two-semester graduate course sequence.

The development of a separate writing and rhetoric department and a new writing and rhetoric major has energized faculty on all levels, and we now find ourselves blessed with a large, professionally engaged, cohort of part-time faculty special lecturers who, in addition to enrolling in graduate study in the field, have begun enrolling in our writing project's Invitational Summer Institute, attending national conferences like NCTE and CCCC, presenting at state and local conferences, participating in Michigan State's Writing in Digital Environments summer workshop (WIDE Paths) and Ohio State's Digital Media and Composition (DMAC) institute, partnering with the Digital Archive of Literacy Narratives project, creating an online archive of local oral histories, leading professional development workshops for our faculty, hosting bi-weekly brown-bag lunches, serving on the committee to evaluate our writing excellence awards, and coordinating our fall and winter festival of writers. Marshall Kitchens, who worked with us to revise the major proposal, is now department chair, and he has been the driving force behind most of our department's curricular changes and has found ways to support our most significant professional development initiatives. By tying the first-year course together with our major, we have been able to unite the two strands of our departmental mission, and the result has been, we believe, the strengthening of both the first-year course and the degree program. Bringing part-time and full-time faculty together for professional development events, brown-bag lunches, conferences, committees, and graduate classes has also strengthened the writing and rhetoric department's sense of community and shared goals.

WHAT WE WISHED WE KNEW THEN

There are, of course, a number of things we wished we knew when we began re-envisioning our major proposal a little over two years ago.

While we revised the existing major proposal, the three of us also wrote proposals for six new upper-level writing courses to complement and strengthen the major: Literacy, Technology, and Civic Engagement; Issues in Writing and Rhetoric; Advanced Writing; Introduction to Professional Writing; Composing Audio Essays; and Digital Storytelling. We believe that attaching these new course proposals to the major proposal strengthened our argument for the major by providing curricular context for the more conceptual arguments we constructed. Upon reflection, we also suspect that many of these courses may have been approved, in part, because of their connection to one another and to the major proposal. In other words, the individual courses made sense to our College's Committee on Instruction because they were able to understand those courses within the context of the proposed program and vise versa. Another possibility for the seemingly easy passage of these courses is that there may be a honeymoon period associated with a new major during which time courses that might raise resistance from other departments can work their way through governance without much objection. For example, when we proposed a web design course the following year—a course that we had included in the original proposal as part of the new media track but did not provide a full course proposal for—we met with resistance from studio art and the department of communication and journalism. This resistance lasted nearly a year before our College Committee on Instruction finally settled matters and approved both a new Writing for New Media course and the original Rhetoric of Web Design course. While proposing those first six new courses as a part of the major proposal added a significant amount of work to the proposal process, we believe now that it would have been beneficial and time saving in the long run to include as many new course proposals as we believed were necessary for the program during the major proposal process to avoid as much political posturing over academic turf as possible.

We also wish that we would have taken time during the proposal revision to think specifically about program assessment. As with most colleges and universities, ours is pushing assessment. Had we considered the assessment piece of the proposal more during the proposal revision process, we would most likely be more prepared now to enact our assessment procedures, and this probably would have saved us from retracing many of the steps we initially took during the proposal process. Our first formal assessment of the degree will begin in August of this year, using the procedures we established in our major proposal just two short years ago. And while we received valuable feedback from the university senate's assessment committee that helped us to refine our proposed assessment protocol during the proposal review process, we wish now that we had also met with Office of Institutional Research and Assessment (ORIA) to confirm the viability of our assessment plans. Over the last two years we have focused our efforts, in large part, on developing our new courses and recruiting majors, two time-consuming yet important occupations for any new degree program. But in retrospect, had we met with ORIA during the proposal process, we probably would have also immediately established a department committee charged with guiding us through our first assessment of this new degree.

We were also unprepared for the amount of student interest we received. In the original proposal, we predicted we would have five majors by the end of the first year and ten after the second. Halfway through our second year, we have fifty declared majors with more signing up monthly. Had we been prepared for this, we might have set up our administrative structures a little differently to account for the added departmental and administrative work that accompanies a fast growing program. For example, we initially proposed separate independent studies courses for the thesis and internships, allowing students to choose one or the other. But as the number of majors increases, we have come to realize that administering independent studies for a dozen or more theses and internships every semester will put an undue burden on our already over-taxed faculty. So, just two short years into this new major, we're revising our requirements to include, instead of separate independent studies for the thesis or internship, a single capstone course that will meet regularly while students pursue the thesis or internship option. This will transfer the work of mentoring these senior projects to a variety of faculty every semester and permit faculty who would have done this type of advising work as an added responsibility to do so as a part of their usual course load.

Maybe the final lesson we learned from this process is a lesson that is only now taking shape: in developing and implementing the writing major, it's important to remain flexible and open to change. The more we work with majors, the more we learn about what they need. And the more we learn about majors around the country, the more critically we evaluate our own curriculum. Over the course of the next few years, we will begin to discuss changes to the core curriculum. For example, we are beginning to wonder if a required research methods course might help to focus the various methods students learn in their required and elective writing classes. We are also thinking about adding electives in technical editing, courses in the history of composition in the university, courses in classical rhetoric, and additional 300-level courses in new media.

Developing a new BA program provided our faculty with an opportunity to re-think every aspect of our program, from the ways we conceptualize first-year writing to the ways we interact with, support, and professionalize our part-time faculty. Working with majors has, in many ways, helped us to refigure how we work together, and the result has been a uniting of our department's mission, goals, and values.

Coda: Where We Are Now

When this program profile was written in 2010, we had just concluded our second year of the major. We had experienced a period of growth that exceeded our expectations and had begun to graduate strong and motivated students. We had yet to see our numbers level off, then decline, then level off again. We had yet to really grapple with the logistical challenges of course scheduling, advising, assessment, governance, etc.

On rereading this profile, we can't help but appreciate our own optimism and idealism, like new parents talking about the bright futures awaiting their children. Like familial relationships, there have been times of growth, happiness, and discord for our major and our program. Change has also been a constant in our first five years.

The editors of this volume provide a theoretical structure for it by framing each program profile, as well as the volume as a whole, as an ecological system, "envisioning writing as bound up in, influenced by, and relational to spaces, places, locations, environments, and the interconnections among the entities they contain." While we used a simplistic biological framework of familial relationships between programs and those who design and implement them, the effect is essentially the same. Indeed, as we look back upon the first five years of our major, we see more clearly the interconnectedness of our program and the material circumstances within which it exists. We more thoroughly understand and embrace the fluctuations our program continues to go through as it, and we, mature. We both encourage and work to keep under control the complexities of the program. And we embrace the emergent nature of the program through minor and major revisions that become necessary as we learn more about what the program is and should be and the students that it is designed for.

One of the more significant revisions that we made to our curriculum was in regards to our Composition II course, WRT 160. Much of our original profile articulates why this service course was included as a core course in the major. To those of us familiar with writing's history in the academy, the justification sounded reasonable. However, when faced with the realities of running a program and designing a meaningful curriculum, we had to face up to the fact that including WRT 160 in the core made little sense given our local circumstances and the fact that it did little to improve our status in the University, as envisaged in the profile.

While we stand by the value of the course and our FYC program (the program was awarded a 2012 CCCC Certificate of Excellence), we realized that most of our majors were not actually taking the course with us but were transferring the credits in from other institutions. So we eliminated WRT 160 from the core and developed a new course to address a long acknowledged gap in our curriculum: the history of rhetoric.

We have also worked to fill curricular gaps within the major's tracks with several new courses, such as Global Rhetorics, Video Game Culture, and Research Methods. Based on a request from the dean's office, we more clearly defined the major's three tracks by identifying each upper-level course we offer with only one track. We have redefined the capstone course on a couple of occasions, most recently to accommodate the needs of program assessment.

The department's Committee on the Major is currently working on our first large-scale promotion and recruitment campaign in effort to grow our numbers (which have leveled off at around 50 majors). We are also conducting a survey of the faculty to get their feedback on the current state of the major and to gauge where we should be headed. Whether this leads to substantial revisions or minor tweaks to our major program, it remains clear that our program is growing up, maturing in ways we had and had not anticipated, and that change has been a consistent force for good in our first half-decade.

Appendix

Requirements for Major in Writing & Rhetoric

Requirements for a Liberal Arts Major in Writing and Rhetoric, B.A. degree program.

The major in writing and rhetoric requires a minimum of 40 credits. A maximum of 8 credits may come from areas other than the writing and rhetoric rubric with the permission of the department chair.[4]

Students who earned college credit for the AP writing course and those who have received credit for the equivalent of WRT 160 at other institutions are not required to take WRT 160. Students who have been exempted from WRT 160 for submitting a portfolio as described under the General Education Program in the Undergraduate Degree Requirements section of the Undergraduate Catalog do not need to take WRT 160 and can instead choose an additional elective course to complete their 40 credits of course work.

Only courses completed with a grade of 2.0 or higher will be counted for the major. Students must complete the following:

1. Core Courses (12 credits): Majors will complete the following Core Courses.

- WRT 160 Composition II (or equivalent)
- WRT 340 Issues in Writing and Rhetoric

- WRT 394 Literacy, Technology, and Civic Engagement

2. Major Track (16 credits): Students will choose one of the following major tracks for their course work and complete both the required course and three of the elective courses from that track. One of the elective courses may be chosen from another track with the permission of the WRT department chair:

Writing for the Professions (16 credits)
- WRT 331 Introduction to Professional Writing (required)
- WRT 305 Advanced Writing: Various Themes
- WRT 332 Rhetoric of Web Design
- WRT 335 Writing for Human Services Professionals
- WRT 341 Rhetoric of Professional Discourse
- WRT 350 Service Learning Writing
- WRT 380 Persuasive Writing
- WRT 382 Business Writing
- WRT 381 Scientific and Technical Writing
- WRT 460 Writing across the University: Language and Disciplinary Culture

Writing for New Media (16 credits)
- WRT 330 Digital Culture: Identity and Community (required)
- WRT 305 Advanced Writing: Various Themes
- WRT 232 Writing for New Media
- WRT 231 Composing Audio Essays
- WRT 233 Digital Storytelling
- WRT 320 Peer Tutoring in Composition
- WRT 332 Rhetoric of Web Design
- WRT 364 Writing about Culture: Ethnography
- WRT 381 Scientific and Technical Writing

Writing as a Discipline (16 credits)
- WRT 320 Peer Tutoring in Composition (required)
- WRT 305 Advanced Writing: Various Themes (new course)
- WRT 341 Rhetoric of Professional Discourse
- WRT 342 Contemporary Rhetorical Studies
- WRT 350 Service Learning Writing
- WRT 360 Global Rhetorics
- WRT 364 Writing about Culture: Ethnography

- WRT 365 Women Writing Autobiography
- WRT 380 Persuasive Writing
- WRT 414 Teaching Writing
- WRT 460 Writing across the University: Language and Disciplinary Culture

3. Two Electives (8 credits): Electives are chosen from additional WRT courses numbered 200 or above. Students may substitute appropriate courses from other departments with permission of the WRT department chair.

4. Capstone Course (4 credits): The capstone includes either WRT 491 Internship or WRT 492 Senior Thesis. The internship should demonstrate grounding in the discipline and application of disciplinary theory. In addition to evaluation by the internship supervisor for the course grade, the student will produce a reflective research project on the experience to be presented in an annual public research forum (e.g., Meeting of the Minds or a special program colloquium) and evaluated by a committee of the tenured/tenure-track faculty using the evaluation criteria in the assessment plan.

As an alternative to the internship, a student may elect to complete a senior thesis project under the supervision of a Writing and Rhetoric faculty member (exceptions can be made for a mentor outside of the department) within whose professional discipline the subject of the project lies. The thesis project should bring together the student's knowledge and skill in his or her specific track of the major/minor (writing for the professions, writing for new media, or writing as a discipline). The senior thesis, like the internship, should demonstrate grounding in the discipline and application of disciplinary theory. In addition to evaluation by the thesis supervisor for the course grade, the student will present the results of the project in an annual public research forum (e.g., Meeting of the Minds or a special program colloquium) to be evaluated by a committee of the tenured/tenure-track faculty using the evaluation criteria in the assessment plan.

Notes

1. The complete proposal as submitted to college and university committees and eventually approved by the OU Board of Trustees is available online. Included with the proposal are memos documenting the objections, questions, and suggestions posed by each committee that reviewed the proposal and our

memos in response to those objections, questions, and suggestions. To see these materials, go to http://www.oakland.edu/?id=7837&sid=230.
2. For course descriptions, visit: http://www2.oakland.edu/wrt/courses.cfm.
3. For course descriptions, visit: http://www2.oakland.edu/wrt/courses.cfm.
4. Oakland University uses a 4-credit course system.

WORKS CITED

Andersen, Wallace May. "Outside the English Department: Oakland University's Writing Program and the Writing and Rhetoric Major." Giberson and Moriarty. 67–80. Print.

Balzhiser, Deborah and Susan H. McLeod. "The Undergraduate Writing Major: What Is It? What Should It Be?" *College Composition and Communication* 61 (2010): 415–33. Print.

Baker, Lori and Teresa Henning. "Writing Program Development and Disciplinary Integrity: What's Rhetoric Got to Do with It?" Giberson and Moriarty. 153–73. Print.

Beard, David. "The Case for a Major in Writing Studies: The University of Minnesota Duluth." *Composition Forum* 21 (2010). Web.

Delli Carpini, Dominic. "Re-writing the Humanities: The Writing Major's Effect Upon Undergraduate Studies in English Departments." *Composition Studies* 35 (2007): 15–36. Print.

Giberson, Greg A., and Thomas A. Moriarty, Ed. *What We Are Becoming: Developments in Undergraduate Writing Majors*. Logan: Utah State UP, 2010. Print.

Howard, Rebecca Moore. "Curricular Activism: The Writing Major as Counter Discourse." *Composition Studies* 35 (2007): 41–52. Print

Shamoon, Linda K., Rebecca Moore Howard, and Sandra Jamieson, Ed. *Coming of Age: The Advanced Writing Curriculum*. Portsmouth: Boynton/Cook, 2000. Print.

Yancey, Kathleen Blake. "Made Not Only in Words: Composition in a New Key." *College Composition and Communication* 56 (2004): 297–328. Print.

10 The Case for a Major in Writing Studies: The University of Minnesota Duluth

David Beard

Introduction

This essay profiles the department of writing studies at the University of Minnesota Duluth to explicate the arguments for a major in writing studies. In one sense, it picks up work done in the forthcoming *What We Are Becoming: Developments in Undergraduate Writing Majors* (edited by Moriarty and Giberson). The title intentionally echoes the title of an essay by Charles Bazerman ("The Case for Writing Studies as a Major Field of Study"), from which this project took a great deal of direction. Therein, Bazerman defines writing studies by stipulation, claiming that "Writing Studies is the study of writing—its production, its circulation, its uses, its role in the development of individuals, societies and cultures" (32). As such, writing studies is a discipline structured like other disciplines with similar titles: American studies, women's studies, library and information studies, communication studies, cultural studies—with writing as the object.

In other essays, Bazerman defines writing studies by differentiation. Within the larger umbrella of English studies, for example, Bazerman differentiates writing studies from literary studies: Writing studies differs from literary studies in part because it does not engage "the traditional historical work of rhetorical and literary studies in recovering, editing, and interpreting major texts" (as well as new additions to the canon) ("Theories of the Middle Range in Historical Studies of Writing Practice" 298). It may be helpful to think through "writing studies" as a broad term; in paraphrasing it to colleagues, I often appeal to the distinction between literary studies and literacy studies. Literacy studies entails reading and writing activities of great variety—multiple genres, multiple contexts—and so is clearly distinct from literary studies, which keeps its eye on a narrower range of writing and reading activities. Further, because we are interested in a broader array of writing and reading activities, we can focus on a greater variety of sociocultural effects of writing. We can be interested in the ways that a variety of writing forms sustain institutions, generate communities, and enable (or domesticate) individual and social cognition. Bazerman is not the only scholar to advance claims for the disciplinarity of writing studies, but at Minnesota Duluth, we found these claims particularly persuasive.

The arguments that Bazerman advances in these essays, however, were transformed when planted in the soil of the department of writing studies at UMD. The intellectual resources and curricular raw material available in that context engaged the core claims of Bazerman's arguments in uniquely local ways. What emerges, then, is something new and different—a curricular innovation.

This profile traces the intellectual arguments that created space for a department and major in writing studies at UMD. Those arguments included a differentiation from the contested spaces of other disciplines (literary studies and communication studies). We needed the good will of colleagues in other departments—their support made curriculum development and approval easier. We also took this opportunity to differentiate the old department of composition (the service unit that offered primarily first year writing courses from 1988–2008) from the new department of writing studies (the unit which we became). In becoming the department of writing studies, we found an opportunity to rebrand our department in terms of its intellectual project rather than only its service mission.

Our innovation, then, stems from our arguments about the object of the discipline of writing studies at UMD. We teach and research writing, defined as a practice, as a tool both for cognition and for social action, and as a force for sociocultural change.

These claims about the nature of writing are manifest in the core curriculum of the major (sixteen credits in six courses across all four years of the student's coursework). The following six core classes (see table one below), required of all majors in writing studies, represent the common core of intellectual work in our department. Following Bazerman's claims, we see the study of writing from a variety of perspectives. We study writing with a critical eye toward its role in the development of individuals, societies and cultures (WRIT 1506; WRIT 2506; LING 2506). We examine its contemporary systems of circulation and uses (WRIT 2506; JOUR 3700). And across the curriculum, culminating in the portfolio class (WRIT 4506) and in the new media writing class (WRIT 4250), our students master its production—becoming skilled writers in their own right.

Table 1. Six Core Courses in Writing Studies at the University of Minnesota Duluth

Course Number	Course Title	Course Description
WRIT 1506	Literacy, Technology and Society	Historical survey of cultures without writing systems and cultures with writing systems and then later with printing, telegraph, radio, telephone, television, computers as well as other forms of technology. Survey of attitudes toward technology from Thoreau to Gandhi and beyond.
WRIT 2506	Introduction to Writing Studies	Considers writing itself as both a practice and an object of study. Drawing on composition, journalism, linguistics, literary studies, and rhetoric, the course offers a survey of historical, critical, and theoretical issues in writing studies. Writing assignments ask students to apply a writing studies framework to produce and analyze specific texts.

Course Number	Course Title	Course Description
LING 2506	Language and Writing	Different from a traditional linguistic approach, language and its system will be examined with emphasis on writing, as opposed to speech. Based on the formal theoretical foundations of language and linguistics, three main topics are discussed in detail. First, world's major writing systems and a short history of writing are introduced. Second, the English sentence structures are studied from a contemporary theoretical and historical linguistic perspective. Third, language use in writing is discussed in various genres.
JOUR 3700	Media Law and Ethics	Examines laws, regulations and major court decisions that affect journalists and news organizations. Topics include First Amendment principles of press freedom, libel, invasion of privacy, prior restraint, access to information, and the regulation of electronic media content.
WRIT 4250	New Media Writing	Combines the theory and production of new media writing—digital, verbal practices in converged media—through the application of readings and discussion to five projects that progress from written, print-based genres to new-media presentation.
WRIT 4506	Capstone Course: Senior Portfolio Preparation	Required capstone course for all writing studies majors. Portfolios for multiple purposes will be prepared under the guidance of the student's adviser.

In addition to the completion of this core, students elect an emphasis in journalism or in professional writing. Those emphases explore the theoretical, practical and sociocultural issues in writing in two contemporary contexts (the media and the workplace).

In the first section of this profile, I describe how our local context yielded specific arguments for the major in writing studies—arguments that have implications for larger scholarly discussions concerning the disciplinary distinctions between and intellectual status of rhetorical studies, composition studies, rhetoric and composition, or writing studies. In the second section, I rehearse the arguments that located writing studies among its sibling disciplines at UMD; in the third section, I examine the arguments that defined the proper object of study of our new discipline and manifested that definition in our core curriculum.[1]

THE LOCAL CONTEXT AND THE SCHOLARLY CONVERSATION

The new major in writing studies was born of two decades of evolution in a freestanding department of writing studies, formerly department of composition. The department of composition separated, administratively, from the department of English in 1988. At that point in time, the split was largely amicable, rooted in a largely budgetary desire to separate the costs of the first-year composition program from the costs of the large English major. (The very large department was allotted a single budget to serve the needs of a large number of English majors as well as a composition program serving 10,000 students with two semesters of required courses. Too often, courses in literature intended to serve the English majors were cancelled so that another liberal education writing course could be added to the schedule. The solution, it seemed, was to disentangle the two programs: to move the liberal education writing courses into a freestanding academic department.)

Generally speaking, collaboration characterized the relationship between the then newly-formed department of composition and the department of English. Their joint role in maintaining the terminal MA in English necessitated the collaboration, as teaching assistantships remained allotted to the first-year writing program (and so administered by the department of composition). Faculty in both departments were appointed with graduate faculty status in the English MA program; they were equally able to direct MA projects, teach MA courses, and advise MA students. Over time, the collaboration has become a strong point in positioning the MA competitively across the region. Together, we now offer two traditional curricula in English (with emphases either in a narrow form of literary studies or in a broader form of English studies, composed of courses in literature, language, and writing studies). We also offer an emphasis in publishing and print culture, a track designed to blend an emerging subfield of English (print culture studies) with the practical skills to enter a career in writing and editing.

Nonetheless, at the undergraduate level, the two departments developed independently. Left to its own devices, over time, the department of composition grew as it entrepreneurially came to house the following academic programs:

- First-year composition (an integral part of the liberal education program)
- Advanced composition (required courses in professional writing serving multiple majors for accreditation purposes, etc.)
- An undergraduate minor in Linguistics, the first formal minor available in the department and the only graduate-level minor available in the department
- An undergraduate minor in professional writing and communication, developed to advance the status of the professional writing courses from "merely" service to a designated area of study for students
- An undergraduate minor in information design (a selection of courses in web design and digital culture studies)
- An undergraduate minor in journalism

Some of these programs developed from faculty strengths (the minors in linguistics and information design). One program developed from collaboration with another unit (the minor in professional writing and communication, developed with the department of communication). One program was inherited as a legacy (as the journalism program migrated from the department of English to the department of communication to the department of composition over the span of three decades). Growth came like the root system of a tree—spreading in multiple directions over time.

Because the split was primarily administrative and the growth in programs within the department over the years was primarily entrepreneurial, there was little opportunity to engage, deliberately and as a department, some of the scholarly conversations that have come to define the questions of disciplinarity for rhetoric, composition and writing.

When Margaret Strain tells us that "we recognize when an area of professional inquiry has achieved the status of a discipline [when] its members begin to write its history" (57), she refers to the histories that focus on standard, intellectual measures of disciplinarity. Kitzhaber, Berlin, Connors and dozens of other scholars in composition studies write histories that focus on intellectual measures and measures of disciplinary prestige:

> a defined subject of study; a canon of texts which theorizes and historicizes that subject; the establishment of a research community; the creation of apparatuses that insure the field's continued visibility such as scholarly journals, presses, and professional organizations; and, the ability to authorize and reproduce practitioners through the establishment of graduate programs. (Strain 57)

As a freestanding department at a regional, MA-terminal university dedicated primarily to teaching, we had little place in the development of the discipline by these measures. As the number of programs grew and diversified, there was even less coherence in the object of study, it seemed. While our faculty published occasionally, there was not a vital position taken in the development of a canon of texts or a research community to study those texts. And while the graduate program would occasionally produce students who would pursue a PhD in English, it did not produce practitioners of composition studies.

Louise Wetherbee Phelps articulates a disciplinary history of rhetoric and composition that simply does not map onto the narrative of our then department of composition:

> The formation of Rhetoric and Composition as a contemporary discipline is conventionally dated to around 1963, but founders saw it as reconnecting writing to a history of Western rhetoric stretching back to ancient Greece. In the sixties and seventies, scholars from various disciplinary backgrounds brought this humanistic tradition together with the methods of social and behavioral sciences to develop a new field focused on studying written language. (1)

Phelps's narrative is designed to justify a full and clean break from English as an intellectual umbrella (for purposes of National Research Council classification). She is "opposed [to] classifying Rhetoric and Composition as a subfield of English studies. . . . [S]uch a classification [is] historically inaccurate and misleading because of this field's multiple source disciplines and the varied intellectual configurations and institutional locations of its doctoral programs" (5). Phelps needs to articulate rhetoric and composition as a unique intellectual enterprise to justify its administrative independence from English.

In a very real sense, because we were created as an administrative expedience, we achieved departmental independence before we had disciplinary coherence. We were the freestanding department of composition before we had a coherent intellectual mission beyond the service courses in first year composition and advanced composition that defined the bulk of our teaching. We existed outside the disciplinary narratives that have defined the field, in many ways, as a freestanding department with autonomy in funding and tenure decisions, but without a place within the larger scholarly conversation about the intellectual work of rhetoric and composition. And the entrepreneurial patterns of our growth, one academic minor at a time, only increased the diffusion of our voices.

By 2007, the department was nearing the largest total number of tenure-line faculty it had ever held: one full professor of composition (noted Ong scholar Tom Farrell), one full professor of linguistics (sociolinguist Michael Linn), three associate professors of composition (Kenneth Risdon, Craig Stroupe, and Jill Jenson, who serves both as department head and director of composition), and five assistant professors (three in composition: David Beard, Kenneth Marunowski, and Juli Parrish; one in linguistics: Chongwon Park; one in journalism: John Hatcher). The department is also staffed by eight teaching assistants and a dozen adjunct faculty and one full-time administrative assistant. There was a critical mass of scholars willing to collaborate toward a common vision—toward articulating a common intellectual project.

In 2008, we sought to be renamed the department of writing studies. "Writing studies" would replace "composition" as a term to differentiate us from our past (as a department largely defined by the service composition courses that were the justification for our independence). But more importantly, "writing studies" better collected the various strands of research extant in the department. "Writing studies" would pull first year composition and professional writing, information design and journalism, and even some elements of linguistics into a coherent intellectual project that would be the backbone of our new, first major. This name change was important both for the culture of the department and for the public face of the department within the university. "Writing studies" was to become our new public face.

Finding Writing Studies among the Disciplines

To craft that public face, we had to be able to define writing studies (for ourselves and for others) through a variety of strategies. We had to be able to define what writing studies *is not* (to differentiate a new public face from the old public face of the department as a service unit; to differentiate the mission of the newly renamed department from extant departments at the university). We had to define what writing studies *is* (as a positive statement aligned with the research, teaching and service goals of the university). And, lastly, we needed to identify our complementarity—our points of intersection with other units.

Writing Studies Is Not Identical with Literary Studies at Minnesota Duluth

The longest-standing academic discipline for the study of written texts in the United States is the discipline of English. And English as a disciplinary formation has come to include a wider and wider range of objects of study: film, popular culture, visual communication, creative writing, linguistics and much, much more.

But, during the last 100 years, English as a departmental (as opposed to disciplinary) formation at Minnesota Duluth has become more and more narrow. Facets of language and literacy have split into other departments. Speech and theatre scholars divided from English in 1937 at UMD. When the "communications" minor in English moved into UMD's communication department in the 1960s, so did the study of mass communication and media texts. And when the faculty member who pioneered cultural studies at UMD saw his tenure line moved to sociology, so did the range of courses in that area migrate with him. As a result, in many ways, the scope of English studies at Minnesota Duluth has narrowed to a strictly defined conception of literary studies.

Writing studies, as a disciplinary practice, may use methods that overlap those developed in literary studies—drawing on the same bodies of critical theory, for example. But those methods are put to different political, intellectual and cultural ends. Scholars in writing studies do not suffer from what John Schilb has called "canonoia" (131)—here, Schilb means the contortions that followed the canon wars in literary studies at the end of the last century and that define the literary curriculum. By contrast, the nonliterary texts under analysis from a writing studies perspective are intentionally a-canonical, but no less important because they circulate and do work in the world. Such writings include workplace writings, political and civic writings, and student writings (the traditional domain of composition studies).

Writing Studies Is Not identical with Composition

Composition is a term that has only recently come into crystallization. Arguments for the disciplinary status of composition have multiplied in the last twenty years, from Steve North's monograph *The Making of Knowledge in Composition* to Louise Wetherbee Phelps' systematic arguments for rhetoric and composition as an "emerging field" (1). At UMD, composition, as a term for an intellectual enterprise, has generally become shorthand, however, for a particular component of writing studies: the pedagogical component. When David Bartholomae claims that composition is the "institutionally supported desire to organize and evaluate the writing of [student] writers and to define it as an object of professional scrutiny" (327), he is clarifying the particular commitment inherent in the term "composition" to the pedagogical project that defined our department. We were a department of composition for two decades, and the pedagogical project was foremost in the department's mission for much of that time. Part of the transformation in the change of our name to a department of writing studies was an embrace of an intellectual project larger than just that of the best pedagogy for the first-year course. "Composition," as a term for the public face of the department within the Minnesota Duluth context, was inextricable from the first-year course. Rather than attempt to redefine the term for our colleagues, we struck out for a new term.

Writing Studies Is Not Communication Studies

Writing studies is, in many ways, roughly analogous to communication studies. Both draw deeply but not exclusively from the rhetorical tradition. We both use rhetoric as a field and method of inquiry, but the term was both bigger and smaller than what each discipline has been within the 20th century. And like the compositionists, the communicationists initially built their claims to disciplinarity in claims to pedagogy. (The first national association for communication faculty was the National Association for Academic Teachers of Public Speaking, as discussed in Keith.)

Despite our clear similarities (common roots in pedagogy and common sources in the rhetorical tradition), writing studies is not identical with communication studies. Whole ranges of human communication overlap in these two fields (as, for example, scholars from both disciplines may be interested in print magazines as an object of study). But whole ranges of human communication do not overlap (as speech-communication scholars continue to explore areas like proxemics and nonverbal communication).

This differentiation was essential in the context of Duluth because of the relative size of the communication faculty and the communication major at Duluth. Within the college of liberal arts, the communication faculty is the 800-pound gorilla, and we needed to articulate our mission as parallel to, rather than competing with, their mission. A sharp focus on writing helped make our argument that our major complemented, rather than competed with, the major in communication.

In differentiating ourselves from our siblings (from literary studies and from communication studies) and from our own past (composition studies, to the extent that the term "composition" reflected only the service mission of the department), we had not yet articulated what it meant to be a new department of writing studies. We needed to advance those positive arguments.

Writing Studies, Defined

Writing studies, once differentiated from other disciplines, must be articulated on its own ground, one defined by a newly recognized and important object of study. Similar disciplines include American studies, women's studies, and ethnic studies (African-American studies, Asian-American studies, etc.). The appearance of these disciplines was an act of legitimating their objects of study and an act of facilitating interdisciplinary inquiry.

It is not the case that no one studied America prior to the rise of American studies departments. However, the centrality of American culture as an object of study was established with departments and the discipline bearing that name. American studies departments include literature scholars, historians, sociologists, art historians and other scholars, united by their interest in the common object of study.

Our argument is that *writing* is now poised to take its place alongside those objects of study at the core of a discipline. In making that claim, we are heavily dependent upon and build from the claims made by Charles Bazerman, transformed to understand writing in three ways.

These three ways shape our curriculum and define our claims to disciplinary status at Duluth.

1. We research and teach writing as a practice (with its own theoretical grounds).
2. We research and teach writing as a tool (used in a variety of human activities).
3. We research and teach writing as a historically embedded phenomenon that has transformed human socio-cultural structures.

The Study of Writing as a Practice

Our roots in composition studies mark our commitment to the study of writing as a practice. By far, the highest number of courses offered in the department are still courses in writing (first-year composition; professional writing; and the courses in journalism, professional writing and publishing that emphasize writing as a practice). Additionally, the single most vital contribution to the MA curriculum remains the course in teaching college writing.

The major curriculum is built upon the study and execution of the practice of writing. In the core curriculum (required of students in both the journalism and professional writing emphases), students must complete New Media Writing (WRIT 4250) and the production of a Professional Portfolio (WRIT 4506).[2] These courses put students through their paces, generating a body of writings that can demonstrate their mastery of the skill of writing. They signal our basic commitment to teaching writing as a practice.

Courses in journalism (in broadcast media production and online journalism, especially) and courses in the information design program (in visual rhetoric, in document design and in web design, especially) appear, to the casual observer, to not be courses in writing. The final product isn't typed, doesn't run through a laser printer, and cannot be photocopied; it must not be writing. But these surface observations fail to see the coherence behind the Latourian perspective on inscriptions that complements Bazerman's perspective in our approach. Latour considers meaningful marks of an immense variety that include the points on a line graph, the halftone dots that constitute a photograph in a newspaper, the pixels on a computer screen, the analogue waves on an audiotape (articulated in *Pandora's Hope*).

Following from that perspective, we define writing very broadly as we design experiences for our students, hoping that each will develop a truly multimedia portfolio (in WRIT 4506). And the universal requirement of WRIT 4250, New Media Writing, demonstrates our commitment to this broad conception of writing—a conception that informs courses across our curriculum.

The Study of Writing as a Tool

When we teach writing as a practice, we teach the skills of rhetorical production—of crafting texts to rhetorical situations. We also research and teach reflection on writing as a tool—a component in both complex human activity systems and in individual human cognition.

Writing is a material object, capable of doing work in the material world. The materiality of text has been mapped in the classical world, for example, by Robert Gaines who articulates the study of classical rhetoric as the study of "anything written using any medium that has survived complete or in fragments . . . [including] original and copied writing on papyrus, wood, wax, or animal skin or writing on or in pottery, masonry, stone or metal . . . man-made objects of aesthetic, practical, religious or other cultural significance" (65). In the contemporary world, that materiality of the text is echoed in the writings of Latour and Woolgar, for example, when they discuss the variety of inscription machines that drive contemporary scientific practice (*Laboratory Life* 48–51). The text is an object, worthy of study, useful in a variety of human activities.

A variety of scholars trained in rhetorical genre theory (Aviva Freedman, David Russell, and others) have advanced the idea that writing is a tool and have made important connections to activity theory. Like these scholars, we want our students to see writing as an integral part of what activity theorists call goal-directed, historically-situated, cooperative human interactions. We want them to see the ways that writing can be the tool to mediate between a subject (a person or persons) and an objective (a goal or common task). *Résumés* mediate between job seekers (and their goal of securing a job) and interviewers (and their goal of finding the best applicant for a position). A Facebook page mediates between the desire of the individual to express himself or herself and the desire of the corporation to target advertising as tightly and efficiently as possible. The core course WRIT 2506, Introduction to Writing Studies,[3] introduces students to this perspective (among others) through readings like John Seeley Brown's and Paul Duiguid's "Social Life of Documents."

We are particularly interested, within our curriculum, in the use of these material pieces of writing as tools for thinking. From the perspective of rhetoric and composition, this follows from adages as old as claims from the expressivist movement that writing enables thinking (roughly paraphrased in the teacherly question, "How do I know what I think until I write it down?") through Patricia Bizzell's claims that writing enables a kind of critical consciousness. But from the important perspective of our collaboration with our faculty in Linguistics, we understand that the features and typifications of language map onto the processes of mind. Linguistics is understood not just as the study of language as spoken practice (the perspective of sociolinguistics, a perspective that has been a part of composition studies since the 1970s). The core curriculum includes a course (LING 2506, Introduction to Language and Writing[4]) that helps majors in writing studies connect diction and syntax to thinking and cognition.

Andrea Lunsford has begun to think through this perspective, as well, in articulating that writing is

> a technology for creating conceptual frameworks and creating, sustaining, and performing lines of thought within those frameworks, drawing from and expanding on existing conventions and genres, utilizing signs and symbols, incorporating materials drawn from multiple sources, and taking advantage of the resources of a full range of media. (171)

Lunsford here provides a fresh articulation of the key insight: writing is a tool for thinking, an insight plumbed to the depths in LING 2506.

Writing and the Development of Human Societies

To be sure that students recognize both the incredible power that comes from effective writing and the awesome responsibility that comes with being an effective writer, we ask them to encounter writing as historically transformative. We ask them to recognize that writing is embedded in human social structures and human cultural institutions. This has been true since the classical period; again, we can turn to Gaines for the most complex articulation. Gaines gives us a survey of what he called "discourse venues," or discourse "places culturally associated with purposive communication, including rostra, legislative assembly areas, courts, theaters, temples, salons, schools, libraries, festivals, and other public and private locations" (65–66). In the contemporary world, we find similar diversity: we find writing acting in politics, in corporate life, in journalism, in education, in online communities, in photocopied 'zines on sale in record shops and in scrapbooks that map the lived experiences of families. Writing is embedded in these human activities and, in some cases, is constitutive of these activities.

Writing has led to immense sociocultural change. We see these claims in the early scholars of literacy in the ancient world (Havelock, Chaytor, and Ong) who noted the transformative power of the written word in ancient Greece. We see these claims in the works of media ecologists like Elizabeth Eisenstein (who noted the transformation of Renaissance culture after the development of printing), Benedict Anderson (who connected printing technologies to the development of nation states), Harold Innis (who connected those same printing technologies to the development of empire), and Bolter and Grusin (who explored the implications of online writing for an "electronically constituted society" at the turn of the 20th century).

Our curriculum reflects both the historical consciousness that stems from these important precedents in media ecology studies and a firm grasp of the contextualized nature of contemporary writing. Students enter the major through a liberal education humanities course called Literacy, Technology and Society (WRIT 1506)[5] that traces the very social impacts of writing upon human sociocultural institutions indicated here. In exploring the complexities of the contemporary context, Media Law and Ethics (JOUR 3700)[6] explores the current context for writing practice. In exploring issues of free speech, for example, students in JOUR 3700 explore the ways that writing can effect personal and political change. In understanding contemporary media institutions, students understand the power of writing to shape popular opinion and cultural norms.

Conclusion

The core curriculum in writing studies at the University of Minnesota Duluth is an expression of our disciplinary and curricular identity. At once, we made clear that we were distinct in our intellectual project from the departments with whom we had collaborated so collegially for decades—and so we could continue to collaborate in the future. We created a new public face—a distinction between the old department of composition (filling a service role in the institution) and the new department of writing studies. We are no longer defined entirely by our liberal education offerings; we have an object of study as coherent as any department on campus.

Reflecting Backward, Looking Forward

This profile is too short to discuss in greater depth the kind of work that students do outside the core curriculum (in electives and in the core courses of the emphases in journalism or professional writing). The goal of this profile has been to articulate the unique innovation of a major in writing studies—a clear break from prior examples of "writing intensive English" and "technical writing."

Personally (and without consultation with any of my colleagues), I take a great deal of pride in the ways that the lower division curriculum of this major differs from, say, the lower division of typical majors in English. "Sophomore" courses in English majors tend to be surveys of genre, of national literature by period (e.g. BritLit I, II; AmLit I, II). Literary criticism and theory, on the other hand, tends to serve as a capstone course in the major—a terminal or exit experience. The emphasis on history and theory in the core curriculum at our first-year and sophomore level (in WRIT 1506, Literacy, Technology and Society; in LING 2506, Language and Writing; and in WRIT 2506, Introduction to Writing Studies) helps students and faculty articulate the ways that our project is about more than just polishing academic essays or drafting feasibility studies. We study writing as an artifact, a tool for human activity and human cognition, and a historical force in human communities. As we move toward collecting portfolios from our graduating students (our first cohort graduates this year) for purposes of assessment, it will be interesting to see whether our students are prepared to reflect on the writing that they produce within those intellectual frames.

Whether this model advances substantially as a model transferable to other institutions or contexts (in which writing faculty maintain appointments in English or other departments) is an open question worthy of further reflection. We needed to demonstrate that we were developing something substantially different from an English major or a communication major. Many writing majors, of course, must demonstrate their consonance with literary studies to survive departmental curriculum meetings. Instead of arguing a difference from literary studies, they must articulate ways that (for example) introductory courses in literary genres are also foundational to writing curricula. The intellectual and political challenge is entirely different.

Precisely for this very reason, we have some concerns about the possibilities for our students in graduate work, as well. As it presently stands, the admission requirements for the MA in English at UMD, for example, require a distribution of work in literature that our students would not meet; if admitted, they would need to take "catch up" coursework in British and American literature. (Of course, this is ironic because the writing studies department houses all of the teaching assistantships in the graduate program in English.) Whether a BA in writing studies serves as a disadvantage relative to a BA in English with a minor in writing for purposes of graduate admissions, for example, is also an open question. These are open questions, to be sure, that only time will tell and that we will revisit as our program is assessed externally.

We are a year and a half from the approval of the major and just one semester in from its first appearance in the catalog—and we are in the middle of the worst budget cuts I've seen in my short career. While it was easy to claim that our major was budget neutral when budgets were flush, those claims will be harder to make in the immediate future. As a department, we are equally committed to the advancement of writing across the university (in the first year writing and advanced writing curriculum) and to the growth of our major. The conditions that led to the split between the English major and the writing program (that led to the development of our department twenty years ago) face us today. At that moment, literature faculty splintered the service curriculum from the major curriculum—a move we will not repeat. How we will be able to sustain the major and the service curriculum is an ongoing challenge in these budgetary times, but a challenge that we welcome. This may very well be what makes writing studies unique: that our commitment to our service curriculum and our major curriculum is equal. How we negotiate these values in this budgetary environment remains to be seen.

Coda: Where We are Now

When I am not a professor of rhetoric in the department of writing studies, I am an educator for a local wildlife rehabilitation center. The opportunity to reflect on this chapter in light of the ecological metaphor is a powerful opportunity to me. The same tensions in my work in wildlife are present in this project and in the struggles in the major ever since.

In wildlife education, there are two positions in tension. On the one hand, there are wildlife rehabilitators who focus on healing individual injured and orphaned wildlife. They devote resources to rescuing and doing medical work on individual wildlife. On the other hand, there are conservationists who focus on questions of habitat—water quality, suburban encroachment, agriculture and landscape degradation.

Successful wildlife work requires both the rehabber and the biologist to move back and forth between the individual organism and the larger habitat, while utilizing scarce, often competing resources. Sometimes, we have to emphasize the water, because water quality determines the vibrancy of an ecosystem. But sometimes, we also need to focus on the fish. This back and forth is immensely challenging, and it makes collaboration difficult for people otherwise united in their passion for wildlife.

It may be no different for scholars in writing studies. Because we are an inherently interdisciplinary field (even to the ampersand in the title of some doctoral programs, "rhetoric & composition," pulling together humanistic and social scientific epistemologies of writing instruction), we, too must be able to move from the fish to the water. We must be able to move from our unique disciplinary differences to the larger ecology in which we work together to teach writing.

The larger ecology that has been the major in writing studies has been phenomenally successful in drawing students as a whole (nearly 100 students in its third year). On the other hand, the bulk (nearly two-thirds) of those students identify with one disciplinary tradition: the journalism emphasis (no doubt because of its recognizable career track). While the ecology has been successful, it has been easy to fix our eyes on one fish (journalism).

The tensions between the ecology of the major and the unique disciplines that constitute the major were not limited to students. Of the nine tenure-line faculty and more than twenty adjunct faculty in the program, few identified as "writing studies" faculty—that is, few identified as members of the writing studies discipline. Historically, the combination of English, composition, rhetoric, linguistics, journalism and information design faculty in the department has always been imagined as an administrative and budgetary maneuver, not an intellectual one—a collection of fish instead of a natural ecology. The ongoing success of minors in these areas encouraged faculty to see their work as independent and within the area of disciplinary autonomy in which they were trained—it encouraged the emphasis on the fish. The identity formation that begins in graduate training plants deep roots. As a result, the collaborative project of the major remains the subject of persistent discussion.

I suppose that, like a convert to a new philosophical movement, some of us placed too much hope in the transformative power of writing studies (in Bazerman's formulation) as providing intellectual focus—a shared ecology in which all of us fish might swim. In five years, I don't know whether the pond will exist in the same way. Perhaps we will be swimming separately, each in one of a chain of adjacent lakes (as UMD might develop majors in journalism, in linguistics, and in writing).

But in the meantime, the exercise has been a powerful one, one that has created a powerful experience for students, too. Whatever the tensions in our collaboration, we are committed to the ultimate goal of success for our students.

Notes

1. This profile is, inevitably, partial in its perspective. It is composed by a faculty member intimately connected to the proposal process and largely responsible for pulling various contributions together for submission to administration, giving the profile something of a synoptic perspective. At the same time, each contributing faculty member would, undoubtedly, inflect the curriculum differently in its explication. This profile is not meant to replace or supersede any of those other perspectives.

2. WRIT 4250 was initially developed in the minor in information design and has been refined by the faculty who have taught in that Minor over the years: Craig Stroupe, Kenneth Risdon, and Rob Wittig. WRIT 4506 is in large part developed under the expertise of department head Jill Jenson, a noted scholar in the uses of portfolio for assessment.

3. This course was developed by Juli Parrish.

4. This course was developed by linguist Chongwon Park.

5. This course was developed by noted Walter Ong scholar Thomas Farrell and re-imagined by Craig Stroupe.

6. This course has been redeveloped and taught by John Hatcher.

Works Cited

Anderson, Benedict. *Imagined Communities: Reflections on the Origin and Spread of Nationalism.* New York: NY: Verso, 1983. Print.

Bartholomae, David. *Writing on the Margins: Essays on Composition and Teaching.* New York: Palgrave Macmillan, 2005. Print.

Bazerman, Charles. "The Case for Writing Studies as a Major Field of Study." *Rhetoric and Composition as Intellectual Work.* Ed. Gary A. Olson. Carbondale, IL: Southern Illinois UP, 2002. 32–38. Print.

—. "Theories of the Middle Range in Historical Studies of Writing Practice." *Written Communication* 25.3 (2008): 298–318. Print.

Berlin, James A. *Rhetoric and Reality: Writing Instruction in American Colleges, 1900–1985*. Carbondale: Southern Illinois UP, 1987. Print.
—. *Writing Instruction in Nineteenth-Century American Colleges*. Carbondale, IL: Southern Illinois UP, 1984. Print.
Bizzell, Patricia. *Academic Discourse and Critical Consciousness*. Pittsburgh, PA: U of Pittsburgh Press, 1992. Print.
Bolter, David and Richard Grusin. *Remediation: Understanding New Media*. Cambridge: MIT, 1999. Print.
Brown, John Seeley and Paul Duiguid. "Social Life of Documents." *First Monday* 1.1 (6 May 1996). <http://firstmonday.org/htbin/cgiwrap/bin/ojs/index.php/fm/article/view/466/387>
Chaytor, Henry John. *From Script to Print: An Introduction to Medieval Literature*. London: Cambridge UP, 1945. Print.
Connors, Robert J. *Composition-Rhetoric: Backgrounds, Theory, and Pedagogy*. Pittsburgh, PA: U of Pittsburgh Press, 1997. Print.
—. "Dreams and Play: Historical Method and Methodology." *Methods and Methodology in Composition Research*. Eds. Gesa Kirsch and Patricia A. Sullivan. Carbondale, IL: Southern Illinois UP, 1992. 15–36. Print.
—. "The Rise and Fall of the Modes of Discourse." *College Composition and Communication* 32.4 (1981): 444–455. Print.
—. "Writing the History of Our Discipline." *An Introduction to Composition Studies*. Eds. Erika Lindemann and Gary Tate. New York: Oxford UP, 1991. 49–71. Print.
Connors, Robert J., Lisa S. Ede, and Andrea A. Lunsford. "The Revival of Rhetoric in America." *Essays on Classical Rhetoric and Modern Discourse*. Eds. Robert J. Connors, Lisa S. Ede, and Andrea A. Lunsford. Carbondale, IL: Southern Illinois UP, 1984. 1–15. Print.
—. "A History of Writing Program Administration." *Learning from the Histories of Rhetoric: Essays in Honor of Winifred Bryan Horner*. Ed. Theresa Enos. Carbondale, IL: Southern Illinois UP. 60–71. Print.
Eistenstein, Elizabeth. *The Printing Revolution in Early Modern Europe*. Cambridge UP, 2005. Print.
Freedman, Aviva and Peter Medway, eds. *Genre and the New Rhetoric*. London: Taylor & Francis, 1994. Print.
Gaines, Robert. "De-Canonizing Ancient Rhetoric." *The Viability of the Rhetorical Tradition*. Eds. Richard Graff, Arthur E. Walzer, and Janet Atwill. Albany: SUNY Press, 2005. 61–73. Print.
Havelock, Eric A. *The Muse Learns to Write: Reflections on Orality and Literacy from Antiquity to the Present*. New Haven: Yale UP, 1986. Print.
Innis, Harold Adams. *Empire and Communications*. 1950. Ed. David Godfrey. Victoria, B.C.: Press Porcepic, 1986. Print.
Keith, William M. *Democracy as Discussion: Civic Education and the American Forum Movement*. Lanham, MD: Lexington, 2007. Print.

Kitzhaber, Albert R. *Rhetoric in American Colleges, 1850–1900.* Dallas, TX: Southern Methodist UP, 1990. Print.

Latour, Bruno. *Pandora's Hope: Essays on the Reality of Science Studies.* Cambridge: Harvard UP, 1999. Print.

Latour, Bruno and Steve Woolgar. *Laboratory Life: The Construction of Scientific Facts.* Princeton: Princeton UP, 1986. Print.

Lunsford, Andrea. "Writing, Technologies, and the Fifth Canon." *Computers and Composition* 23 (2006): 169–177. Print.

Mead, William E. "The Graduate Study of Rhetoric." Report of the Pedagogical Section of the MLA, *Proceedings of the MLA* (1900): xx-xxxii. Print.

Moriarty, T. and G. Giberson. *What We Are Becoming: Developments in Undergraduate Writing Majors.* Logan: Utah State University Press, 2010. Print.

North, Stephen M. *The Making of Knowledge in Composition: Portrait of an Emerging Field.* Upper Montclair, NJ: Boynton/Cook, 1987. Print.

Ong, Walter. *Orality and Literacy: The Technologizing of the Word.* London, Routledge, 1982. Print.

Phelps, Louise Wetherbee. "The Case for Rhetoric and Composition as an Emerging Field: A Report from the Consortium of Doctoral Programs in Rhetoric and Composition." 31 Oct. 2004: 1–9. 6 July 2009 <http://www.cws.illinois.edu/rc_consortium/rhetcompcase.pdf.>

Russell, David. "Rethinking Genre in School and Society: An Activity Theory Analysis." *Written Communication* 14.4 (1997): 504–554. Print.

Schilb, John. "The History of Rhetoric and the Rhetoric of History." *Pre/text: The First Decade.* Ed. Victor Vitanza. Pittsburgh, PA: U of Pittsburgh Press, 1993. Print.

Strain, Margaret M. "Local Histories, Rhetorical Negotiations: The Development of Doctoral Programs in Rhetoric and Composition." *Rhetoric Society Quarterly* 30.2 (2000): 57–76. Print.

Part IV. Interconnected Sites of Agency: Situating Assessment within Institutional Ecologies

As standardized learning outcomes such as the Common Core State Standards attempt to create uniformity across state and regional boundaries, as externally imposed assessment standards and their funding consequences increasingly drive local teaching practices, as Massive Online Open Courses (MOOCs) gain momentum on the promise of extending the geographic and democratic reach of university courses, and as debates rage about the efficacy and ethics of computer scoring of student writing, writing program assessment is increasingly becoming a decontextualized activity, externally imposed, politically and economically driven, and disconnected from writing program needs. The two program profiles featured in this section of the book highlight the ecological attributes of fluctuation and complexity of interaction as writing programs adapt to externally imposed constraints. As the profiles emphasize, assessment must be a situated and interconnected activity, one located within program and curricular goals and existing in dynamic relationship to complex ecologies of curricular development and revision, learning outcomes, teacher training, and student learning. In developing assessment programs that are responsible to externally imposed exigencies and institutional accountability while also responsive to the needs of local writing programs, these profiles serve as powerful examples of how assessment measures can become appropriated as part of the fluid framework of program development, curricular revision, and systems of institutional accountability—what Asao Inoue describes in his profile as a "culture of assessment" that is informed by and informs institutional ecologies.

Asao B. Inoue's "Self-Assessment as Programmatic Center: The First Year Writing Program and Its Assessment at California State University,

Fresno" (published in 2009) profiles CSU, Fresno's writing program and its program assessment initiatives. At the heart of its assessment program is the guiding belief that assessment is not a fixed (or fixing) project but rather a fluctuating and evolving one, driven by a commitment to self-reflection and an on-going process, as Inoue puts it, of "becoming 'literate' about our teaching, our curriculum, and our students' learning." Assessment as literacy work embeds assessment into all aspects of a writing program and its larger context (student demographics, the need to support diverse student discourses and needs, the development of learning outcomes, teacher training and reflection, student portfolios, tools for directed self-placement, student learning and retention, and data gathering), demonstrating the ecological characteristic of interconnectedness. In detailing the piloting and implementation of an assessment program predicated on a culture of self-assessment at all levels, Inoue offers an example of how assessment, rather than an isolated activity, can become an ecological practice that is intertwined with other environmental structures and processes. In that spirit, Inoue's "Where We Are Now" coda points to CSU, Fresno's continued need to develop student assessment practices that honor students' diverse discourses within translingual ecologies.

Approaching assessment work from a similar ecological perspective but with an eye toward supporting structural changes across campus, Paul Walker and Elizabeth Myers's 2011 profile, "Utilizing Strategic Assessment to Support FYC Curricular Revision at Murray State University," reveals the ways in which writing programs are emergent by chronicling a transition period in their writing program's development, when the first-year composition requirement was revised from a six-credit-hour, two-semester sequence to a four-credit-hour, one-semester course. Walker and Myers describe how they used this period of fluctuation, change, and even volatility to shift the emphasis from general skills writing instruction to writing as situated inquiry. In particular, they describe how they used assessment strategically to support and "contribute to changing perceptions about writing on campus and about the role that FYC can play in supporting students' on-going writing development." In this case, we can see how large-scale assessment was used in order to support curricular innovation—how, that is, rather than being externally imposed in order to constrain WPA work, assessment was used actively and *rhetorically* by WPAs to persuade colleagues and administrators across campus of the value of curricular and structural revision of writing instruction. Walker and Myers outline the methods and results of an assessment of the revised course in comparison to the previous course sequence, along with how the assessment guided the instruction, administration, and structural development of writing at the university. In their "Where We Are Now" coda, they describe how such assessment practices have since become connected to instructor self-reflection on student learning and how such reflection has generated professional development projects, both of which have helped the writing program make a case for itself amid budget cuts. The coda also outlines how Myers has applied similar assessment practices to her new high school context. In both profiles, we have illustrations of the complexity of writing programs and examples of assessment as a complex, interconnected site of agency.

11 Self-Assessment as Programmatic Center: The First Year Writing Program and Its Assessment at California State University, Fresno

Asao B. Inoue

> Reading the world always precedes reading the word, and reading the word implies continually reading the world . . . [;] however, we can go further and say that reading the word is not preceded merely by reading the world, but by a certain form of *writing* it or *rewriting* it, that is, of transforming it by means of conscious, practical work. . . . To sum up, reading always involves critical perception, interpretation, and *rewriting* of what is read. (Freire and Macedo 35–36)

Freire's famous rendition of the process of literacy sums up the central philosophy and pedagogy of the first-year writing (FYW) program, which is also a directed self-placement (DSP) program, and its assessment at California State University, Fresno. In fact, Freire defines literacy as a reading and assessment practice. These are at the heart of the writing program's curriculum. We are currently finishing our third year (AY 2008–09) in a five year pilot. My good colleagues (Rick Hansen, Virginia Crisco, and Bo Wang) who designed and proposed our pilot before I was hired formulated much of the philosophy, structure, and curricula; however, nothing had been done to assess the program yet—one of the reasons they hired me. Outcomes had been developed, but there were too many to measure, and they weren't associated closely with any of the courses students took in the program. Additionally, we had not yet developed a formal mission statement or goals that could be used to support the outcomes we were promoting. So in the process of designing and implementing the program assessment in the first year that I came on board (AY 2007-08), these articulations were developed as well. Much like our students, we as a writing program are continually coming to know ourselves, continually becoming "literate" about our teaching, our curriculum, and our students' learning.

The focus of this profile, then, is this: The important shaping force of our FYW program and its program assessment is *the concept of self-assessment itself as integral to the process of literacy,* much like Freire's account of reading the world and word. Freire's articulation emphasizes two characteristics that define our program and students' learning: (1) reading and writing are joined practices; and (2) self-assessment practices ("interpretation" and "rewriting" in Freire's conception) are equally important to reading-writing processes.

So self-assessment is the center of our writing program's curriculum. It is why we chose a program portfolio as the key artifact to make important pedagogical decisions and programmatic measurements. Self-assessment is a key to how we teach reading practices, and writing practices, and it's the gateway into our program. DSP demands, of course, that students self-assess. Once they're in the program, the portfolio asks them to continue to self-assess their reading and writing practices in order to understand and measure their own progress, and we give them programmatic opportunities to assess others. Assessment of other's work as a reading practice is set up as important in learning how to write better, to write in community, and to understand practice as more than pragmatic, as reflective and theoretical. Additionally, since we're a pilot, and DSP is mostly an untested assessment technology, self-assessment is also the way we approach the program's on-going maintenance, administration,inquiry, and the validity of its course placements.

I have a larger argument in this profile, however. Centering writing programs on the concept of self-assessment as a defining aspect of the processes of literacy helps create an important culture of praxis, a culture of self-assessment, in the classroom and writing program.[1] If we (our students and writing programs themselves) are always *becoming* literate, always coming to understand our own practices, then we are always in the process of self-assessment, making the two processes, literacy and self-assessment, one and the same practice. As in our program, a centering on self-assessment asks that teachers and students turn their energies and time not to perfecting products but articulating reflexive, effective, and flexible reading and writing *practices,* rhetorical activities that are also self-assessment activities.

I should distinguish briefly three elements in our program that may be confusing to some in my discussion. The program portfolio that we use at CSUF is not synonymous with program assessment, nor is DSP synonymous with program assessment. Each of these elements (program assessment, portfolio, and DSP) is separate, but they work together, always serving multiple purposes. Our program assessment may best be characterized as a large set of inquiries and data gathering, while the portfolio is a pedagogical element in classrooms that we happen to use to collect data in order to make program assessments. Meanwhile, the DSP program is ONLY our course placement technology, but it's one that purposefully sets up the curriculum of self-assessment, epitomized in the portfolio. DSP is an "assessment," a self-assessment in fact, but it's only the way in which students get into their writing course(s); the portfolio drives the pedagogy, and the program uses the portfolio as a convenient way to capture student learning. Thus, we feel, the best way to find out whether the DSP is working properly, making appropriate placements (a validation inquiry, which is a kind of program assessment), is to investigate student learning in courses; as a result, most of our direct evidence of student learning comes from the portfolio.

The FYW Program

The program's conception of literacy and its philosophy, mission statement, program goals, and learning outcomes will make clear why we measure what we do and why we measure in the ways we do, i.e., why program assessment and self-assessment are central and defining of our writing program.

Our FYW program is relatively large. In AY 2007-08, we enrolled 1,762 students in ENGL 5A and 844 in ENGL 10.[2] Historically, many of CSUF's students have lacked confidence in their reading and writing abilities, as many first year students do. Our students, however, come with a variety of home languages, from schools that often do not ask them to read or write much or write in the ways we ask of them, from working-class backgrounds, and as first-generation college students. As many of our long-standing writing teachers have attested to, our students need particular kinds of reading and writing practices in the right kind of environment. The DSP and the FYW program's philosophy were designed to respond to at least three important historical issues shaping the FYW program in the past:

- the need to reduce the program's reliance on an outside, standardized placement test because it is not valid enough for our writing placement purposes (e.g., the EPT, SAT, etc.), and because in spite of our students' scores, the vast majority complete successfully their writing courses when given the right educational atmosphere, pedagogies, curriculum, and responsibilities;[3]
- the need that students have to place themselves and gain agency and responsibility over their educational paths in the university; that is, research shows that when students feel responsible for their own choices, when they've chosen their classes, they tend to be more invested in them, and succeed in higher numbers;[4]
- the need to give students credit for all of the writing courses they take since university credit acknowledges their work, does not penalize students for wanting extra practice in writing, and reduces the institutional and social stigma of "remedial" writing courses.[5]

Table 1. CSUF's Student Populations by Race

Race	Percentage
White	32.6%
Latino/a	31.6%
Asian Pacific Is.	18.5%
Unknown/Other	7.8%
African American	7.5%
International	1.1%
American Indian	1.0%

Table 2. Languages Spoken by CSUF's FYW Students

Languages	Percentage
English	61.3%
Spanish	21.87%
Hmong	8.31%
Armenian	2.96%
Punjabi	1.72%

Languages	Percentage
Tagolong	0.67%
Vietnamese	0.67%
Chinese	0.57%
Lao	0.57%
French	0.48%
Portuguese	0.48%
Khmer	0.38%

These historical exigencies helped us rethink our assumptions about literacy and its teaching in the program. Our FYW program promotes a philosophy that understands literacy as social practices sanctioned by communities (Volosinov 13); thus, learning "academic literacy" (however one wishes to define the concept) effectively will demand social processes that call attention to the way the local academic communities and disciplines value particular rhetorical practices and behaviors when reading and writing. The "academic literacy" we promote in our program, then, is a self-conscious process of discursive practices and strategies that acknowledge their roles in societal power structures, very akin to James Paul Gee's conception of "powerful literacy" (Gee 542). Gee calls these power structures "dominant" or "secondary" discourses, acquired typically in institutions, such as schools (Gee 527–28). However, our program's concept of literacy is also informed by Lisa Delpit's important critique of Gee. She argues that students of color and other groups who come to our classrooms from discourse communities distant to the academic discourse taught are not socially predetermined to never quite "get in" to the club (Delpit 546). Delpit discusses several examples that disprove Gee's claim, such as Bill Trent, a professor and researcher who came from a poor, inner-city household in Richmond, Virginia, and whose mother had a third grade education and father had an eighth grade education (548). Additionally, Delpit argues against Gee's claim that someone "born into one discourse with one set of values may experience major conflicts when attempting to acquire another discourse with another set of values" (546–47), which often leads teachers to not teach dominant literacies, in our case, conventional academic literacy found in texts like Graff and Berkenstein's *They Say / I Say* textbook.[6] Delpit's criticism of Gee is important to our program's concept of academic literacy because CSU, Fresno is a member of the Hispanic Association of Colleges and Universities (HACU), with large populations of Latino/a and Asian Pacific Islanders who come into our classes from discourse communities distant from the one we promote.

Thus the program consciously attempts to acknowledge three things that Delpit highlights in her critique: one, that what we offer our students are academic literacy practices that are connected to dominant discourses and economic power, which are associated with white, middle-class academic English; two, that all who enter have equal claim to these literacy practices, and can learn them; and three, that subordinate discourses (those not taught in the expressed, formal curriculum) are not just a part of the classroom but may transform the curriculum, literacy practices, and the discourse we promote (Delpit 552). This is idealistic, yes, but it's crucial to the articulation of our program not just because of who enter and reside in our classrooms, but because, as Stanley Aronowitz and Henry Giroux have argued, dominant literacy is "a technology of social control" that "reproduces" social and economic arrangements in society (50). We do not want to blindly reproduce discourses and social arrangements from the assessments of those discourses that may be inequitable and unfair to some, while privileging others.

Our classrooms are primarily working class students of color, many of whom are first-generation college students. Since we do not wish to simply reproduce hegemony by promoting a dominant academic literacy, the process of academic literacy begins with course placement, constructing an environment for engaging students reflectively with the ways power moves through discourses that "legitimat[e] a particular view of the world, and privilege, a specific rendering of knowledge" (Aronowitz and Giroux 51). This critical notion of literacy as reflexive practices that are both hegemonic and counter-hegemonic means our DSP program is vital to the curriculum and our students' learning. It introduces them, through self-assessment of their own reading and writing practices, to their learning paths in our program and begins to orient them to the new role they must play, one more active, more participatory, and one with more power than they may be used to. From this understanding of literacy, four elements make up the writing program's philosophy, which centers on literacy, reading, and writing practices:

- Literacy learning is social.
- Reading and writing are connected processes.
- Reading and writing are academic practices of inquiry and meaning-making.

- Reading and writing practices are shaped by and change based on the academic discipline.

The central activities in the program's philosophy are reflection and self-assessment of reading and writing practices, which are crucial to developing one's literacy *practices*. To be a good reader and writer means ultimately that a writer must be able to see her writing as a product of a field of discourse(s), which means she can assess its strengths, weaknesses, choices, and potential effects on audiences for particular purposes in particular rhetorical situations. It also means she can situate her practices in relation to a community of other readers and writers engaging in similar activities, which mirrors Graff and Birkenstein's "conversation" metaphor used to describe academic literacy. This additionally means that all students must practice assessment chronically—their literacy processes are not punctuated with self-assessment and reflection but are interlaced with these practices. In a nutshell, *good writers are always good self-assessors of their practices.* At all levels in our program, (self) assessment is fundamental to the way agents—and the program itself—make decisions, move through the program, understand literacy, and situate, understand, and define themselves in courses through the portfolio.

You can hear this central theme in the program's mission statement:

> The FYW program is committed to helping all students enter, understand, and develop literacy practices and behaviors that will allow them to be successful in their future educational and civic lives. In short, our mission is to produce critical and self-reflective students who understand themselves as reader-writer-citizens.
>
> More specifically, through instruction, community involvement, and a wide variety of other related activities, the FYW program's mission is to:
>
> - teach and encourage dialogue among diverse students (and the university community) about productive and effective academic reading and writing practices (i.e., academic literacy practices), which include on-going self-assessment processes of students;
> - assess itself programmatically in order to understand from empirical evidence the learning and teaching happening in the program, measure how well we are meeting our program and course outcomes, aid our teachers in professional devel-

opment, and make changes or improvements in our methods, practices, or philosophy.

Our mission is not just to help students learn academic literacy practices. Our mission is also to learn about our program, its students, and our practices. So as a program, we take to heart the practice of self-assessment. It's central to the way literacy is theorized, taught, and practiced in the classroom, and it's central to the way we approach ourselves as an evolving program also learning about itself.

Finally, there are three primary learning goals for the entire program, and eight learning outcomes. During the pilot, we measure only five of these outcomes, central ones. They are articulated as follows:[7]

- READING/WRITING STRATEGIES: Demonstrate or articulate an understanding of reading strategies and assumptions that guide effective reading, and how to read actively, purposefully, and rhetorically;
- REFLECTION: Make meaningful generalizations/reflections about reading and writing practices and processes;
- SUMMARY/CONVERSATION: Demonstrate summarizing purposefully, integrate "they say" into writing effectively or self-consciously, appropriately incorporate quotes into writing (punctuation, attributions, relevance), and discuss and use texts as "conversations" (writing, then, demonstrates entering a conversation);
- RHETORICITY: Articulate or demonstrate an awareness of the rhetorical features of texts, such as purpose, audience, context, rhetorical appeals, and elements, and write rhetorically, discussing similar features in texts;
- LANGUAGE COHERENCE: Have developed, unified, and coherent paragraphs and sentences that have clarity and some variety.

THE STRUCTURE OF THE PROGRAM

The FYW program consists of three courses: English 5A, English 5B, and English 10. One either takes ENGL 5A-5B or ENGL 10 to fulfill the university writing requirement. These three courses amount to three course paths, when adding LING 6 (discussed below), taught in another department. Students place themselves into FYW through a directed self-placement (DSP) model.[8] According to Daniel Royer and

Roger Gilles, the first to design, administer, assess, and publish results on directed self-placement ("Directed Self-Placement"), DSP "can be any placement method that both offers students information and advice about their placement options (that's the 'directed' part) and places the ultimate placement decision in the students' hands (that's the 'self-placement' part)" (Royer and Gilles, "Directed Self-Placement" 2). In our model, students make the placement decision based on two kinds information that we provide: One, a set of course outcomes, and two, prompts that ask them to consider their own reading and writing histories and behaviors. Students place themselves into one of three options:

- Option 1: English 10, Accelerated Academic Literacy. This is an advanced class, and students who choose this option typically are very competent readers and writers, ready to read complex essays, develop research supported analyses, and complete assignments at a faster pace. Generally, these students have done a lot of reading and writing in high school and feel comfortable with rules of spelling, punctuation, and grammar. This course starts with longer assignments (5–7 pages) and builds on students' abilities to inquire, reflect, compose, revise, and edit.

- Option 2: English 5A/5B, Academic Literacy I & II. These courses "stretch" the reading and writing assignments over two semesters and have the same learning outcomes as English 10. Students in this option often do not do a lot of reading and writing (in school or outside of it) or may find reading and writing difficult. Students get to be with the same teacher for both courses (a full year). The first semester (ENGL 5A) starts with shorter assignments, focusing on reading practices, and moves toward more complex reading and writing at semester's end. The second semester (ENGL 5B) builds on work in 5A and leads students through longer and more complex reading and writing tasks. Finally, these courses focus more on researching, citing, and including sources correctly in writing, with more direct instruction in language choice, sentence variety, and editing.

- Option 3: Linguistics 6, Advanced English Strategies for Multilingual Speakers, then English 5A & 5B. The first class (LING 6) assists multilingual students with paraphrasing and summarizing while also providing help with English grammar. Students who take this course usually need instruction that addresses the challenges second language learners face with academic reading, writing, grammar, and vocabulary. Students in this option tend to use more than one language, avoid reading and/or writing in English, and/or have a hard time understanding the main points of paragraphs or sections of a text. This course, LING 6, focuses on increasing reading comprehension while developing a broader English vocabulary. This class builds language skills through short readings and writing assignments that prepare them for English 5A and 5B.In the AY 2007-08, the majority of our students selected option 2, the English 5A-5B sequence (the stretch program). Of the 2,606 enrollments in the year, 1,762 (68%) enrolled in ENGL 5A, while 844 (32%) enrolled in ENGL 10. Only a handful of students each year choose option 3, usually after enrolling in option 2. Because LING 6 is not taught in our program or by our teachers, I will not speak at all about this course. We gather little data on it. LING 6 is very small, and not managed by our FYW program.

Program Portfolio As Pedagogy and Program Assessment Device

The central device we use to guide the program's pedagogies and assess directly student learning along program outcomes is the program portfolio. Portfolios have been shown in many places to emphasize student agency, control, selection, development along dimensions/outcomes, and reflection/self-assessment (Hamp-Lyons and Condon; Elbow and Belanoff), each of which are important to our program.[9] Drawing on Sylvia Scribner's "literacy metaphors," Belanoff argues that portfolio pedagogies can provide a richer environment for literacy practices. They can account for "literacy as adaptation" to a dominant discourse, "literacy as power" through practicing dominant forms, and "literacy as grace" by participating socially in human meaning-making activities (13), each of which is an important practice for our program.

Because the central component of the assessment technology that seeks to understand student learning in our program is the program portfolio and the activities and agents that surround it, the portfolio is required of all students in all three courses and has a set of common requirements dictated by the curriculum [A reflective letter (approximately 2–4 pages); ten pages of formal, polished, revised writing; completed assessments of peers' portfolios (usually two)]. All courses require common sets of assignments (projects) by course, which are used to fill the program portfolio. While documents, assignments, and prompts may vary in terms of readings and details, each project must center on a specified set of outcomes (listed above in the previous section). In fact, we provide all teachers with "template" assignment sheets that articulate the main features and outcomes of each project, which allows the teacher to modify and adjust invention and warm-up assignments.[10]

At midterm and final, a sampling of portfolios is gathered by the program as direct evidence of student learning. The evidence from portfolios is three-fold:

- Independent Ratings: Over each summer, we rate portfolios along the five outcomes listed previously (i.e., reading/writing strategies, reflection, summary/conversation, rhetoricality, and language coherence). These blind ratings are done by teachers in the program, but not the teacher of record.

- Teacher Ratings (Competency Measures): We capture teachers' ratings (in the semester) from these same portfolios by the teachers of record. I call these "competency measures" because the idea is to gather a more contextually sensitive rating of the portfolios and writing competence in the course—a contextual judgment. Since teachers know the students well, how others are performing in the class, and have a sense of where each portfolio as a presentation of learning fits into their particular class, these ratings, I argue, are competency measures and may be different from the independent ratings.

- Student-Peer Ratings: Additionally, since self-assessment, student agency and control, and critical understanding of literacy practices are important to the philosophy, mission, goals, and outcomes of the program, all students engage in peer portfolio assessments at midterm and final, which ask them to do two things: (1) read and write descriptive assessments along each program outcome that begin with evidence building from the reading of peer portfolios and end by making descriptive judgments along each outcome; and (2) rate each portfolio along each program outcome with the teacher in class after the descriptive assessments are finished.

The program uses all three measures from portfolios to get a rich sense of how well students are learning from three important stakeholders: outside teachers who teach in the program, the teacher of record, and the students themselves. This year is the first year we've tried the student-peer ratings, so we are looking forward to seeing what this tells us about the other ratings gathered. What we hope to find is that students' ratings come close to those of their teachers and/or outside teachers, which would suggest that they are able to reflect upon and self-assess their own writing practices. And if they diverge noticeably from teachers' ratings, then this provides important pedagogical information for building curricula and lessons, rethinking our own judgments, etc. Additionally, folding this third layer of evidence into our program assessment also builds the construct and content validity (and fairness) of the portfolio by allowing students to become agents in the assessment of their own learning through practicing assessment as a Freirian reading practice. Teachers also use student ratings in a variety of ways in their classes to discuss the midterm portfolio assessments, program outcomes, portfolio documents, self-assessment as a practice, among other things. For instance, one teacher, Matthew Lance (a graduate student), recently shared with other teachers his practice of radar graphing the portfolio ratings (using Excel) by students and the teacher as a tool to discuss disagreement and variance in multiple readings of portfolios in follow-up student conferences. Each graph forms a pentagon of various sizes, depending on the ratings of that one reader (see Figure 1). Larger pentagons show higher ratings, while misshaped ones suggest the reader sees weakness in one or more areas and strength in others. All three pentagons are mapped together to show graphically difference among readers along outcomes. The radar graphing simplifies the judgments, so it is mostly useful for starting and focusing discussions about the descriptive assessments of each reader. It points out variance and the degree of variance in a simple but provocative way, a way that students and teacher can use to begin conversations about ratings, self-assessments, and outcomes.

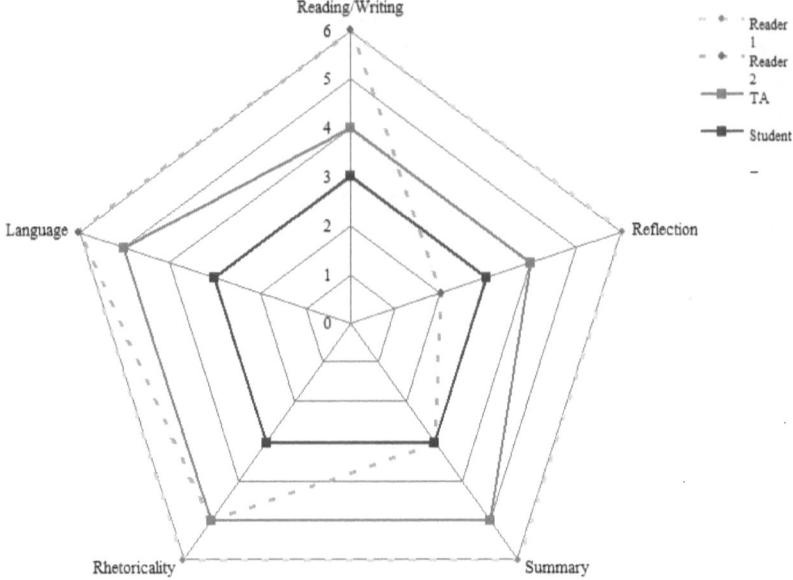

Figure 1. Example Radar Graph Used to Discuss Difference in Portfolio Ratings with Students

Assessment Research Methods

Ultimately, the program assessment that I designed is robust, mainly because I wanted to make sure that we had plenty of data, from multiple sources, to be able to argue the effectiveness of our pilot DSP, the validity of its course placements, the appropriateness of decisions to pass students through the program, and I wanted to understand the learning occurring in the new writing curriculum that we instituted at the same time as the pilot DSP. In a nutshell, the program gathers the following information each semester from a sample of students (approximately 25 percent of the total students enrolled in our program):

- Entry and exit course surveys that ask about satisfaction and course placement accuracy;
- Independent Portfolio ratings that are conducted in the summers and by teachers teaching in the program (discussed above);
- Portfolio competency measures that take ratings from the teacher of record (discussed above);

- Student-Peer portfolio ratings that gather ratings from students, allow them to exercise assessment practices purposefully, and provide voice in the program assessment (discussed above);
- Student progress measures in courses that gather basic information at midterm and final about each student's general progress in the course, identifying simply "passing," "not passing," or "borderline passing";
- Course grade distributions in all FYW courses that provide an indirect measure of student learning, as well as a way to see how the portfolio and our curriculum are affecting students' chances for continuing education (persistence/retention in the university) and their educational paths in the university (degree of success);
- Passing rates in all FYW courses, which are a function of grade distributions;
- Frequent ongoing projects/measures of various kinds that look at particular aspects of the program that beg for inquiry; these are often one year or semester, short-term data gathering projects.

All of the above evidence that we gather and I consider when assessing the program and DSP decisions are first separated by DSP option (i.e., ENGL 5A-5B and ENGL 10), then in most cases analyzed along three lines: race, gender, and generation of student (i.e., first generation or continuing generation).

The final category of ongoing special projects in the program has varied over the last year or so. In the spring 2008 semester, for instance, a group of graduate students (most of whom were TAs) and I gathered teacher commenting data on student midterm portfolio drafts to measure empirically the kinds, quality, quantity, and frequency of teachers' comments on various racially identified student drafts. This began in my composition theory seminar and culminated in a CCCC roundtable discussion in San Francisco in 2009. Since then, it has spawned three theses projects, which also will be folding into our future program assessments. The first of these follow-up projects is an empirical look at the rhetoric of reflection in student portfolios in our program. The second is a closer look at the construction of error in teacher commenting practices on Latino/a writing in the program.[11] The third project is an experiment that attempts to test a hypothesis that was induced from the findings in the initial commenting project: race of students seems to affect the commenting practices of teachers. Additionally, this year,

I've begun gathering simple survey data and grade distributions from students in the program who are in classes that use grading contracts, something I use and introduced to teachers about a year and a half ago. Our grading contracts guarantee each student a "B" grade in the course as long as she meets certain criteria (negotiated with the entire class). These criteria are based on work done in the spirit it is asked, regardless of quality, which allow teachers and students to deemphasize grades and focus on the more descriptive and formative assessment practices in the classroom, providing more agency and control for students. About twenty-five teachers use grading contracts in our program each semester, so this is a significant factor that shapes the learning of our students.[12]

I detail these special projects because they highlight one very important aspect of our FYW program: self-assessment as program culture. Self-assessment is not just a student learning outcome, but it's half of our program's mission statement (listed above). This promotes assessment as a set of teacher and student practices that are pedagogical and help us understand ourselves and our growth. It also makes for a robust, dynamic, and pedagogically sound curriculum because it develops a rich corps of teachers who know well the learning in the program and are continually reflecting upon their practices and those of the program. Similar insights about using assessment as a way to build culture in Washington State University's writing programs have been discussed in *Beyond Outcomes* (Haswell 2001).

It's not easy to summarize our findings, so the summary below of last year's data (AY 2007-08) is by no means complete, but it gives you a general idea of what we learned in the first full year of our program assessment efforts (AY 2007-08). (For a more complete discussion of all the data collected, see my annual program assessment report listed in note 7).

Assessment Research Findings

In its most basic form, the question our program assessment asks is: "What are our students learning in our program?" or "How well are they learning along our program outcomes?" We also ask: "How appropriate are the DSP placements?" Overall, it appears our students are developing along all of the outcomes we measure in both options 1 (ENGL 10) and 2 (ENGL 5A/5B), meet our expectations in high numbers in their final portfolios, and are highly satisfied with their DSPs. Additionally, students pass at acceptable rates, persist in the university after our courses, but there is some room for improvement.

First, the direct evidence from *independent portfolios ratings* suggest that students are learning along all outcomes. In option one (ENGL 10) portfolios,

- all students achieved an overall rating of "adequate quality" or better (100%) in final portfolios.[13]
- on average, portfolios improved by .68 points (out of six points) in all five of the outcomes measured, with the most improvement between midterm and final portfolios occurring in "Reflection" (.94 out of six points), with the second-most improvement occurring in "Summary/Conversation" (.89).

In option two (ENGL 5A/5B) portfolios,

- the average student met all outcomes, and 95.3 percent of all students achieved an overall rating of "adequate quality" (rating of three) or better in final portfolios.
- only 4.7 percent of all students' final portfolios were rated of "poor quality" (1 or 2), a 25.3 percent decrease from 5A midterm.
- students improved by .52 points in all five of the outcomes measured, with the average student score resting in the "proficient" category for each outcome, making the most improvement in "Rhetoricality" (.59) and the second-most improvement occurring in the outcome of "Summary/Conversation" (.55).

So based on both the resting points for final portfolio ratings and the development students made over the course of their studies, our students, who place themselves in their courses in both options, appear to learn along all dimensions and meet the program expectations in high numbers (see Figure 2).

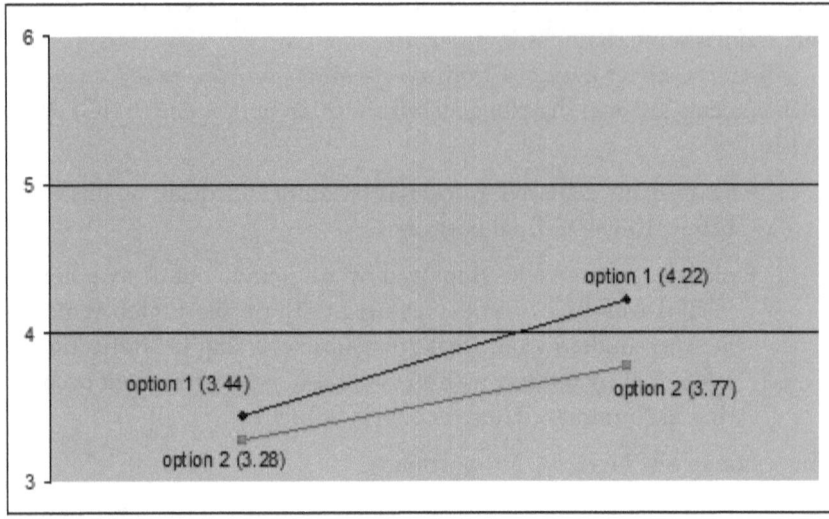

Figure 2. Average Overall Ratings for Option 1 and 2 Student Portfolios in Independent Ratings Increase During Their Time in the Program

Second, the portfolio competency measures (direct evidence of learning) that we gather directly from teachers in the semester suggest similar conclusions. Option one portfolios yielded the following results:

- Competency measures for all students showed a very modest improvement of 3.2 percent, from 83.2 percent at midpoint to 86.4 percent at final. This finding appears NOT to agree closely with the numerical portfolio ratings (100%) mentioned above.

- Interestingly, females in option 1 attained higher measures overall, with 85.1 percent judged competent at midterm and 89.2 percent at final. This is 7.6 percent higher than males.

- Black females (and perhaps all Blacks) appear most at risk in our program, achieving overall competency in the fewest numbers at ten final (72.73%), while Latinos start with higher competency (83.3%) but end similarly as Black females (72.2%).

Option two portfolio results were as follows:

- Students showed continual overall improvement (22.7%), from 73.9 percent at 5A midterm to 96.6 percent at 5B final.

- Their overall competency measures (96.6% at final) appear to agree closely with the numerical portfolio ratings of the same portfolios (95.3%).
- Unlike option 1, gender DID NOT play a significant factor in competency measures, with males receiving only .5 percent fewer overall competent judgments than females at 5B final.
- Black males, however, are most at risk in option 2: Black males achieved overall competency in the fewest numbers at 5B final (88.89%), and showed 7.11 percent fewer overall competent judgments than the next closest group (Latinos).
- Latinos (and Latinas to a lesser degree) may be at risk as well, starting with lower competency measures (60%) but ending comparable to most other groups (96%).

Table 3. Overall Proficiency Ratings on Portfolios by Option and by Race

Race	Option 1: 10 Midterm	Option 1: 10 Final	Option 2: 5A Midterm	Option 2: 5B Final
Asian/Pacific Islander (API)	84.85%	93.94%	68.89%	97.78%
Black	56.25%	81.25%	86.67%	93.33%
Latino/a	88.89%	83.33%	69.77%	96.51%
White	94.29%	85.71%	77.27%	96.59%

Overall competency rates (ratings generated by teachers of *record* from portfolios) are more evenly distributed at final in option two, suggesting that the stretch program creates a racial leveling effect in terms of competency read in portfolios. Meanwhile Blacks and Latinos/as perform the worst in option one (see Table 3).

Third, our student survey results (indirect evidence of learning) suggest that, generally speaking, students are highly satisfied with their chosen writing courses, which makes for more active and engaged learners. In option one,

- students grew in satisfaction, from 91.2 percent at midpoint to 96 percent at final.
- the vast majority of females (94.6%) started satisfied and slightly more of them ended satisfied (97.3%).

- Whites (97.2%) and APIs (97%) ended up the most satisfied with their DSP choice of all racial groups.
- as perhaps expected, Black males ended up being the least satisfied (80%) racial group.

In option two,

- students grew in satisfaction with their DSPs during the course of their studies, from 83.9 percent at 5A midterm to 87.7 percent at 5B final.
- males achieved higher levels of satisfaction (93.8%) than females (84.5%) by 5B final.
- females remained constant in their satisfaction with their DSPs (84.5%).
- interestingly, 91.1 percent of all students of color felt mostly or completely satisfied in their DSPs in option two, with APIs being the most satisfied group (97.8%).
- White females, however, had the lowest satisfaction at most points in the year (5A midpoint: 82.5%; 5B final: 75.4%).

Our survey results suggest that, if student satisfaction is linked to learning and more deeply engaged students (as Reynolds suggests), then our program appears to be cultivating high levels of learning and engagement because of satisfaction levels, with a few groups, particularly Blacks, needing some additional attention.

Fourth, passing rates (indirect evidence) of students suggest indirectly that students are learning adequately along the outcomes we promote, and they are choosing appropriately their courses in our DSP. For option one,

- 5.1 percent fewer students passed compared to the previous writing course (Engl 1A) in 2005, a course students were placed into by standardized test scores.
- 6.2 percent fewer APIs (72.5%), 13.9% fewer Blacks (69%), and 9.5% fewer males (70.4%) passed than did in the previous writing course (Engl 1A).
- regardless of racial group (except for APIs), passing rates appeared to be consistently lower by an average of 10 percent than overall competency measures, meaning grades seemed to represent stu-

dents' competencies consistently regardless of race or gender if competency is considered a criterion validation measure.

Option one appears to be more difficult to pass than the previous one semester writing course. For option two

- roughly the same percentages of students passed overall (82.4%) as compared to the previous writing course (ENGL 1A) in 2005.
- both males and females passed at similar rates (81.1 percent for males, and 83.3 percent for females) as the old Engl 1A.
- 4.5 percent more APIs (83.2%) passed than did in the earlier writing course.
- 9.2 percent fewer Blacks (73.7%) passed than did in the earlier writing course.
- 13.1 percent fewer Native Americans (78.6%) passed than did the earlier writing course.
- regardless of group (except for Blacks), passing rates appeared to be consistently lower by an average of 10 percent than overall competency measures, meaning grades seemed to represent students' competencies consistently regardless of race or gender in option two.

Similar to option one, option two is more challenging than the previous one semester course. Our passing rates suggest internal consistency in both options, and suggest that option two passes just as many students as the old Engl 1A course, while option one fails a few more students. Option two also appears to reduce more effectively negative racial and gender formations in passing rates, which agrees with the direct evidence in portfolios mentioned above. While Blacks fail more frequently in both options, they do end highly satisfied with their DSP (see Table 4 for a comparison of passing rates and satisfaction rates by race and option).

Fifth, the office of Institutional Research, Assessment, and Planning (IRAP) at CSU, Fresno recently conducted a one semester comparison of the current writing program to the old Engl 1 course, using retention and passing rates(indirect, comparative evidence of learning). Most interesting in their brief report is the apparent effect that our option two (ENGL 5A/5B) has on retention rates in the university. The ENGL 5A/5B program appears to have a positive effect on student retention rates, especially for those who are designated as needing remediation

Race	Option 1: 10 Passing	Option 1: 10 Entry	Option 1: 10 Exit	Option 2: 5A Passing	Option 2: 5A Entry	Option 2: 5A Exit	Option 2: 5B Passing	Option 2: 5B Entry	Option 2: 5B Exit	Engl 1A: 1A Passing
All	76.9%	91.2%	96.0%	82.7%	83.9%	89.8%	82.4%	82.2%	87.7%	79.6%
APA	72.5%	97.0%	97.0%	86.4%	84.4%	91.1%	83.2%	86.7%	97.8%	78.7%
Black	69.0%	87.5%	93.8%	75.6%	86.7%	93.3%	73.7%	86.7%	93.3%	82.9%
Latino/a	78.0%	86.1%	94.4%	76.9%	81.2%	92.9%	80.9%	83.5%	89.4%	78.1%
White	82.6%	91.7%	97.2%	86.2%	85.1%	85.1%	83.7%	77.0%	80.5%	81.4%
CGS	no data	90.2%	97.6%	no data	84.0%	85.3%	no data	80.0%	85.3%	no data
FGS	no data	91.6%	95.2%	no data	83.4%	91.7%	no data	82.8%	88.5%	no data
All females	80.5%	94.6%	97.3%	84.9%	84.5%	87.1%	83.3%	80.0%	84.5%	83.50%
APA female	no data	100.0%	100.0%	no data	90.0%	90.0%	no data	83.3%	96.7%	no data
Black female	no data	90.9%	100.0%	no data	83.3%	100.0%	no data	83.3%	83.3%	no data
Latina	no data	83.3%	94.4%	no data	83.3%	90.0%	no data	83.3%	88.3%	no data
White female	no data	100.0%	95.7%	no data	82.5%	80.7%	no data	73.7%	75.4%	no data
All males	70.4%	88.0%	94.0%	79.4%	82.7%	95.1%	81.1%	86.4%	93.8%	79.9%
APA male	no data	92.9%	92.9%	no data	73.3%	93.3%	no data	93.3%	100.0%	no data
Black male	no data	80.0%	80.0%	no data	88.9%	88.9%	no data	88.9%	100.0%	no data
Latino male	no data	88.9%	94.4%	no data	76.0%	100.0%	no data	84.0%	92.0%	no data
White male	no data	83.3%	100.0%	no data	90.0%	93.3%	no data	83.3%	90.0%	no data

Table 4. Comparison of passing rates and satisfaction levels in options 1 and 2 and Engl 1A. Abbreviations: APA=Asian Pacific-American; CGS=Continuing Generation Student; FGS=First Generation Student

by the university—that is, our DSP is working *better*, helping retain university designated remedial students better than the old ENGL 1LA (with lab).[14] IRAP's results compared Fall 2005 student cohorts (the old ENGL 1LA) with our new program's ENGL 5A/5B cohorts starting in fall 2007:

- The old course obtained a 52.3 percent retention rate among "remedial" designated students who failed their courses; however, this retention rate *increases* to 74.1 percent for the ENGL 5A/5B cohort of students who failed. So even when our "remedial" designated students fail the stretch program (option two), they stay in the university in greater numbers—by 21.8 percent.
- When our 5A/5B students pass our courses, they retain at 96 percent.

When comparing these findings to our high satisfaction numbers among all groups of students, one could make the argument that when students do well in our writing program, one that gives them choice and agency through the DSP, they stay in the university longer, especially when choosing option two. Does this mean that DSP actually helps retain "remedial" and at risk students? Do their abilities to make choices in the DSP help them stay in the university, even when they fail? All of our data so far suggest yes. Additionally, Greg Glau, the Director of the Writing Programs at Arizona State University, shows similar patterns in their ten year old stretch program; however, their retention rates are at just over 90 percent when students pass their stretch courses (42–43).

Finally, these findings are tentative. This is just one year's data, and the independent portfolio measures discussed above were too small of a sample to make firm conclusions. The present year's data will allow us a much better sense of how well our students are learning and whether race and gender are factors for success and learning in each option. We'll also be able to confirm or nuance the positive findings we have already. All this helps us make the argument to keep our program as it is (once our pilot is over in two years) and present our program (and the DSP) as a possible model for other CSU campuses.

These findings tell me that while we are doing a lot right, that most (if not all) teachers are not teaching on islands—and this itself is a big accomplishment given how many teachers and classes we have—there are still cracks in the program, things that need fixing. We have strong indicators, along every data stream, that tell us our program, from the DSP to our curricula, is working for our students, but we also have evidence from just about every source that shows Blacks are most at risk, least satisfied, and fail most often. And yet, it appears that our DSP encourages retention, even when students fail their courses.

What I've Learned

I've learned quite a bit about program assessment, data gathering, and of course, our writing program at CSU, Fresno from designing and managing our program assessment efforts. Mostly, I've learned two things. First, the culture of assessment is crucial if you want a dynamic and robust program assessment, one that affects curriculum and teacher practices, one that students themselves are agents in and not just stakeholders of, one that professionally develops staff, teachers, and, yes, students, and one that's understandable (in order to change) as a living environment that (re)produces not just academic dispositions but particular social and racial arrangements in the university and community. This last element of our program assessments is the most important for me, not just because it happens to be crucial to my research agenda, but because it is what a university like CSU, Fresno, a member of the HSACU, needs. Additionally, creating such a programmatic culture was not my original goal, but it happened. It's not uniform across all teachers, and perhaps not as strong as I make it out to be here, but we're working on stronger training and communication lines, and it appears to be an effective and educative culture. Teachers continually bring and find ways to make assessment pedagogical.

Second, I've learned that one cannot conduct good, on-going writing assessment alone. One needs a corps of willing and informed teachers and administrative assistants who see clearly the benefits of a culture of assessment. And that culture of assessment fosters teachers who care about and reflect upon (self) assessment as pedagogy. I have many that work alongside me: the program's administrative assistant, Nyxy Gee; my colleagues, Rick Hansen, Ginny Crisco, and Bo Wang; and the many good, hardworking graduate student teaching assistants and adjunct teachers, such as Meredith Bulinski, Megan McKnight, Jocelyn Stott, Maryam Jamali, Holly Riding, Matthew Lance, among many others. We still have work to do, of course, but it is exciting work, environment-changing work, and maybe, if we're lucky, social justice work.

Coda: Where We Are Now

Since the publication of the program profile more than four years ago, our pilot DSP program, as part of larger writing program ecology, has continued to fluctuate and change. First, it was approved, and we became an established program on campus. We stopped asking teachers of record to provide their portfolio ratings during semesters, as well as stopped asking for students' ratings of peers' portfolios. We simply didn't learn enough from that information, and it was too much work for teachers. Additionally, we are moving to a more focused, qualitative orientation in our program assessment efforts, where we look at particular outcomes more as goals in order to understand all the consequences that our program's curriculum and various writing assessments produce in student writing and writers. This means the data we collect is changing each year. Our program assessment goals, then, are fluctuating and changing as the curriculum, portfolio, and outcomes of the program become stable. Consequently, as numerically complex as our assessment arguments were before, now I foresee them becoming more qualitatively complex, complex in the ways they represent the languages and practices of our students, thus representing them as more complex social and racial formations themselves.

In the profile, my tacit argument was that all writing programs need to find ways to incorporate and access (not necessarily assess) the diverse student discourses native to the students in their programs—that subaltern discourses can help classrooms critique the dominant discourse of the classroom and make it more meaningful, vibrant, and critical. I thought the processes of peer and self-assessment would help do this work, giving more power to students in and through judgment—I still do. But I'm less convinced that as a program we've accomplished this goal. I don't think that students' discourses, the Hmonglish (Hmong English), Spanglish, and Black English Vernacular, have in any real way become discourses that classrooms have routinely used to examine, say, our expectations for portfolios, individual assignments, or the "templates" in Graff and Birkenstein's *They Say / I Say*. Then again, I'm not sure we've found good ways to capture this kind of critical work in the program—to see it—and assess its effectiveness. And so our efforts now move to finding ways to see these things in the writing practices and documents of our students, to see more complexity. Through seeing this complexity, my hope is that what emerges is a valuing in real programmatic ways new competencies and multilingual practices. Ratings and surveys just don't do this work. Ultimately, to grow and improve learning, our program must find ways to develop student assessment practices that do translingual work (see Horner et al.; Young and Martinez).

I also mentioned (in passing) the grading contract (based on Danielewicz and Elbow's and Shor's contracts) just beginning to be used by the program at that time. The contract is now a mandatory part of all ENGL 5A courses, a course that also changed to a pass/fail course. Additionally over the last four years, just over 82 percent of all ENGL 5B courses use grading contracts (by choice of the teacher), and about 57 percent of all ENGL 10 courses use a contract as well. The ENGL 5B final portfolios have shown a slight decrease in their mean score, as seen in Figure 3. I don't know if these changes are related, but since AY 2010–11, I have not been able to train teachers on the grading contract as I did before then (I stepped down as co-director and took on the role of WAC Director). Interestingly, student satisfaction rates with course selections have remained consistent and high overall in both two-year periods. From 2008 to 2010, average overall student satisfaction was 86.3 percent (ENGL 10) and 81.3 percent (ENGL5B), while from 2010 to 2012, it edged up to 89.1 percent (ENGL 10) and 81.7 percent (ENGL 5B).

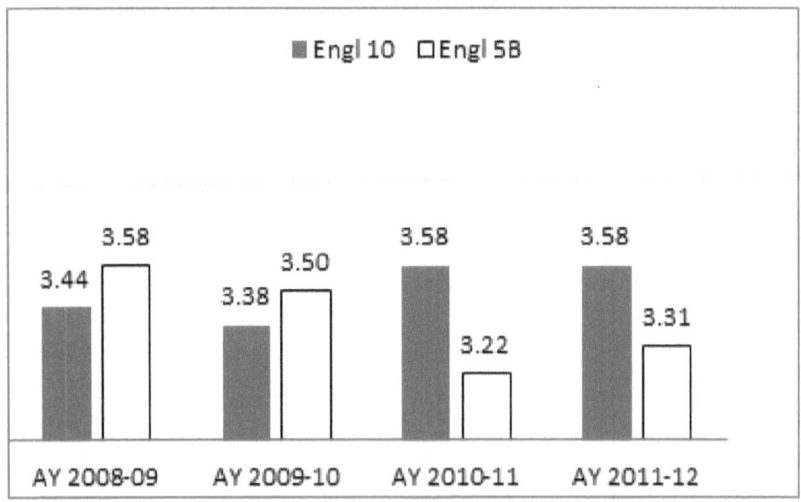

Figure 3. Portfolio Overall Ratings

Perhaps the changing enrollment patterns explain part of the story. From 2008 to 2010, almost 70 percent of all students chose ENGL 5A/5B, but from 2010 to 2012, 5A enrollment dropped to 50 percent of all students. Again, there is more fluctuation, only now it is based on students' choices. There are other factors involved in these changes beyond what students desire to take, but right now, I am more interested in revealing the trend downward in overall portfolio ratings of ENGL 5B portfolios, which match closely the program's increased use of grading contracts, its decrease in training on those contracts to teachers, its increase in ENGL10 enrollment, yet consistent satisfaction rates among students in both options (5A-5B and 10). The data seem contradictory. Why would more students choosing ENGL 10 result in higher mean portfolio ratings for ENGL 10? Are our first-year students changing? Are the slight changes in curricula (e.g., change in textbooks) affecting portfolio ratings? It's hard to know this without looking qualitatively and more carefully than we have at student writing.

As a program that is constantly assessing and understanding itself, we have lots of work ahead of us, and more change. But it is clear to me that one thing is still working well: (self)assessment at all levels is still central to what teachers and students do and defines the program.

Notes

1. In his discussion of "common sense" as separate from "philosophy," Antonio Gramsci says that a philosophy of praxis "must be a criticism of 'common sense,'" and should "renovat[e] and mak[e] 'critical' an already existing activity" (332). "Praxis" then is a dialectic, self-assessment: practices that are theorized and theorizing that become practices.

2. As I'll explain shortly, we have essentially two options that fulfill the university writing requirement, which students choose from: (1) ENGL 10, a semester long "accelerated" writing course; or (2) ENGL 5A followed by ENGL 5B, a year long stretch program.

3. The preponderance of research and scholarship in writing assessment shows that standardized tests for writing placement and proficiency are usually inadequate for local universities' varied purposes and stakeholders (Huot; White; Yancey). The two largest national professional organizations in English instruction and writing assessment (i.e., The National Council of Teachers of English, NCTE, and the Council of Writing Program Administrators, CWPA) have jointly published a "White Paper on Writing Assessment in Colleges and Universities" that acknowledges this research and promotes instead a contextual and site-based approach to writing assessment, which includes placement (NCTE-WPA Task Force on Writing Assessment).

4. DSP systems provide students with more control, agency, and responsibility for their writing courses and their work because they make the placement

decision. Reviewing and analyzing several decades of research in the field of social cognitive learning, Erica Reynolds argues: "students with high-efficacy in relation to writing are indeed better writers than are their low self-efficacious peers" (91). In other words, when students have confidence about their writing course placement, they perform better as writers in those courses. Additionally, as the "Assessment Research Findings" section shows below, our DSP, particularly those students choosing option 2 (the stretch program) and identified by the university as "remedial," achieve high retention rates.

5. Students get credit for the GE writing requirement through taking and passing English 5B and English 10. They get elective credit for English 5A and Linguistics 6. For a full explanation of these writing courses, see "The Structure of the Program" section.

6. Graff and Birkenstein use "conversation" as the metaphor that drives the templates given in this small textbook, each of which illustrate a rhetorical "move" that writers make in academic prose. The authors describe their textbook's understanding of academic discourse, which our program adopted: "The central rhetorical move that we focus on in this book is the 'they say / I say' template that gives our book its title. In our view, this template represents the deep, underlying structure, the internal DNA as it were, of all effective argument. Effective persuasive writers do more than make well-supported claims ('I say'); they also map those claims relative to the claims of others ('they say')." (xii) This is, effectively, Graff's argument in other discussions of his, for instance, *Beyond the Culture Wars*.

7. For a complete list and explanation of the program's goals and outcomes, see my annual program assessment report at: http://www.csufresno.edu/english/programs/first_writing/ProgramAssessment.shtml.

8. To see the literature we provide students, parents, and advisors for course placement, see the first year writing information page on the CSU, Fresno English department web site: http://www.csufresno.edu/english/programs/first_writing/index.shtml.

9. See Section I of Belanoff and Dickson's edited collection, Section II of Black, Daiker, Sommers, and Stygall's collection, and Section II of Yancey and Wieser's collection for accounts of how portfolios can be used to assess a program and provide rich learning environments for students.

10. You can find template syllabi and assignment sheets by course at the CSU first year writing program's web site (http://www.csufresno.edu/english/programs/first_writing/index.shtml).

11. To view the video of student responses to commenting practices that we presented in our roundtable at CCCC in 2009, see: http://www.youtube.com/watch?v=-LA6nBFkNb8 (part 1) and http://www.youtube.com/watch?v=J_MqmBqgWqg&feature=channel (part 2).

12. See Ira Shor *When Students Have Power* for an early discussion of grading contracts; Bill Thelin's "Understanding Problems in Critical Class-

rooms" for a description of one grading contract; Jane Danielewicz and Peter Elbow's forthcoming essay on grading contracts for a discussion of the general, common features of most grading contracts, which can vary; and Cathy Spidell and Bill Thelin's article, "Not Ready To Let Go: A Study Of Resistance To Grading Contracts," for a qualitative study on student attitudes towards grading contracts.

13. The scale used for all ratings of outcomes in student portfolios is described in our training and norming literature as: (1) consistently inadequate, of poor quality, and/or significantly lacking; (2) consistently inadequate, of poor quality, but occasionally showing signs of demonstrating competence; (3) adequate or of acceptable quality but inconsistent, showing signs of competence mingled with some problems; (4) consistently adequate and of acceptable quality, showing competence with perhaps some minor problems; (5) consistently good quality, showing clear competence with few problems, and some flashes of excellent or superior work; (6) mostly or consistently excellent/superior quality, shows very few problems and several or many signs of superior work. The following visual in Figure 4 is used to help raters make decisions:

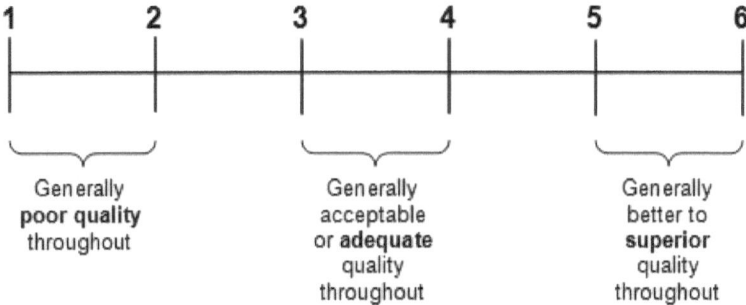

Figure 4. Rater's Scale

Raters are asked first to identify which larger conceptual bucket the portfolio fits into along the outcome in question (i.e., poor, adequate, or superior quality), then they decide if the portfolio is high or low bucket. This produces the rating.

14. IRAP's report on passing and retention rates in our DSP program can be found on our English department's web site. The report identifies "remedial students" at CSU, Fresno as: "students who tested into remediation based on their EPT status. In pre-DSP English, these students enrolled in ENG 1 along with a lab (ENG 1LA or 1LB). In the DSP program (beginning fall 2006), they can enroll in either ENG 10 or the ENG 5A5B sequence."

Works Cited

Aronowitz, Stanley, and Henry A. Giroux. *Postmodern Education: Politics, Culture, and Social Criticism.* Minneapolis, MN: U of Minnesota P, 1991. Print.

Black, Laurel, Donald A. Daiker, Jeffrey Sommers, and Gail Stygall, eds. *New Directions in Portfolio Assessment: Reflective Practice, Critical Theory, and Large-Scale Scoring.* Portsmouth, NH: Boynton/Cook, 1994. Print.

Belanoff, Pat, and Peter Elbow. "State University of New York at Stony Brook Portfolio-Based Evaluation Program." Pat Belanoff and Marcia Dickson, eds. *Portfolios: Process and Product.* Portsmouth, NH: Boynton/Cook, 1991. 3–16. Print.

Danielewicz, Jane, and Peter Elbow. "A Unilateral Grading Contract to Improve Learning and Teaching." *College Composition and Communication* 61.2 (2009): 244–268. Print.

Delpit, Lisa. "The Politics of Teaching Literate Discourse." *Literacy: A Critical Sourcebook.* Ellen Cushman, Eugene R. Kintgen, Barry M. Kroll, and Mike Rose, eds. Boston, New York: Bedford/St. Martin's, 2001: 545–54. Print.

Elbow, Peter, and Pat Belanoff. "Portfolios as a Substitute for Proficiency Examinations." *College Composition and Communication* 37 (1986): 336–39. Print.

Freire, Paulo, and Donald Macedo. *Literacy: Reading the Word and the World.* South Hadley, MA: Bergin and Garvey, 1987. Print.

Gee, James Paul. "Literacy, Discourse, and Linguistics: Introduction and What Is Literacy?" *Literacy: A Critical Sourcebook.* Ellen Cushman, Eugene R. Kintgen, Barry M. Kroll, and Mike Rose, eds. Boston, New York: Bedford/St. Martin's, 2001: 525–44. Print.

Glau, Greg. "Stretch at 10: A Progress Report on Arizona State University's Stretch Program." *Journal of Basic Writing* 26.2 (2007): 30–48. Print.

Graff, Gerald. *Beyond the Culture Wars: How Teaching the Conflicts Can Revitalize American Education.* New York and London: W. W. Norton, 1992. Print.

Graff, Gerald, and Cathy Birkenstein. *They Say / I Say.* New York: W. W. Norton, 2006. Print.

Gramsci, Antonio. *The Antonio Gramsci Reader: Selected Writings from 1916–1935.* Ed. David Forgacs. New York: New York UP, 2000. Print.

Hamp-Lyons, Liz, and William Condon. *Assessing The Portfolio: Principles For Practice, Theory, And Research.* Cresskill: Hampton P, 2000. Print.

Haswell, Rich. Ed. *Beyond Outcomes: Assessment and Instruction Within a University Writing Program.* Westport, CT: Ablex Publishing, 2001. Print.

Horner, Bruce, Min-Zhan Lu, Jacqueline Jones Royster, and John Trimbur. "Language Difference in Writing: Toward a Translingual Approach." *College English* 73.3 (2011): 303–321. Print.

Huot, Brian. *(Re) Articulating Writing Assessment for Teaching and Learning.* Logan, UT: Utah State UP, 2002. Print.

NCTE-WPA Task Force on Writing Assessment. "NCTE-WPA White Paper on Writing Assessment in Colleges and Universities." Online posting. Council of Writing Program Administrators. 31 Aug 2008 <http://wpa-council.org/whitepaper>. Web.

Reynolds, Erica. "The Role of Self-Efficacy in Writing and Directed Self-Placement." *Directed Self-Placement: Principles and Practices.* Eds. Daniel Royer and Robert Gilles. Cresskill, NJ: Hampton P, 2003. 73–103. Print.

Royer, Daniel J., and Roger Gilles. "Directed Self-Placement: An Attitude of Orientation." *College Composition and Communication* 50.1 (1998): 54–70. Print.

Royer, Daniel J., and Roger Gilles. *Directed Self-Placement: Principles and Practices.* Cresskill, NJ: Hampton P, 2003. Print.

Shor, Ira. "Critical Pedagogy Is Too Big To Fail." *Journal of Basic Writing (CUNY)* 28.2 (2009): 6–27. Print.

—. *When Students Have Power: Negotiating Authority in a Critical Pedagogy.* Chicago: University of Chicago Press, 1996. Print.

Spidell, Cathy and Willam H. Thelin. "Not Ready to Let Go: A Study of Resistance to Grading Contracts." *Composition Studies* 34.1 (Spring 2006): 36–68. Print.

Thelin, William H. "Understanding Problems in Critical Classrooms." *College Composition and Communication* 57: 1 (September 2005): 114–141. Print.

Volosinov, Victor N. *Marxism and The Philosophy of Language.* Trans. Ladislav Matejka and I. R. Titunik. Cambridge and London: Harvard UP, 1973. Print.

White, Edward M. *Developing Successful College Writing Programs.* San Francisco and London: Jossey-Bass, 1989. Print.

Yancey, Kathleen Blake. "Looking Back as We Look Forward: Historicizing Writing Assessment as a Rhetorical Act." *College Composition and Communication* 50.3 (February 1999): 483–503. Print.

Yancey, Kathleen Blake, and Irwin Weiser, eds. *Situating Portfolios: Four Perspectives.* Logan, UT: Utah State UP, 1997. Print.

Young, Vershawn Ashanti, and Aja Y. Martinez. Eds. *Code-Meshing As World English: Pedagogy, Policy, Performance.* Urbana, IL: NCTE, 2011. Print.

12 Utilizing Strategic Assessment to Support FYC Curricular Revision at Murray State University

Paul Walker and Elizabeth Myers

In fall 2009, the first-year composition (FYC) requirement at Murray State University transitioned from a 6-credit-hour, two-semester sequence (ENG 101/102) to a 4-credit-hour, one semester course (ENG 105). The initial decision to make this revision stemmed from the labor conditions for contingent faculty in our department and the sense that the FYC curriculum could be invigorated with a fresh look at objectives, outcomes, and structure. Yet the context for these revisions, as we explain later, afforded us a chance to do something more: to contribute to changing perceptions about writing on campus and about the role that FYC can play in supporting students' on-going writing development. In particular, the revision enabled us to foreground to a greater degree the critical inquiry of ideas *through* reading and writing, avoiding overemphasis on "general skills" that can impede rhetorical sensitivity to writing's complexity. Since the revision was adopted in 2009, students and FYC instructors have expressed satisfaction and praise for the changes. Those responses testify to the success of the course objectives, which have been strongly influenced by current writing theories and recommendations by the Council of Writing Program Administrators (WPA) and the National Council of Teachers of English (NCTE). We are heartened by the pedagogical results of the curricular revision, and the process for its approval across the university points to interesting junctures of disciplinary expertise, collegial expectations, administrative objectives, and the "culture" of assessment locally and nationally. Assessment didn't drive the curricular and structural changes, but our designed assessment comparing our old and new course design—and comparing student writing under the previous and revised curricula—was useful for confirming to others in the university that our efforts to address "best practices" from a disciplinary perspective are well founded.

Writing, as any rhetoric and composition scholar knows, evokes strong emotions from people, especially among faculty who complain about students' writing abilities. Those outside our field sometimes distrust our expertise in teaching academic writing, offering up their own ideas for the best ways to ensure quality student writing—ideas that are often current-traditional in nature. For example, despite numerous faculty workshops and a body of research that clearly shows the detriments of using grammar as a central teaching focus, the first solution to less-than-stellar student writing proposed by many outside our discipline is: "Teach the students grammar." Direct assessment of student writing, in our situation, provided a way to justify our theory- and context-based curricular and structural revision. Although we are uncomfortable with the need to "prove" our expertise in writing instruction pedagogy and practice to outside interests through direct assessments, the results of our assessment confirmed our decision to change the structure of our first-year composition course. The process for revising our course, despite being necessarily contextual to our institution, nevertheless may hold value for other institutions that may be considering a similar structural revision based on similar circumstances.

We know that Murray State's writing program is not the first to offer a one-semester FYC requirement for the general population of students; yet describing the change allows us to illustrate the value of thoughtful response to institutional realities that might otherwise hinder efforts to conform to proven models of writing programs. Further, comparing the two-course sequence to the one-course revision shows how assessment can be valuable within an institution as well as of interest to others who are affected by writing assessment's administrative influence. Comparing results of a "traditional" and a "revised" curricula is a common assessment practice, but comparative assessments measuring the efficacy of distinctive structural changes such as we did with our ENG 105 course, if they exist, are not readily available.

Our assessment builds on the gains that have been made in writing assessment, guided by at least two decades of holistic scoring research (e.g., Williamson and Huot; O'Neill, Moore, and Huot; Freitag-Ericsson and Haswell). Calibrated holistic scoring by writing teachers has so far defended writing assessment from widespread machine scoring that keeps being offered as a solution to the inefficiencies and biases of humans. Nonetheless, we recognize that any large-scale writing assessment can oversimplify the complexity of the act of writing or detrimentally transcend the context in which a written document was produced. Overreliance on calibrated holistic scores can presume, like machine-essay-scoring, that individual student writing is effectively represented through a quantified categorization. This is why utilizing writing assessment to justify curricular changes or developments can be problematic—not all practices can be shown to be effective through statistical measures, which, of course, raises all sorts of questions about reliability of human judgment and evaluation. This program profile will leave most of those questions for another time; the objective of our narrative is to show how a thoughtful curriculum change was supported by a local assessment that justified the change to others on campus.

Background and Institutional Context

Murray State University is a public comprehensive university in Murray, Kentucky with an enrollment of approximately 10,000 students. As one of Kentucky's regional universities, it serves the mostly rural population of far western Kentucky and nearby portions of Missouri, Illinois, Indiana, and Tennessee. Murray State has received recognition for being a good value and has consistently been listed among the top twenty-five public comprehensive universities in the south. Incoming students each fall number close to 1,100, and with a twenty-five student cap on the university's required (revised) first-year composition course, 40–45 sections are offered each fall. Of note is that Murray State does not have a campus-wide writing program. The curriculum of first-year composition and scheduling of part-time instructors, along with other administrative tasks, are the responsibility of the composition coordinator, a faculty member in the department of English and philosophy. The coordinator is assisted by a composition committee, which he/she chairs. Developmental writing is taught in a separate office on campus, led by a basic writing coordinator, who also contributes writing assistance to a learning center by training undergraduate writing tutors. A writing across the curriculum (WAC) program was brought back to life in 2010, with the appointment of a WAC coordinator. Also in 2010, a writing center opened, directed by a faculty member in the department of English and philosophy and staffed by trained graduate students or advanced undergraduate students. Thus, although most of the elements of a program are in place, the work is divided among four coordinators who work together sporadically without a real unifying structure. Eventually, we trust this situation will change; we provide this information here to help contextualize the steps we have taken with FYC to perhaps enable those changes.

In the summer of 2008, before the WAC and writing center coordinators were appointed, Paul was appointed the composition coordinator. Paul's curricular responsibilities included ENG 101: Composition, and ENG 102: Composition and Research. At the time, ENG 101 required three papers that generally covered narrative, analysis, and basic argument, and ENG 102 followed with emphasis on analysis, research, and multiple-sourced argumentative papers. The focus on academic inquiry in the second semester worked effectively as a scaffold for developing academic writing, but as the revision was being designed and implemented, that two-semester timeframe did not seem to be a necessity for student success. A well-designed program, focused more intensively on inquiry, might provide effective writing development in one semester. Importantly, Paul didn't come into the position with the intent to change the curriculum, but during an impromptu meeting with the dean of the college of fine arts and humanities, the one-semester idea arose from a sit-down discussion about possible changes to address two issues. First, some faculty members in the department of English and philosophy had expressed interest in "writing seminar" versions of FYC that were conducive to their respective specialties. The other issue concerned the number of FYC sections taught by part-time instructors, which did not align with MLA or NCTE recommendations. During this meeting, we broached the idea to revise the FYC structure from a six-credit-hour, two-semester sequence to a four-credit-hour, one semester course and agreed to begin the process.

Administratively, the key to this revision was that the four-credit-hour course would count as two courses for English faculty loads (normally 4/4), making it more appealing for full-time faculty in rotation to teach it, with the understanding that enhanced individualized writing instruction would be enabled by the extra two credit hours of unassigned time. In the new one-semester course, full-time faculty could design the course with attention to theme and specialty, accomplishing the common objectives and outcomes without worrying whether the material would cohere with the second-semester course. The other key was to front-load FYC sections in the fall to allow students to complete the requirement and enroll in courses where FYC is a prerequisite in the spring. Under our plan, with full-time faculty teaching more sections than obligated to, and fewer sections needing to be offered in the spring semester, money typically used to pay part-time instructors in the spring would be saved to provide "seed money" for eventually transitioning those instructors into full-time lecturers. This goal arose out of the concern (provoked by various outside statements on course load and conditions for contingent faculty)[1] that too many of our sections were taught by adjuncts, preventing the impetus for improving the conditions for those adjuncts. We recognize that the department's and university's mutual investment in full-time faculty positions is beneficial to students because of the permanence of the position. No matter how willing part-time instructors are to improve teaching practice, conduct teacher or other research, and participate fully in departmental and university service and programs, the incentives and rewards for doing so are limited if nonexistent.

At the time, a university committee was revising the university studies, or general education, requirements, and a major-specific, writing-intensive requirement was being considered. If a revision of FYC was going to happen, the ideal time was during this gen ed revision period. Shortly thereafter, with approval from the chair of the department of English and philosophy, the proposal was presented to the department and various university committees until it was approved for the new university catalog for fall 2009. To state the process succinctly hides the amount of work in developing the proposal and defending it among several groups of university faculty. As the readers of this journal might suspect, the proposal met with some opposition, though not from within the English and philosophy department. The opportunity to teach one less course (even though ENG 105 is not a full course release) and the increased flexibility in meeting the common objectives appealed to English faculty. The opposition came from elsewhere. For the university studies committee, the proposal was brought forward late in the process of revising the general education requirements, which meant that the nearly approved revision now would be short two credit hours. This concerned several programs because they had been insisting on a broad general education program and had resisted reducing the number of hours required. Other programs thought it beneficial because they already felt limited in what they could require of their majors with a broad general education program and the state-imposed 120-hour requirement for any bachelor's degree. Eventually, these concerns were worked out without any change—the general education requirements were adjusted to fit the reduced FYC credit hours.

Rationale for the Curricular Revision

The most vehement opposition to the course revision came in various forms of the same sentiment: "Our students don't know how to write; we should be requiring more writing, not less." A few faculty members expressed concern that students would no longer be required to take two semesters of composition prior to taking other classes. We did not ignore this concern, for we acknowledge the importance of writing instruction. However, we felt that our revision was equal in rigor and content to the two courses, no matter what the credit hours indicated. Also, the concerns of other faculty in this regard seemed based on what Joseph Petraglia has called general writing skills instruction (GWSI), which our field has tended theoretically to resist, from the rise of "writing to learn," to WAC/WID, to the post-process understanding of writing's complexity and situatedness. The preparatory function of FYC presumes that writing skills can be taught generally, and that such skills must be taught early on for students to write effectively later on in their "more important" major courses. One problem with these presumptions, as we know, is that GWSI dismisses the complicated and heterogeneous contextual factors involved in writing. Challenging the need to teach once-and-for-all general writing skills in the first year is the largest obstacle our discipline faces in any generation, because arguments against GWSI are esoterically theoretical and counter the general-skills approach common in other "introduction-to-the-discipline" courses. Yet the concern that students must have preparatory writing instruction overlooks two realities of university students: 1) students enroll in general education courses that often require writing concurrently with FYC; 2) due to transfer policies, program requirements, and human tendencies, a significant number of students enrolled in ENG 101 and 102 after their first year. Many seniors, in fact, have historically enrolled in ENG 102 in their last semester, somehow fulfilling all other requirements without this supposedly "preparatory" course. Our revised ENG 105, while not completely averse to such issues, makes it much more difficult for students to attain sophomore status without completing the course.

Therefore, our response to this concern was to assure members of the faculty that the new four-credit-hour course would not only "prepare" students for academic writing and thinking during their first year, but it would also at least match the rigor of its two-course predecessor. Our assurance was based on three "marketable" notions:

- More full-time faculty interested in and actually teaching composition would enhance the students' contextual understanding of and performance in academic writing.
- A comprehensive, intensive writing course in one semester would keep the connections between writing skills and academic inquiry fresh on their minds without a break between an introductory 101 course and a research-focused 102 course.
- Further writing instruction and practice would take place within students' major programs with the implementation of designated Writing Intensive courses required for every major.

The third notion, which was already decided by the university studies committee, we recognize as an adequate, though not ideal, application of WAC and WID research. Since our idea for this revision, as well as the benefits for students and part-time instructors, stemmed from the recommendations of the NCTE and WPA Council on the use of contingent faculty in the composition classroom,[2] our original intention was for this plan to not only invigorate our instruction, but to also enable our adjuncts to apply for soon-to-be-added full-time lecturer positions. Because of budget reduction factors common to most public universities around the country during the last few years, we have been unable to offer these lecturer positions yet, but our full-time faculty representation in the composition classrooms has indeed increased. In fact, the course revision has made the teaching of composition a priority for faculty in our department—they are choosing to teach FYC rather than accepting it as their obligation.

The revised course, pedagogically and theoretically, emphasizes critical reading, writing, and inquiry. The revision committee looked closely at the WPA Outcomes Statement,[3] making minor adjustments to ENG 105's objectives to match the document's language and intent more closely. The finished revision is essentially an expanded ENG 102 course, with additional early writing assignments building toward researched argument papers.[4] Most instructors now design the readings, writing assignments, and classroom activities with acknowledgement of the complexity of writing, or the varied situations and contexts from which writing emerges and to which it responds. Many instructors also choose to design a seminar-type semester with an overriding theme. Examples have included environmentalism, friendship, consumerism, law, veteran issues, globalization, and multiculturalism. With *inquiry* as a part of the course title, students are reminded of the necessity to ask questions of theme-based or other texts, discovering answers through research and writing, and are encouraged to delve deeper into topics through the focused and nuanced study of ideas. Practically, the coherence of the assignments in only one semester has been an important element in the course's success among faculty and students. Meeting frequently and for longer each week seems to enable more connection among assignments, allowing for revision, further development of ideas, and inclusive skill-building without a complete change of classroom dynamic as happened with a two-semester sequence.

Rationale for the Curricular Assessment

The enthusiasm and commitment shown by the faculty to the revised curriculum have been adequate evidence that the changes invigorated the teaching of FYC, but such intangible measures are less persuasive to some audiences. Therefore, the purposes of this study included justifying our curriculum change to our own campus colleagues, while recognizing that the results may be beneficial to the field of composition at large. To our knowledge, there are no studies directly comparing two-semester FYC sequences with one-semester FYC courses. We were able to point to other institutions that have four-credit-hour FYC courses, but we weren't able to locate any research that stated whether there was any difference in student performance. The dearth of research is expected, we suppose, as writing programs should be contextually designed, and the needs of different universities vary. For example, many universities recruit graduate students by providing funding through teaching FYC; Murray State funds only a few graduate teaching assistants per year, which requires many other sections to be staffed in other ways. Some schools don't have any graduate students, and so both historical practice and institutional needs determine how FYC is structured in any context. Therefore, to compare across institutional contexts could set up false ideals that may cause problems when applied elsewhere. We were confident that our revision would address the needs at Murray State; and while the impetus for our comparative assessment was local in nature, stemming from the opposition to our course revision from outside the department, we recognize from our own experience that such research should be valuable to others in composition studies.

At the beginning of fall 2009, with the debut of ENG 105, we were prepared to assess student writing that emerged from the course as a condition for going through with the revision of FYC. If, after one year, we found a significant decline in the quality of student writing from the four-credit-hour course compared to the six-hour sequence, then we would reevaluate the revision, make adjustments, and assess again after three years. If at that point, the quality of student writing in the four-credit-hour course remained significantly lower than the student writing from the six-hour sequence, then we would return to that model. Because of that promise, we developed an assessment plan and carried it out, as described in the next two sections.

Assessment Methods

Elizabeth, a senior at Murray State majoring in history and English education, led the data collection (of final student papers submitted in ENG 102 and 105) under Paul's direction. She developed a permission waiver with assistance from the IRB, which was signed by the majority of ENG 105 students, who understood that their participation would not affect their course grade. The forms were distributed to approximately forty sections of ENG 105 classes and around ten ENG 102 sections (102 sections that were retained for students who enrolled under the jurisdiction of the old catalog). These forms were not necessary for the primary comparative purpose of the assessment, but in the event that the content of the papers might be used, we wanted permission to do so. Because past collection of ENG 102 final papers was strictly for programmatic assessment purposes, we did not have permission forms for past ENG 102 students, making those papers off limits under IRB rules if passages are excerpted.

After we collected the permission forms and categorized them by section number and instructor, we requested that each course instructor provide the students with an e-mail address utilized specifically for storage of the papers. At the end of the semester, all ENG 105 and ENG 102 instructors were asked to have their classes submit their final papers to this e-mail address. For several years, the final paper in ENG 102 has been a researched-argument assignment of at least eight pages in length, and this served as a pattern for ENG 105's final paper. Because of this, we were able to compare papers with similar aims and structure and that served, in the case of each class, as the culminating assignment for the semester.

Upon receiving the papers, we downloaded the approximately 300 student papers to a removable drive and numbered them. We used a random number generator to select two sets of 75 samples from the ENG 105 group (we had more usable 105 papers, and one set would be used for another institutional use). From the past collection of ENG 102 papers, and the recently received ENG 102 papers, we randomly selected one set of seventy-five papers. After the student papers were selected, we printed the papers and eliminated any identifying information for the students by first whiting out their names and section number, as well as marking over them with a black sharpie to ensure total anonymity.

Over the course of several months, a holistic scoring team, formed in conjunction with a university-wide effort to enhance written communication for Southern Association of Colleges and Schools (SACS) accreditation, developed a six-point scoring guide (see Appendix 1) to evaluate student papers and calibrated their scoring of sample papers accordingly. The team consisted of around ten instructors of required composition and humanities courses, and the scoring guide reflected general writing attributes but was particular to elements that work in English studies, though not aligned explicitly to either ENG 102 or ENG 105 objectives. At the end of the 2009 spring semester, the team spent one week evaluating the composition papers using the scoring guide. As typical for this type of calibrated scoring, each paper was evaluated by two readers. If there was more than one point between the evaluator's scores, a third reader evaluated the paper.

As is understood in our field, writing quality is a heterogeneous variable—there are several competing elements in its make-up. If writing assessment intends that results culminate in a single score, then weighting of variables is easily manipulated, adjusted, or altered (Gladwell). Weighting of variables is always subjective. In writing assessment, even the most carefully calibrated, multi-reader holistic scoring cannot reduce internal weighting of elements by individual scorers. The calibration merely elicits *that group's* collective weighting of a rubric's elements relative to a numerical scale. This understanding made us careful in our use of holistic scoring to compare students' writing samples. Since we were not attempting to extend the student scores beyond the study, we felt that we did not overreach, claiming to accurately measure the individual students' overall writing ability. The paper scores simply represented a scoring group's assessment of each paper at that specific time, outside the classroom context but still within the context of a composition program. From that programmatic perspective, the results avoid, importantly, assessing the methods of individual teachers of the composition courses.

Assessment Results

In Table 1, the results of the assessment are shown. As indicated in the table, the assessment included more papers from ENG 105 than ENG 102. The reason for this was that the ENG 105 papers were also being used in another assessment project in conjunction with the SACS Quality Enhancement Plan. However, to avoid picking and choosing among the 150 papers to compare with the seventy-five ENG 102 papers, we believe comparing the mean of the 150 papers is more honest than using half of them. The holistic scorers did not at any time know which paper was for either class, nor did they know the specific assessment purpose of the papers they would be evaluating. The most elementary comparison is between means, shown in Table 1; the ENG 102 papers show a slightly higher, though statistically insignificant, mean (3.23) than all of the ENG 105 papers (3.05). The average score of three for all papers, on a scale of six points, indicates an expected performance range of first-year writing students, though over time we hope for an average in the four range.

Table 1. Average scores for course papers (on a six-point scale)

Course	ENG 102 (n=75)	ENG 105 (n=150)
Mean	3.23	3.05

Enhancing the Foundation for Writing Instruction

The results of our assessment confirm that our efforts to revise our FYC course to better fit the situation of our university were successful—our planning and implementation maintained the level of student writing ability through a one-semester course. We feel we are moving in the direction of meeting the best practices of writing programs, despite the chance that other realities, most likely financial, might in the future undermine our efforts to increase the number of full-time faculty in the composition classroom. In addition to the quantifiable comparison of the student writing, the revision of the course produced a few qualitative results as well, which we hope will help to maintain the overall quality of writing instruction at our institution. These results include increased cooperation between the coordinators of composition and basic writing; a starting point for developing a culture of writing across campus; and an increased interest in and enthusiasm for teaching writing by faculty members.

As mentioned earlier, Murray State lacks a WPA; instead, the work is currently divided among four coordinators, one of whom is a tenured faculty, two of whom are tenure-track, and one of whom is non-tenure-track. The revision of FYC enabled stronger collaboration between the composition coordinator and the basic writing coordinator, as the change to ENG 105 affected when students placed into the developmental writing courses would be able to enroll, as well as the preparation required. The one semester course covers topics and skills quickly; most classes meet four days each week, and students produce a lot of informal and formal writing. The need for support for struggling writers not technically classified as basic writers has been integral to the collaboration between the two coordinators. Now that a writing center has been established, the support for students in FYC continues to grow. With the resurgence of a WAC program at Murray State in 2010, students and teachers outside FYC are recognizing the importance of writing in learning and teaching. By initiating this revision and the discussions that followed, the composition program claims some credit for the increased "culture" of writing on campus. The catalyst for most of the recent developments has been SACS accreditation requirements, but by proposing the revision to FYC, defending it, and promising to assess its effectiveness, issues of writing were foregrounded on campus before the WAC program was implemented, preparing the ground, so to speak, for the valuable theories of writing to disseminate beyond the English department. Paul has had several opportunities in hallways and stairwells to casually share the philosophy behind the ENG 105 revision with fellow faculty members, and many of these conversations have carried into subsequent discussions of criteria for writing intensive courses. The WAC coordinator shares our emphasis on the role of academic inquiry in writing instruction, which is evident in the new WAC program as "writing ambassadors" from each college address writing issues from cross-disciplinary perspectives.

Within the Department of English and Philosophy, the course revision has caused more faculty members to be involved in professional development and orientation sessions. One course that covers what two courses covered previously has simplified hiring of adjunct instructors, placement testing, and the glut of students trying to enroll in full sections of 101 or 102 in the fall or spring semester. Further, composition teacher orientation sessions are slightly more focused because everyone is teaching the same course, which increases the applicability of specific professional development meetings during the semester. The workshops on writing involve most of the faculty members in the department, including philosophy faculty, who are not "credentialed" under accreditation rules to teach English courses. Those who have participated enjoy discussing writing instruction and evaluation, and other issues of writing not unique in any way to English studies. Additionally, the new WAC program has extended these discussions on writing even further across campus.

Furthermore, the revision has increased enthusiasm within the department for teaching composition, not only because of the course-release equivalency, but also because the department was and continues to be involved in ongoing discussions and professional development regarding writing on campus. It's not that they weren't interested in writing before; the revision and subsequent writing-related developments have awakened to a larger degree faculty's interest and intellectual effort in teaching writing. This is evident from voluntary attendance at professional development programs, volunteers for the composition committee and subcommittees, and the lack of complaints about teaching composition. In fact, an unanticipated problem that has occurred is full-time faculty are asking to teach composition, which has caused administrative difficulties in offering and staffing other department courses.

Where We Will Go From Here

A concerted, thoughtful effort in any curriculum will likely show some measure of success, and since this revision reflects the best model for the context in which writing is taught at Murray State, we believe that the comparison assessment confirms to our university colleagues that our efforts have been rightly placed. We recognize that much more can be done to increase our ability to facilitate writing improvement in our students, including, though not limited to, effective assessment of students meeting course objectives and outcomes. Influenced by the Dynamic Criteria Mapping of writing qualities pioneered by Bob Broad, we are examining the newly released Framework for Success in Postsecondary Writing[5] for ideas on understanding student "habits of mind" as they relate to what teachers value in student writing. Teacher accountability is a politically charged phrase, and we are uncomfortable with many of the intentions behind such efforts. One of the principles that we hold in our composition program is that teachers perform best when allowed to do what they do best. Aside from broadly described common objectives, teachers of FYC at Murray State are free to design their courses to best fit their own interests and specialties. Instructors are not left alone, however; guidance and suggestions are always available, in addition to regular professional development workshops and a pre-semester orientation session.

The autonomy we provide reflects our belief that teachers' understanding of the classroom context should be valued and thus protected against overbearing outside efforts to measure accountability or student performance. Because of that belief, our current and future assessments of student performance and teacher effectiveness rely heavily on individual teacher impressions of their own students' accomplishment of course outcomes—acknowledging the importance of in-context assessment of learning. As alluded to previously, the study described in this article was useful for the specific purpose of confirming our department's specialty in first-year writing instruction to our colleagues. The assumption that such a confirmation is necessary is troubling, and to avoid the impression that we will continue to conduct similar assessments, we are purposefully assessing our curriculum by relying on individual teacher impressions of their own students' collective work. Basically, we ask each instructor to participate in this assessment two times during the semester by responding to an online questionnaire after completing the grading of a set of their students' papers (see questionnaire in Appendix 2). The questionnaire asks the teacher to rate on a scale whether his/her students are competent in each of the course objectives. The results show which objectives students are struggling with the most, and thus we can develop workshops for teachers to address these struggles. In this way, the assessment of student performance is able to immediately enhance the curriculum, rather than us trying to interpret quantified, holistic scores of a sample of student papers and to understand which of the holistic-rubric categories affected the students' overall scores. We also think that involving instructors in the assessment builds their instructional expertise as they reflect on the collective performance of their students—a benefit that is more often limited to holistic scoring teams rather than all instructors.

At this point in our writing program's progression, our comparative study confirms the foundational structure of writing instruction at Murray State University as we move forward in all our efforts to enhance our program. We recognize that there are additional ways to increase instructional expertise and student learning, including reducing our course caps in FYC courses, and we hope to address such enhancements as we do our part to contribute to a stronger culture of writing across campus.

Coda: Where We Are Now

Since the publication of our program profile, much of what we anticipated in the article has indeed occurred. Under the banner of SACS reaccreditation, the pressure to formulate student-learning outcomes and attendant assessment methods for student writing increased. In response to this pressure, Paul, as composition coordinator, developed an assessment model for ENG 105 that emphasizes instructor contextual expertise that reflects the ecological aspects of writing programs highlighted in this collection. In our new model, each instructor contributes to the program's collective culture of writing by identifying what we value in student writing. As part of this volume, we recognize that as a program we are collectively more than the sum of our individual selves—this emergence taking place through a seemingly simple process that belies a complex and wide range of teacher expectations that bear on the assessment of student writing. Each semester, following a facilitated conversation to generate and adjust what we value in writing, instructors evaluate their classes by completing a brief impressionistic survey, which provides a sense of how students taught in our program are meeting these values as well as the course objectives. This model—centered on the natural variance of teachers' expertise and experience as well as the pressures and demands on the students' attention and motivation—affords ecological opportunities to see difference and disagreement as symbiotic with the collective strength of our writing instruction.

At a basic level, the results of this survey provide a general understanding of where our FYC students struggle, leading to sensible professional development in these areas. But more importantly, our acceptance of difference enables instructors to utilize expertise in a variety of ways: sometimes conforming to others, sometimes deliberately rejecting traditional program practices. But all of this takes place within an ecology that naturally finds balance within its own complexity because its members are fully trusted to gauge their own students' accomplishments, reinforcing the emergence of continuing meaningful knowledge from the self-organizing model of assessment.

Similarly, we are working to produce more interconnected collaboration among areas of writing across campus, so our efforts continue to create an encompassing writing program that includes first-year writing, WAC/WID, the writing center, and basic writing. For now we enjoy significant support and trust in the composition program's efforts to enhance writing instruction and integration on campus. By proactively designing and implementing a meaningful assessment model for our ENG 105 course, Paul has so far been able to deflect potential changes to the course and program because of budget cuts and questions of faculty loads. Further, because the model yields prescient areas of concern in our students' writing and a noticeable culture of writing is growing among composition faculty and spreading elsewhere, the ecological balance we have found for instruction and assessment seems intriguing to those who are not teaching FYC but who assign multiple writing assignments.

For Elizabeth, who is now a sophomore English teacher in Tennessee, her participation in conducting and reporting on this study has influenced how she values well-implemented assessments and the role assessments play in shaping an effective curriculum in the high school classroom—enabling the ecological balance in a complicated and often frustrating environment. In making the transition from Tennessee's previous standards to the Common Core State Standards (CCSS), her experience utilizing assessments has been integral in helping her fellow teachers prepare students for new intellectual demands. Under a pilot program for schools hoping to begin CCSS implementation a year early, Elizabeth's school conducted four writing assessments throughout the school year, each assessing one of the three modes of writing within the CCSS—narrative writing, informative writing, and argument writing—as well as the regularly scheduled Tennessee Writing Assessment for juniors.

Similar to the newly developed ENG 105 assessment model, Elizabeth's school assessment shows specific areas that could be addressed; for example, students could write proficiently and use adequate grammar in doing so, but are weak in their abilities to elaborate, synthesize, and logically connect thoughts and ideas. In other words, areas of writing that require depth of thinking are areas where the students struggle. Because of the information from these assessments, Elizabeth's English department has begun the process of increasing the rigor within the school, including re-mapping the curriculum to focus upon depth rather than breadth in the study of various literary texts and eliminate texts that are no longer academically rigorous. Combining summer reading with modes instruction, they are focusing on structuring lessons that require students to make multiple connections, explain points thoroughly, and cite textual evidence to support and arrive upon answers. Through a well-constructed diagnostic assessment, Elizabeth and her colleagues have been able to arrive upon concrete needs of our students, and make a concrete path upon which teachers can build in the coming years.

Because of the MSU study, she became aware of how assessments shape curriculum in real ways, and gained insight into the ecological aspects of writing instruction at all levels—teachers cannot separate themselves from the system that seems to be the cause of the problems that they are asked to solve. For both Paul and Elizabeth, the environment for writing remains as complicated as ever, but their utilization of ecological principles to address that complexity illustrates how writing programs are indeed ecological: the more diverse an ecosystem is, the more it thrives. If writing programs work to find balance and symbiosis among different ideas and methods, they will successfully adapt to the inevitable changes, surprising themselves with the resulting benefits and innovation.

Appendix 1: Holistic Scoring Guide

Upper Half: Mastery			
Category	consistent	effective	exceptional
	4	5	6
Purpose	Makes major aims clear, demonstrating the conventions of an appropriate genre	and adopts aims central to the topic, demonstrating control of generic conventions	and engages subtleties and nuances of the topic and the generic conventions
Audience	Reflects consistent awareness of desired impact on audience	and effectively appeals to audience expectations	and involves and engages the audience in the topic
Voice	Establishes a voice appropriate to audience and purpose	and effectively maintains control of voice	and may use that voice compellingly
Representation	Develops and represents an idea, experience, and/or text, using logic, reasoning, and/or examples	and analyzes its/their significance effectively	and offers notable insight
Organization	Selects and arranges material to establish a clear focus, providing essential transitions	and supports that focus effectively with purposeful arrangement and purposeful transitions	and may support that focus compellingly through clever but subtle control of arrangement and coherence
Presentation	Controls sentence level features of written language, including grammar, spelling, punctuation, usage	and shows mature command of these features, particularly as regards clarity and precision	and exhibits mastery of these features in an engaging rhetorical style

	Lower Half: Lack of Mastery		
Category	*inconsistent*	*marginal*	*incompetent*
	1	2	3
Purpose	Does not always make clear its major aims nor observe conventions of the implied genre	and confuses the reader about its major aims, perhaps also violating some of the conventions	and persuades the reader that it has neither aims nor understanding of generic considerations
Audience	Reflects inconsistent awareness of desired impact on audience	and may occasionally violate audience expectations	and apparently has no awareness of audience impact; alienates audience
Voice	Does not always establish an appropriate voice for audience and purpose	or uses an inappropriate or inconsistent voice	or creates an image of the writer that undermines and jeopardizes credibility or sympathy
Representation	Does not develop and represent an idea, experience, and/or text, using logic, reasoning, and/or examples	and may partially misrepresent it	or appears to have misunderstood the idea, experience, or text
Organization	Fails to arrange material effectively	and may show a lack of focus	and may confuse readers
Presentation	Loses control of one or more elements of written language at the sentence level	or reveals only rudimentary knowledge of the conventions of standard written English	or fails to acknowledge these conventions

Appendix 2: New Assessment Questionnaire

Piloted Fall 2010

(Coincidentally, the state of Kentucky had begun a process of university and community college articulation with student learning outcomes in 2010, and a few of those outcomes are listed here in addition to the ENG 105 objectives.)

Introduction/Description

All ENG 105 sections list common objectives for the course on the syllabus. As a teacher of ENG 105, you are the best source of determining whether students meet those objectives during the semester, since you are involved in all class activities aiming toward those objectives. The composition committee has designed this survey as a way to determine whether the ENG 105 curriculum helps students accomplish its stated objectives and student learning outcomes. Based on your responses, the composition committee can identify specific areas where faculty workshops or revision of those objectives may be beneficial to the program as whole.

The following survey should be completed after you have finished evaluating a set of student papers. This survey will be for *one formal paper during the semester OR the final paper*. Your responses to these surveys are anonymous. Please do not include any identifying information.

Again, complete this form after you have finished grading a full set of student papers.

Survey

On a scale of 1 to 5, with 5 representing freshman-level competence, answer the following questions about your class as a whole. Try to think in terms of trends: most students demonstrate . . . or most students don't demonstrate . . .

For this assignment, the majority of students:

Demonstrate the ability to write clear and effective prose in response to the writing prompt.
    1   2   3   4   5
Demonstrate their knowledge of and ability to use conventions ap-

propriate to audience, purpose, and genre.
    1   2   3   4   5

Demonstrate the ability to identify, analyze, and evaluate statements made by others, and integrate those views into their own work.
    1   2   3   4   5

Demonstrate the ability to construct informed, sustained, and ethical arguments in response to diverse points of view.
    1   2   3   4   5

Demonstrate effective planning, organization, and revision in developing this paper.
    1   2   3   4   5

Demonstrate the knowledge of rhetorical elements, methods, and aims of expository or persuasive writing.
    1   2   3   4   5

Competently examine complex ideas and situations in developing cohesive arguments.
    1   2   3   4   5

Demonstrate the ability to cite sources correctly and thoughtfully within their work.
    1   2   3   4   5

The following questions are not related to specific objectives or outcomes, but instead address how we as teachers read our student work. For this assignment, the majority of students:

Write in a way that I value as a reader.
    1   2   3   4   5

Exhibit writing that is compelling:
    1   2   3   4   5

    offers insight: 1   2   3   4   5
    enthusiastic: 1   2   3   4   5

As a teacher, what do you value in student writing? (open-ended)

Upon Completion: Thank you for participating in this survey. The data compiled will help in the development of future composition workshops.

Notes

1. We were most influenced by statements on contingent faculty conditions from the AAUP and MLA. These statements respond to the decline in

tenure-track positions and the concern that universities are exploiting part-time faculty without providing adequate professional conditions for teaching. For further information, see the AAUP statement on "Contingent Appointments and the Academic Profession" (http://www.aaup.org/AAUP/pubsres/policy-docs/contents/conting-stmt.htm) and the MLA "Statement on Non-Tenure Track Faculty Members" (http://www.mla.org/statement_on_nonten).

2. See the NCTE "Position Statement on the Status and Working Conditions of Contingent Faculty" (http://www.ncte.org/positions/statements/contingent_faculty) and "Part-time Faculty in English Composition: A WPA Survey" (http://wpacouncil.org/archives/05n1/05n1mcclelland.pdf).

3. See the Council of Writing Program Administrators' "Outcomes Statement for First-Year Composition": http://wpacouncil.org/positions/outcomes.html.

4. For details and comparison of the old and new curriculum, see: http://compositionforum.com/issue/24/murray-state-appendices.php#appx1.

5. See the Council of Writing Program Administrators' (with NCTE and National Writing Project) "Framework for Success in Postsecondary Writing": http://wpacouncil.org/framework.

Works Cited

AAUP. "Contingent Appointments and the Academic Profession." November 2003. Web. 31 May 2011.

Broad, Bob. *What We Really Value: Beyond Rubrics in Teaching and Assessing Writing.* Logan: Utah State UP, 2003. Print.

CCCC Committee on Assessment. "Writing Assessment: A Position Statement." November 2006 (Revised March 2009). Conference on College Composition and Communication. Web. 30 April 2011.

Ericsson, Patricia Freitag, and Richard H. Haswell. *Machine Scoring of Student Essays: Truth and Consequences.* Logan, UT: Utah State University Press, 2006. Print.

Gladwell, Malcolm. "The Order of Things." *The New Yorker* 14 Feb. 2011: 68 75. Print.

MLA. "Statement on Non-Tenure-Track Faculty Members." December 2003. Web. 31 May 2011.

NCTE-WPA. White Paper on Writing Assessment in Colleges and Universities. 2008. Web. 30 April 2011.

O'Neill, Peggy, Cindy Moore, and Brian Huot. *A Guide to College Writing Assessment.* Logan: Utah State UP, 2009. Print.

Petraglia, Joseph ed. *Reconceiving Writing, Rethinking Writing Instruction.* Mahwah, NJ: Lawrence Erlbaum, 1995. Print.

Williamson, Michael M., and Brian A. Huot. *Validating Holistic Scoring for Writing Assessment: Theoretical and Empirical Foundations.* Cresskill, NJ: Hampton Press, 1993. Print.
WPA. Outcomes Statement for First Year Composition. 2008. Web. 11 May 2011.
WPA-NCTE-NWP. Framework for Success in Postsecondary Writing. 2011. Web. 7 May 2011.

Part V. Third Spaces: Creating Liminal Ecologies

Operating "outside and alongside" writing programs are institutional "third spaces"—hybrid spaces such as writing centers, studios, summer bridge programs, academic student advising, libraries, student study groups, etc. that exist in-between traditional classrooms and other institutional spaces. Third spaces, reflecting the ecological attribute of interconnectedness, are the liminal, in-between sites that students inhabit and negotiate as they traverse various parts of institutional ecologies. Recent research on knowledge transfer suggests that third spaces function as networks of affiliation within writing ecologies and are important sites for facilitating the reflection and connection-making that can enhance students' learning as well as their ability to transfer and adapt knowledge learned in one context to other contexts. What happens in FYC, WID, and WAC courses is affected in consequential ways by their interaction with and interdependence on the kinds and structures of third spaces available to students. Too often, due to bureaucratic, disciplinary, and budgetary boundaries, third spaces exist in relative isolation, and so their unique ability to network various part of students' academic lives is minimized. The program profiles featured in this section of the book describe efforts at three institutions to cultivate and coordinate third spaces that connect writing to larger institutional ecologies.

In "A Collaborative Approach to Information Literacy: First-Year Composition, Writing Center, and Library Partnerships at West Virginia University" (published in 2009), Laura Brady, Nathalie Singh-Corcoran, Jo Ann Dadisman, and Kelly Diamond highlight the ecological characteristics of interconnectedness and emergence as they profile a partnership between the FYC program, the library, and writing center, and examine how professional and disciplinary boundaries can effectively be redrawn within institutional ecologies of writing to support students' multi-literacies. A network of writing faculty, tutors, and librarians at West Virginia University takes a trans-disciplinary approach to teaching research, reading, and writing as intertwined processes. This collaborative project encourages each member of the team to re-examine professional and disciplinary boundaries and to explore interconnections, resulting in the emergence of new assignments and activities that successfully engage students in researched writing; in addition, by expanding students' understanding of information literacy, the project also invites students to see how their composition courses are integral to other parts of their academic success. As the "Where We Are Now" coda indicates, however, redrawn boundaries and the emergence of new structures can be difficult to sustain within complex, fluctuating systems where multiple variables, such as budget cuts and personnel changes, operate. Yet the value of collaboration as a way to create new ecologies persists, as can be seen in the way librarians have used their partnership with FYC to continually update their research assignments and resources.

Daniel Sanford's "The Peer-Interactive Writing Center at the University of New Mexico" (published in 2012) profiles an innovative, emergent peer-interactive model of tutoring that realigns the goals of writing tutoring to more effectively embody collaborative, process-oriented views of writing and to reposition writing tutors and the writing center as a crucial part of the writing environment and culture of writing. As Sanford describes it, the peer-interactive writing center approach moves away from the one-on-one model and towards a format that encourages interconnections and relationships through genuine peer collaboration, recreates the writing center as a place to actually engage in writing, and encourages students in their intuitions about writing. In this way, Sanford redefines the writing center as a different kind of third space, an emerging and evolving structure that maintains the integrity of writing center work as distinct from writing classrooms but also coordinates the work done in the writing center and the writing classroom in ways that facilitate overlap between them. The profile provides an overview of the fluctuations and changes to the writing center as it moved towards a peer-interactive approach and reports from its assessment. In the "Where We Are Now" coda, Sanford reports on the latest assessment of the program and describes how the pedagogical changes to the writing center have resulted in physical changes in the ways students and tutors interact in this third space.

Published in 2013, the final profile, Jessie L. Moore, Kimberly B. Pyne, and Paula Patch's "Providing Access to College Capital: A Transitional, Summer College Writing Experience for Underrepresented Students," addresses a transitional space within a larger ecology of writing—a summer transition program that prepares non-matriculated students for FYC and access to "college capital." The ecological themes of emergence and interconnectedness are evident in this profile. Emerging from a partnership between a college access and success program (the Elon Academy) and a first-year writing program to reimagine a summer section of a required first-year writing course, the summer transition program worked to develop interconnections and relationships that would improve underrepresented students' opportunities for college success and retention. Importantly, the authors highlight the value of building opportunities for student reflection into transitional spaces, since transitions are consequential developmental periods that can allow students to see and seek connections between prior and new knowledge. The authors' "Where We Are Now" coda points to the value of keeping third spaces uniquely positioned to address transitions. Instead of continuing to offer an abbreviated version of FYC, the authors describe how their partnership with the Elon Academy has resulted in other forms of transitional support and the emergence of new networks of affiliation. In this way, this profile joins the other two in grappling with how to create and sustain third spaces that are both unique and interconnected within institutional ecologies.

13 A Collaborative Approach to Information Literacy: First-year Composition, Writing Center, and Library Partnerships at West Virginia University

*Laura Brady, Nathalie Singh-Corcoran,
Jo Ann Dadisman, and Kelly Diamond*

First-year composition is foundational in that it helps students hone their critical writing and research abilities. However, when we teach composition courses, we seem to teach research as a separate skill. In "Locating the Center: Libraries, Writing Centers and Information Literacy," University of Iowa librarian James Elmborg draws attention to the ways that information literacy fits into the general education of college students and argues that the teaching of writing and the teaching of research should be described and taught as one literacy practice. Members of the writing faculty, tutoring center, and the library at West Virginia University, a large public university, agree with Elmborg and have begun to put his ideas into practice by taking a team-approach to teaching research, reading, and writing as intertwined processes. This collaborative project forced each member of the team to re-examine professional and disciplinary boundaries. By focusing on the connections between and across disciplines (rather than divisions), the project demonstrated that writing teachers, tutors, and librarians could foster new literacies through innovative inquiry and collaboration.

THEORY INFORMING THE PROGRAM: DEFINING INFORMATION LITERACY

At the outset of the project we knew that "information literacy" could be a difficult concept to define. A recent article in the *Chronicle of*

Higher Education traces the term "information literacy" to 1989 "when the American Library Association called it a necessary skill and urged schools and colleges to integrate it into their curricula" (Foster A39). That narrow skills-based definition, however, has been challenged.

Stanley Wilder, associate dean of the River Campus libraries at the University of Rochester, contends that "information literacy remains the wrong solution to the wrong problem" (B13)—and his contention centers on a concern that "information literacy" can too easily focus on skill sets and protocols rather than strategies for engaging with reading and writing and thinking. He does not see students having to scale mountains of new information. In fact, Wilder says that we need to make a conscious effort to characterize students as *more* than "information seekers"; he suggests that we cast them as "'apprentices engaged in a continuous cycle of reading and writing'" (Dow; qtd. by Wilder B13).

As writing teachers and librarians who are collaboratively engaged in teaching research, reading, and writing as intertwined processes, we agree with Wilder's emphasis on a *cycle* rather than a skill set. We also think he is right to question what we collectively and professionally mean by "information literacy" and what is at stake as we continue to circulate that term.

For our collaboration, we started with the definition of information literacy offered in The American Library Association's *Presidential Committee on Information Literacy: Final Report*. The ALA report defines information literacy as a person's ability "to recognize when information is needed and have the ability to locate, evaluate, and use effectively the needed information." The Association of College and Research Libraries' Competency Standards for Higher Education elaborates on this basic definition by emphasizing the way individuals use information within the specific context of the collegiate environment. According to the ACRL standards, "[a]n information literate individual" is able to do the following six things:

1. Determine the extent of information needed
2. Access the needed information effectively and efficiently
3. Evaluate information and its sources critically
4. Incorporate selected information into one's knowledge base
5. Use information effectively to accomplish a specific purpose
6. Understand the economic, legal, and social issues surrounding the use of information, and access and use information ethically and legally.

Our collaboration has taught us that composition instructors play a key role in fostering students' information literacy processes. The act of writing requires students to determine what information they need to support their theses, how to organize that information, and how to present it in such a way as to be relevant to their audience. In other words, the processes of researching and writing are recursive, each one mutually informing the other.

REVIEW OF LITERATURE ON INFORMATION LITERACY

In a 2004 article in the *Journal of Academic Librarianship,* librarian Barbara D'Angelo and composition scholar Barry M. Maid describe "a natural partnership" between the libraries and writing programs (212). Such partnerships, they argue, can be an effective means of implementing information literacy across the curriculum. As we reviewed the existing scholarship on information literacy, we were thus a bit surprised to find little on the topic in the main composition and rhetoric journals. Just as D'Angelo and Maid's essay was published in the *Journal of Academic Librarianship,* university librarians and the professional publications in their field continue to take the lead on information literacy instruction, curriculum development, outreach, assessment, planning, and professional development for both teaching librarians and other faculty members.

In developing our program, we turned to several library publications that address the information seeking behavior of college students, the role of information literacy in rhetoric and composition instruction, and collaborations between faculty and librarians.

Among the more visible works on information literacy collaborations between writing faculty and librarians is the collection of essays edited by James K. Elmborg and Sheril Hook: *Centers for Learning: Writing Centers and Libraries in Collaboration.* As the title suggests, this collection gathers a wide range of case examples that show how writing centers and libraries can support students as they learn to find, evaluate, and integrate a wide range of information from various sources. The cases show how librarians and writing center tutors (individually and collectively) can guide students as they navigate information and come to understand what they do and do not know. The strategies outlined in the cases range from shared principles (Elmborg) and shared practices (Hook), to specific archive projects (Gannett, et al.) and examples of research and writing "clinics" (Boff and Toth). Only one chapter, however, explicitly goes beyond the library and writing center to include the classroom as a space for collaboration among faculty, students, tutors, and librarians (White and Pobywajlo).

Wendy Holliday and Britt Fagerheim, librarians at Utah State University, outline the process of collaboration between writing faculty and librarians in "Integrating Information Literacy with a Sequenced English Composition Curriculum." Their process starts with the assumption that information literacy instruction works most effectively when integrated into a course, and that writing courses that focus on research and argument lend themselves particularly well to this integration because faculty and librarians share similar goals. Holliday and Fagerheim describe their initial assessment of student, faculty, and librarian needs and how they used common themes and priorities to develop carefully sequenced instruction around specific student-learning objectives. Because the librarians recognized that the English instructors took varied approaches, they created several "core" and "supplementary" learning activities that were organized around goals and objectives rather trying to develop activities that corresponded to specific assignments (179–80). This goal-centered, course-integrated approach to information literacy suggests a model that can be adapted to a wide range of programs and disciplines.

In general, the literature on information literacy instruction emphasizes the need for a dynamic and flexible model that goes well beyond print-based bibliographic instruction. As John Perry Barlow, the co-founder of Electronic Frontier Foundation, explains, "Cyberspace is gradually teaching us that information is a verb, not a noun" (Albanese 44). Librarian Troy Swanson extends this idea in his essay, "A Radical Step: Implementing a Critical Information Literacy Model." Swanson encourages instructors—both content faculty and librarians—to make information literacy skills relevant to students' lives. We can do this by recognizing that "students enter our classrooms with their own experiences as users of information" (265). If we start with that common ground, it changes the dynamic of how we teach about information. For instance, source evaluation starts not with the printed page, but with the results screen as students "recognize and choose from the various types of information (scholarly, news, opinion, etc.) to best meet their information need regardless of format" (263). Swanson uses a variety of examples to remind us that students need to think about how information is created, stored, and circulated, and then consider these multiple factors as they select information that will be relevant for a particular audience and purpose.

Edward K. Owusu-Ansah, a reference librarian and coordinator of information literacy and library instruction, provides two excellent reviews of existing library scholarship. In his 2004 essay, "Information Literacy and Higher Education: Placing the Academic Library in the Center of a Comprehensive Solution," Owusu-Ansah examines the relationship between librarians and faculty as he addresses the question of "what role the academic library should play in achieving information literacy on campus" (3). The essay takes for granted the need for information literacy and the American Library Association's definition of the concept and, instead, focuses attention on the debate over the effectiveness of who should teach information literacy: subject-area faculty members or academic librarians? It can be a tense issue on campuses if librarians are seen primarily in support roles rather than as faculty members credentialed to teach their own subject. Owusu-Ansah proposes a comprehensive solution that consists of course-integrated components in combination with an independent credit course in information literacy taught by librarians. He notes that this phased solution builds on existing faculty-librarian collaborations and creates strong alliances across campus (10–11). Owusu-Ansah extends his argument in the 2007 essay "Beyond Collaboration: Seeking Greater Scope and Centrality for Library Instruction." In this second essay, Owusu-Ansah traces the success of and support for cross-campus collaborations that integrate information literacy into discipline-specific courses and contexts but renews his argument for credit-bearing library courses if the library is to take on a central role in education.

For the purposes of the collaboration among librarians, composition teachers, and tutors that we undertook, we end this review of literature with Rolf Norgaard's essay, "Writing Information Literacy in the Classroom." Norgaard is a faculty member in the Program for Writing and Rhetoric at the University of Colorado at Boulder. In his contribution as a "guest columnist" for the spring 2004 issue of *Reference and User Services Quarterly*, he notes that his title is meant to be provocative. He explains that he deliberately avoided the title, "Writing *and* Information Literacy" because he thinks that "both fields might benefit in important ways from eliding that 'and.' Each can and should 'write' the other" if we want to achieve "situated, process-oriented, and relevant literacy" and writing that stays connected to a rich civic, rhetorical tradition (225, 226). We can confirm the value of collectively writing information literacy based on the results of our collaboration among writing teachers, librarians, and tutors.

THE COLLABORATIVE CONTEXT: INFORMATION LITERACY WITHIN A REQUIRED WRITING COURSE

Since our review of literature revealed that librarians have taken the lead on information literacy scholarship over the past decade, it was not surprising that the librarians on our campus initiated our local conversations. In late spring of 2006, the head of the reference department approached the director of the writing program to ask how we might *together* develop a significant information literacy component for the second course in a two-course required writing sequence that focuses on research and argument. Both sides were eager for the collaboration. The composition instructors wished for ways to convince students that writing and researching each build upon the other, and the librarians believed that the best way to put information literacy into practice would be to integrate it into campus programs and student work. As a result, English faculty, writing tutors, and university librarians on our campus have worked together to scaffold assignments, activities, and responses that help students understand and value writing and research as intertwined processes.

To launch this information literacy collaboration, however, we decided that a seventy-five-section course taught to 1,500 students each semester by more than thirty different teachers was simply too large if we wanted to try new approaches across all sections, seek feedback from all those involved (students, tutors, teachers, and librarians), and then revise and seek more feedback. Instead, we turned to a new accelerated academic writing course that had recently been designed to help already-strong writers satisfy the university's two-course sequence in one writing requirement. The new course had the advantage of a smaller scale: eight sections a semester for a total of no more than 160 students, taught by a small group of four teachers who were committed to working together, meeting regularly, and developing a new curricula. It was the ideal size for students, tutors, teachers, and librarians to collaborate on information literacy.

Beyond offering a manageable size for our project, this accelerated writing course shares goals that are similar to the 75-section research and argument class—an important feature if our project proved successful as a model for other writing courses. All of our required writing courses draw on the WPA Outcomes Statement of rhetorical knowledge; critical thinking, reading, and writing; processes; and knowledge of conventions. We translate (and reorganize) the WPA Outcomes Statement just slightly to make the goals a bit more transparent to students:

- Goal 1: Understand Writing as a Process
- Goal 2: Argue Effectively and Persuasively in a Variety of Contexts
- Goal 3: Explore and Evaluate Ideas
- Goal 4: Integrate Research Effectively
- Goal 5: Know the Rules

To meet these five goals, students in our writing courses (probably like students in many composition courses) develop a portfolio of writing that consists of four major papers, informal writing done in and outside of class, and reflective writing. For the accelerated course, the four essays include:

1. a narrative based on personal experience;
2. a rhetorical analysis based on two texts;
3. a critical interpretation;
4. and a researched analysis on a topic of their choice.

The first three essays ask students to think, read, and write about some aspect of their lives and some texts that matter to them. The last essay asks them to think, read, and write about a larger issue while actively engaging with the ideas of others. All but the first essay requires some sort of outside source; each subsequent essay poses new rhetorical challenges and research questions.

Prior to the accelerated writing course, no formal partnership existed between the composition program and the library. In past semesters, individual instructors contacted librarians and requested their assistance in teaching students about available resources. They sought librarian expertise at the end of the semester when students were doing more extensive research for their final papers. This informal arrangement limited the instructors' and students' concept of information literacy to *locating information.* Evaluating and integrating sources were taught as separate concepts.

LIBRARIANS AS COLLABORATORS

In the pilot course (English 103), students start incorporating outside information by the time they are writing the second of four essays for the semester. The first information literacy session in the library thus focused on quick methods to get information; the emphasis was not so much on how to retrieve information, but how to *evaluate* the information. The teaching librarians showed students a few tricks for searching Google—such as restricting searches by using phrases and required terms. They also demonstrated how to find government documents on the web using Google's advanced search. A discussion of Wikipedia was also included in this session. To illustrate the way that anyone can edit Wikipedia, for instance, one librarian showed a clip of *The Colbert Report* which discussed the Word of the Day: "Wikiality." To illustrate the concept of "wikiality," Colbert explained, "I'm no fan of reality, and I'm no fan of encyclopedias. I've said it before: Who is Britannica to tell me George Washington had slaves? If I want to say George Washington didn't have slaves, that's my right. And now, thanks to Wikipedia, it's also a fact." Colbert typed in some of his own entries to Wikipedia to alter reality and encouraged others to log on and do the same. The example let the students see how fluid online information could be and why it's important to think critically about what's considered a "real" or "wikial" source. The librarians concluded the information literacy session by discussing how to evaluate information on the web and how to use Network Solutions to find the owner of a site.

To assess how well students grasped the session's concepts, and also to help them gather needed sources, the sessions taught by the librarians initially required two research logs: Research log one required an evaluation of a website and an article on Wikipedia. Research log two asked them to develop a focused research question. Eventually, students were required to complete a total of four Research Logs. All four comprised a total of ten percent of each student's final grade.

Before each of the library sessions, the 103 instructors introduced their students to the essay assignments. The rhetorical analysis (and by fall 2008, the persuasive essay) required limited external research. Students had several days to prepare for the session and bring a focused topic or several potential research questions. Both the instructors and the librarians stressed the importance of the library session in locating and evaluating potential sources early in the writing process. The second library session was held during week eight, immediately after instructors introduced the third essay. Students were given about a week to complete the research logs and submit them to their instructors to be mailed to the librarians.

The librarians who taught the sessions also asked to evaluate these logs. By participating as respondents to student writing, the librarians could see for themselves whether or not their sessions were effective (and then fine tune them for future students). The process of responding to student writing also reinforces the collaborative aspect of the course because it keeps the librarians in conversation with the writing teachers and, from the students' perspective, it underscores the authority and voice that the librarians have as part of the instructional team.

The students' research logs provided some key insights. For instance, some students found a disconnect between the library segment and their writing course; they didn't understand why they had to do two distinct assignments asking them to focus their research topic. Some found the research log questions asking them to evaluate websites and articles on Wikipedia "tedious" and "redundant." Other students were offended thinking that we were spending a class period teaching them how to search Google. Students also expressed a desire for handouts so they could replicate certain searches using Network Solutions.

Overall, however, the library sessions received positive responses. After each information literacy session, students were asked to complete a survey via Zoomerang, an online survey tool. Here's what we found in the first semester of the pilot (fall 2006):

- Seventy-eight percent said the information presented in the information literacy session was the right amount.
- Ninety-four percent said they clearly understood the benefits and risks of doing research on the Internet.
- Seventy-three percent said they gained a better understanding of strategies for evaluating websites.
- Sixty-three percent also said they now understood how to find a focus for their research.

After comparing the surveys to the students' research logs, the librarians decided to make some changes to the research log assignments for the spring semester 2007. The first revised information literacy session condensed the Google searching tips and added Lexis-Nexis; this database is a good choice for getting quick information as it is full-text and is easy to use. The revisions also changed wording in the lesson plan so the session focused more on *evaluating* web sources and less on searching skills. (For example, constructing searches in Google were presented as "quick tips.") Google's Advanced Search was shown as an easy way to find government documents on the Web. Librarians discussed evaluation of Web sources as a necessary strategy that college students should develop.

As for research log assignments, the librarians combined logs one and two by spring of 2007 and then designed new activities for two more sessions. The logs are now integrated more fully into student prewriting for their researched essays. The revised activities also streamline the website evaluation section and include a new question that asks students to find a newspaper article in Lexis-Nexis and evaluate its usefulness *for their essay*. In addition, the librarians now pass out handouts at the beginning of the research sessions to emphasize points during that class period and to provide a guide that students can use later in their own work outside of class. This handout includes the grading rubric for the revised first research log. In all, the course now has three research logs that have each been fine-tuned over four semesters in response to student comments and to support the key assignments in the course. The librarians focus the current research activities as follows:

- Research log 1: Evaluating Internet Resources/Using Lexis-Nexis (Appendix 1)
- Research log 2: Finding Books (Appendix 2)
- Research log 3: Finding Articles (Appendix 3)[1]

These materials are available online to our students and instructors. When the librarians who designed, taught, and evaluated the three research activities were surveyed at the end of the two-year pilot, all observed that the amount of information presented in each of the library sessions seems appropriate and that the pilot had helped them establish a model that should work well for other students in the larger research and argument class.

Writing Centers as Collaborators

Writing centers have often been perceived as spaces for remediation and/or grammar fix-it-shops. The information literacy pilot project provided an opportunity to alter these historical perceptions of writing center work by adding another layer to the collaboration. The students in the pilot course, Accelerated Academic Writing, were already strong writers. When strong students such as these are encouraged to use the writing center, we shift the emphasis away from remediation and grammar correction and instead emphasize that *all* writers—regardless of their writing proficiency—need good readers and responders.

The pilot also presented an opportunity for meaningful collaboration. Writing centers are often characterized as supplemental to writing instruction: Places designed to serve or back-up writing intensive courses. In practice, however, writing centers play a significant role in college writing as one of the few places where writing instruction happens beyond first-year composition. Centers are much more than adjuncts to the curriculum. Certainly, they support writing-intensive courses, but more importantly, they are partners in writing instruction.

The information literacy pilot emphasized the writing center's partner role. The writing center coordinator and a graduate student tutor both attended curriculum and development meetings, offered input on the information literacy component as writing center professionals, and revised the model assignments to reflect the expertise they provided as both teachers and tutors.

In this important partnership, our university made a conscious choice to follow the model of other campus-integrated tutoring spaces like the Stanford Writing Center, whose aim is to influence the campus culture of writing. The Stanford Center is significantly involved in first-year composition and general education courses. The center is recognized as a place where university instructors and professors can learn how to teach writing effectively. Because of the center's involvement at multiple levels, it plays a role in how students envision writing. According to Stanford's five-year longitudinal study of writing, students at the end of their undergraduate careers did not see writing as simply a means of recording ideas but rather "as a way of managing and making sense of enormous amounts of information and as a way of creating new knowledge" (Stanford Study of Writing). Because the Stanford Writing Center is instrumental to university-wide writing instruction, it played a sizable role in shaping the students' conceptions.

The English 103 Information Literacy pilot is one way through which our local writing center has had a hand in helping students re-envision writing and research. Before the semester began, three tutors were trained in information literacy. The librarians introduced these tutors to databases and search engines and offered the tutors advice on how to help students discover appropriate sources. In the first few weeks of the semester, tutors visited and promoted the center in each pilot section of English 103. No student was required to visit the center, but during the tutors' class visits, they emphasized how they could help students with their writing *and* research processes. Thereafter, each time an instructor introduced a new assignment or collected a draft, she would remind students of the writing center services. The students who took advantage of the center responded positively. For example, in an end-of-the-semester assessment of the course, one English 103 student spoke about what it means to construct an argument for a real audience and what it means to locate, evaluate, and incorporate information:

> [My fourth essay] was an argumentative paper about whether we should drill for oil in national preserves. After I researched both sides of the issue, I decided that "drilling in national preserves should not be allowed because the short-lived economic benefits do not outweigh the permanent environmental damage." This paper evolved from a peer-reviewed introduction to a nine page essay, mostly with the help of the Writing Center. At the Center, they advised me to include more of the opposing point of view, rather than just dismissing their points, which was a good idea.

In the fall 2006 semester, the center saw about sixteen of the 120 students enrolled in English 103 students. About half of the students asked for help with their persuasive research papers, and about a third asked for help with their portfolios. The total number of English 103 students who visited the writing center represents just over ten percent of the total course population. These averages are consistent with the percentage of first year writing students (those students who take English 101 and 102) who use the writing center each semester and are also consistent with writing center usage on a national average (*Writing Center Research Project*).

The writing center continues to be part of the ongoing collaboration with the librarians and the writing teachers. By the end of the first pilot year (spring 2007), one of the reference librarians teamed up with the writing center coordinator to implement a Writing and Research Clinic—an idea borrowed from the Bowling Green Writing Center. The Writing and Research Clinic combined two existing WVU services, the writing center and the Term Paper Clinic. Students who have questions about how to find resources and how to integrate sources can meet with a librarian at the Term Paper Clinic (held in the main campus library). The writing center also helps with research, but tutors often refer students to the clinic when the students need more resources than are available at the center or when they would further benefit from the knowledge of an information literacy expert. When the library and writing center partnered to offer the Writing and Research Clinic, one tutor and one librarian jointly met with students. This allowed students who had writing and/or research questions to meet with a librarian and a writing center tutor in one location and at the same time.

As we offer more sections of Accelerated Academic Writing, the writing center will continue to broaden its tutoring base to include information literacy and will thus be prepared to support the larger population of students in the standard research and argument class as that class benefits from the information literacy model that we are developing. In addition to some of the writing assistance already offered—like helping students in brainstorming research topics and integrating sources, the writing center will increasingly need to help first-year composition students locate and evaluate sources and use those sources effectively and ethically.

Beyond and outside of English 103, the pilot collaboration with librarians also suggests ways to revise the writing center mission statement to include the ACRL standards for information literacy, especially as information literacy becomes a university-wide focus.

It is equally important that the writing center continue to support not only students, but instructors and faculty, too. The writing center is already in the habit of teaming up with the undergraduate writing program to offer teacher workshops on topics such as "how to avoid plagiarism in the writing class" and "how to respond and evaluate student writing." The writing center, the undergraduate writing program, and the libraries will collaborate again to offer an information literacy workshop for those who teach writing and writing-intensive classes. And the Writing and Research Clinic jointly sponsored by the libraries and the writing center will certainly continue. Such partnerships reemphasize critical writing, reading, and researching as an integrated literacy practice.

The Importance of Feedback: Student-Guided Revisions to the Course

Instructors, tutors, and librarians all made a point of seeking student comments through the course in a variety of ways. In addition to the online surveys following the library sessions, we also invited sustained reflections at the middle and end of the term. The students let us know what was working. This first statement comes from a freshman engineering major—intense, anxious to learn, and full of academic drive. He says: "The library sessions have shown me more proper, reliable methods of researching my interests through the utilization of accredited, professionally-reviewed literary databases . . . [in order to] back my ideas with credible resources, rather than only well worded personal opinion."

A sophomore biology major had this to say about the first information literacy sessions: "The library sessions revealed a wealth of new resources, such as Google's advanced search and Network Solutions that have proven extremely helpful in gathering legitimate information on non-scientific topics." By semester's end, she reflected on her newly acquired research skills: "In high school I had no qualms with turning immediately to Wikipedia or any site that *looked* reputable for information to use in a paper. Now I go straight to EBSCO-host, JSTOR, or my personal favorite, the bookshelves."

As a result of the revisions to the library sessions, the student comments now reflect increasingly positive responses. In the student survey, most students now say they find the library sessions "informative." No one made any negative comments about being "taught how to use Google." The comments about revised research log one were also positive; many students commented that the activities helped them focus their topic. Our student survey this semester shows that our changes are yielding some strong improvements when compared to the first semester's numbers (fall 2006). Here's what the numbers tell us at the end of our two-year pilot program:

- One hundred percent of spring semester 2008 students say that amount of information presented in the library sessions is "the right amount" (which is up 23 percent compared to the response we received during the first term in fall 2006).

- One hundred percent of spring semester 2008 students say they understand the benefits and risks of doing research on the Internet (up 6 percent from fall 2006).

- One hundred percent of spring semester 2008 students say they understand strategies for evaluating websites (up 27 percent from fall 2006).

- Ninety-seven percent of spring semester 2008 students say they understand how to find a focus for their research problem (up 34 percent from fall 2006).

When students see a direct correlation to their researching and writing goals, they appear to embrace their expanded audience of librarians. They recognized the librarians as instructors and gatekeepers of electronic access and delivery systems, and as experts who could help them narrow a topic, find legitimate support for their ideas, and even serve as a sounding board for those ideas. As one student put it, "The library sessions revealed a wealth of new resources, such as Google's advanced search and www.netsol.com that have proven extremely helpful in gathering legitimate information on non-scientific topics."

Now that we have completed the two-year pilot, the instructor-librarian relationship is very solid and we have, as one of the librarians put it, established a model that we can now expand to other writing courses. As we think about scaling up from eight sections to eighty sections, not every instructor or student will know the librarians as personally as those who participated in the pilot, but the feedback provided by the smaller cohort has helped fine-tune activities, questions, and assignments that will ensure that any web-based tutorials will be responsive to specific student needs. We also plan to continue using student and instructor surveys to be sure that the crucial feedback component remains strong as we continue to revise and expand.

Reflecting on What We've Learned

The instructors teaching the pilot sections of English 103 provided feedback on the effectiveness of the information literacy sessions in teaching journals, surveys, and on-going conversations with the librarians. The instructors have been consistently positive, but their survey responses at the end of the two-year pilot are overwhelmingly enthusiastic:

- One hundred percent of instructors found the library instruction sessions and the research logs useful to students for completing their class assignments.
- One hundred percent noticed that the research sources their students selected are more on-target.
- One hundred percent noticed improvement in their students' research papers due to the library instruction the students received.

Equally compelling are the instructors' written comments over the two years of the pilot. Here's a quick sample: "My students and I collectively enjoyed the information literacy sessions. I believe the library sessions were well thought out and well executed, so kudos to the librarians." In terms of revisions, an instructor from the first year of the collaboration noted the need to think about how to use shared class time well: "The instructors need to prepare their students for the [library] sessions and keep their assignments relevant to the research plan introduced by the librarians." As a plan of action, yet another instructor acknowledged the role of the writing center in this project: "Next semester I will take my students to the writing center early on and push students to use the center as a way of staying on track."

One suggestion consistent among the instructors in the first semester of the pilot (fall 2006) was the need to relate the library sessions' research log activities directly to the students' topics for their major essay assignments. Instructors were aware that some students were unable to pursue the topic they used to complete their first two research logs, in effect creating an assignment that served no immediate purpose. Instructors realized that they needed to prepare students for the information literacy sessions by introducing the next major assignment, incorporating class discussion and inventing exercises *before* the students met with the librarians.

This collaborative project encouraged writing faculty, tutors, and librarians to re-examine professional and disciplinary boundaries, and resulted in new assignments and activities that successfully engage students in the intertwined processes of research, reading, and writing. We're glad we did it. We want to wrap up by sharing four lessons we've learned.

We are just completing the two-year pilot collaboration. By fall 2009, we hope to extend the faculty-librarian-writing center collaboration beyond the small number of accelerated academic writing sections to reach many sections of the required research-and-argument course that serves over 3,000 students a year. By fall of 2010, we hope that all sections of this large-enrollment course will benefit from a close association with the university libraries as well as our writing center.

At the outset of this project, the librarians and the writing instructors shared the hope that students would come to see the processes of research and writing as recursive and intertwined. This last instructor comment from spring 2008 suggests that we have made progress toward this shared goal:

> Students gained confidence from being able to gather good sources. This improved writing. And they also used research at several stages of the writing process, which was invaluable. Faced with organizational problems, problems with an argument or explanation, or just being stuck in a rut with a draft, they could go BACK to doing more research at this later stage.

This comment reflects a consistent observation made by both the librarians and the instructors: students research more thoroughly, evaluate sources more critically, and integrate that research and critical thinking as part of their writing process.

Conclusions and Recommendations

This collaborative project encouraged writing faculty, tutors, and librarians to re-examine professional and disciplinary boundaries, and resulted in new assignments and activities that successfully engage students in the intertwined processes of research, reading, and writing. We're glad we did it. We want to wrap up by sharing four lessons we've learned.

1. Collaboration depends on communication.

We learned that we needed several channels to keep communication lines open among all team members: a combination of emails, meetings, web-based resources, and regular surveys helped give everyone a voice.

Even so, it has been easy for the writing center to fall out of the loop this year when we no longer had a member of the writing center teaching one of the sections. We still advertise our services to the writing students in the pilot sections, but we lost the close collaborative connection. We are already thinking of things we can do in the future to make sure that someone from the writing center remains integral either as a teacher of one of the sections, or (on a larger scale) by implementing a writing fellows program where tutors visit course sections at strategic points during the semester to offer their perspectives on the writing and research processes, offer advice on particularly challenging assignments, assist with peer review, and so forth.

We are also aware that communication will remain our biggest challenge as we increase the number of teachers, librarians, and tutors involved in this collaborative approach to research and writing.

2. Collaboration depends on teamwork.

Instructors, tutors, and librarians must keep in mind that they are a team. What one person does affects the other members of the team. We learned this lesson the hard way when one instructor changed an assignment and didn't let the librarian know ahead of time. When the librarian delivered her lesson, it was no longer relevant for the students. Everyone was frustrated.

3. Collaboration depends on feedback—and lots of it.

Student course evaluations alone are ineffective for evaluating a program such as this one. The surveys of students, instructors, and librarians (and clearly we should add the tutors) help assess the effectiveness of the information literacy sessions and their ultimate impact on the work submitted for final portfolio. Instructors are also asked to keep a teaching journal and then write a reflection at the end of each semester. This is where even the most autonomous teachers tend to become aware of why teamwork has to take priority.

4. Collaboration can be time-intensive.

We admit that the communication and coordination that are so central to the success of this program both take time. We are lucky to have a very committed group, but it also helped to have some summer support to offer the initial team members, and we have kept course sizes as low as we can (about sixteen per section) as another way of supporting innovation and collaboration.

To sum up: the lessons we've learned tend to overlap. Communication builds community and teamwork; feedback helps solve problems that may pose challenges to communication and teamwork. The collaboration can be time-intensive, but it is also very rewarding. Would we do it all again? Absolutely.

Coda: Where We Are Now

Four years later, the collaborative information literacy project at West Virginia University remains largely successful and productive despite a few changes and challenges.

Fluctuations in Collaboration

As the editors of this collection note, change is inherent to ecological systems. Internal and external forces act on writing programs, and they must "adapt, transform, and evolve" if they are to endure. Our information literacy project has changed over time. One of the key members of the initial collaboration retired two years ago, another changed jobs, and sabbaticals and family leaves have created other breaks in continuity. We were, however, fortunate to have one writing instructor and several librarians continue with the project to help us maintain our momentum as well as our focus. Careful documentation also helped convey the purpose and significance of the project to new participants. We faltered, however, in our continued inclusion of the writing center. We were aware of this challenge in 2009, but retirements and leaves made it difficult to follow through on plans to make sure that someone from the writing center remained integral as a teacher of one of the sections. Budget and staff limitations have also made it impossible to implement plans for a writing fellows program. As a result, the writing center's presence diminished in the minds of librarians, the writing instructors, and students. The writing center nonetheless continues to promote its services to English 103 students each semester, recruits undergraduate peer tutors from the course, and remains committed to helping students re-envision writing and research as part of its ongoing tutoring work. The Writing and Research Clinic, renamed Research Help and Consultation, also remains a viable option for collaboration between the library and the writing center.

Continuity and Emergence

Our initial information literacy pilot was an emergence in and of itself; it was something new generated from its component parts: first-year composition, the writing center, and the library. However, because ecological systems are in a constant state of flux, emergence can happen at multiple points in time. While some facets of our collaboration remain the same, new pieces have come into being as our program has evolved. The team of writing instructors, librarians, and writing center representatives continues to meet at least twice each semester. The librarians have proven to be particularly dynamic collaborators. In addition to adapting their information literacy lessons as instructors updated their writing assignments, the librarians reviewed several sets of student papers to analyze student citation practices. As a result of their findings, they removed Google and Wikipedia as pre-search activities because the citation analysis showed that students were relying too heavily on these sources instead of exploring the proprietary reference resources that the librarians were also demonstrating. The new model now presents more expert and discipline-specific sources such as CREDO and Sage to expand students' knowledge base. Based on student and instructor feedback, the librarians also revised the initial short-answer research logs into research notebooks that asked students to do more drafting and reflecting. To remain mindful of the workload expected of the students, the librarians streamlined three research logs into two (slightly longer) notebooks. Along with this shift, the librarians also changed their response practices: they now hold individual conferences with each student. These conferences inform the librarian's evaluation of elements such as students' sources, research questions and search process, summary paragraphs, and participation. Students, instructors, and librarians are all responding positively to the conference system, which has given the librarians an active and visible role in these sections. While this is one of the most positive changes in the last four years, it is also the change that may prove most challenging to implement on a larger scale.

Looking Forward

We remain committed to a collaborative approach to information literacy even as we recognize the need to renew our partnership with the writing center, anticipate further changes in participants, and tackle the challenges of sustaining this model in a climate where resources remain limited.

As we now look beyond our own program to the complexities articulated in the profiles of other liminal ecologies in this section, we are reconsidering questions of agency. How might a "third space" allow WPAs to initiate change rather than simply reacting to new conditions? What new constructions or collaborations does it allow? These other third space program profiles are already helping us identify the conditions and collective actions that we want to promote and protect in our own environment. For instance, the peer-interactive tutoring model at the University of New Mexico suggests ways that we might reintegrate our writing center as a way of facilitating *groups* of writers very early in their writing processes—not unlike the way that the librarians currently work with groups very early in the research process. Likewise, the summer college writing experience at Elon reminds us of the importance of shared goals and shared reflections, two components that have proved central to the information literacy project. While our program's specific goals and environmental conditions differ from those at Elon and the University of New Mexico, all three profiles in this section illustrate how place affects what and how and when and why we write—as well as who writes to whom. The profiles also demonstrate how the writers shape the places where their writing occurs. This volume's emphasis on writing program ecologies reminds us to pay attention to shared genealogies and shifting conditions; it helps us analyze the causes and effects of change; and it encourages us to develop alliances for collective action.

Appendices

Appendix 1: Research Log 1: Evaluating Internet Resources / Using Lexis-Nexis

Name:
Section and Professor's Name:
Date:
I. Getting Started with Google (http://www.google.com/)

1. Write down as many words, phrases, or images you associate with education (think about grade, middle, and high school or college life). Consider how education is portrayed in the media (e.g., in movies, television, music). Also think about how your family and friends describe or talk about education. Consider ideas both inside and outside of the classroom.
2. Based on your lists above, what aspect of education would you like to research the most? Write a down a narrow and focused research question based on this topic.
3. What are the important keywords and phrases in this research question?
4. Come up with as many synonyms as you can for your key words and phrases and write them down.

5. Use Google's Basic Search to research your question and find a website *for your next essay.* Copy the search terms below exactly as you entered them into Google.
6. Browse through a few sites on the first page. Which site is the most relevant to your research question? Why? Be specific.

II. Evaluating Websites

1. Title and Address

Pick one web site among your Google search results that you'd like to use for your next essay.

- What is the title of the site and its URL (address)?

2. Authority and Affiliation

- Who is the author or sponsoring institution / organization listed on the webpage?
- Is this website a government site? That is, does it have a URL that ends in .gov?
- If it isn't a government site, go to Network Solutions (http://www.netsol.com) to find out who owns the site. Write down the owner of the site and the contact information.
- Do you think this person or organization is a credible source of information? Why or why not? Be specific

3. Timeliness

- When was the page first published or last updated?
- Does the date matter? Why or why not? Be specific.

4. Coverage

- What questions do you have about the topic that the website does not answer? Be specific. ("No questions" is not an acceptable answer.)

5. Final Analysis

- Do you think your professor would want you to use this page? Why or why not? Be specific.

III. Lexis–Nexis

At the WVU Libraries' homepage (http://www.libraries.wvu.edu), under Popular Databases select Lexis-Nexis. Make sure you use Guided News Search for this exercise.

For this exercise, you'll find a newspaper article for your next essay.

1. Write down your search terms exactly as you entered them into the Lexis–Nexis search.

2. How many records are found?
3. Are these results too few, too many, or just right? Explain why.
4. Skim through your results and choose a news article you'd like to use for your essay. Write down its information:

   Name of the newspaper:
   Issue Date:
   Section:
   Length:
   Headline:
   Byline:

5. Why would this be a good article for your essay? Be specific.

APPENDIX 2: RESEARCH LOG 2: FINDING BOOKS

Name:
Section and Professor's Name:

Research Question:
For this exercise, you'll search MountainLynx Catalog, go to a WVU Library, find two books, and evaluate their usefulness for your next essay. Go to the MountainLynx Catalog at http://mountainlynx.lib.wvu.edu

1. Select Assisted Search and write down the words/phrases exactly as you typed them in the blanks:
   How many books did this search retrieve?
2. From your list of results, select one book that is "Not Checked Out" and write down the following information:
   Author(s)/Editor(s):
   Title:
   Publisher:
   Date & place of publication:
   Call Number:
3. Go the Library and find the book. Look over the table of contents, the index, and skim the Preface or Introduction chapter; this method is a good way to get an understanding of a book's content.
4. Now write a well-developed paragraph, using specific reasons, why or why not this book would be a good choice for your essay. Discuss how the information in the book will or won't help you answer your research question.

5. Browse the other books shelved nearby. Write down the information of one additional book that might be a good source for your essay:
   Author(s)/Editor(s):
   Title:
   Publisher:
   Date & place of publication:
   Call Number:
6. Review the book as you did in question 3. Now write a well-developed paragraph, using specific reasons, why or why not this book would be a good choice for your essay. Discuss how the information in the book will or won't help you answer your research question.

Appendix 3: Research Log 3: Finding Articles

Name:
Section and Professor's Name:

Research Question:
For this activity, you'll search 3 databases for articles, choose two articles, and evaluate their usefulness for your next essay.
1. Choose 3 databases to search. (Remember that EbscoHost isn't a database as such, but a collection of databases. You can pick databases within EbscoHost, like Academic Search Premier.) What 3 databases did you choose and why did you choose them? Please be specific; think about how the scope of the databases relate to your research question.
2. What search terms did you use for each database?
3. Where you satisfied with your results? Why or why not? Please be specific.
4. Which database seemed to be the most helpful? Please be specific.
5. Choose two articles from your results. Write down the full citation information for each article.
   ARTICLE 1
   Author:
   Title of the article:
   Name of the journal/magazine:
   Volume number: Issue number:

Page number: Date of issue:
Today's date (when the article was retrieved):

Read or carefully skim the article. Now write a well-developed paragraph, using specific reasons, why or why not this article would be a good choice for your essay. Discuss how the information in the article will or won't help you answer your research question.

ARTICLE 2
Author:
Title of the article:
Name of the journal/magazine:
Volume number: Issue number:
Page number: Date of issue:
Today's date (when the article was retrieved):

Read or carefully skim the article. Now write a well-developed paragraph, using specific reasons, why or why not this article would be a good choice for your essay. Discuss how the information in the article will or won't help you answer your research question.

Note

1. To see the grading rubrics for Research Logs 1, 2, and 3, see: http://compositionforum.com/issue/19/west-virginia.php

Works Cited

Albanese, Andrew Richard. "Cyberspace: The Community Frontier." *Library Journal* 127.19 (2002): 42–44. Print.

The American Library Association's Presidential Committee on Information Literacy. *Presidential Committee on Information Literacy: Final Report.* 2005; 2007. American Library Association. Web. 11 June 2008.

Association of College and Research Libraries (ACRL). "Information Literacy Competency Standards for Higher Education." 2000. *American Library Association.* Web. 11 June 2008.

Boff, Collen and Barbara Toth. "Better-Connected Student Learning: Research and Writing Project Clinics at Bowling Green State University." Elmborg and Hook 148–57. Print.

Colbert, Stephen. "The Word: 'Wikiality.'" *The Colbert Report*. Comedy Central. 30 July 2006.

D'Angelo, Barbara J., and Barry M. Maid. "Moving Beyond Definitions: Implementing Information Literacy across the Curriculum." *Journal of Academic Librarianship* 30.3 (May 2004): 212–216. Print.

Elmborg, James K. "Locating the Center: Libraries, Writing Centers and Information Literacy." *Writing Lab Newsletter* 30.6 (February 2006): 7–11. Web. 11 June 2008.

Elmborg, James K., and Sheril Hook, eds. *Centers for Learning: Writing Centers and Libraries in Collaboration*. Chicago: Association of College and Research Libraries, 2005. Print.

Gannett, Cinthia, Elizabeth Slomba, Kate Tirabassi, Amy Zenger, and John C. Brereton. "It Might Come in Handy. Composing a Writing Archive at the University of New Hampshire: A Collaboration Between the Diamond Library and the Writing-Acoss-the Curriculum/Connors Writing Center, 2001–2003." Elmborg and Hook 115–37. Print.

Holliday, Wendy, and Britt Fagerheim. "Integrating Information Literacy with a Sequenced English Composition Curriculum." *portal: Libraries & the Academy* 6.2 (Apr. 2006): 169–184. Print.

Norgaard, Rolf. "Writing Information Literacy in the Classroom." *Reference & User Services Quarterly* 43.3 (Spring 2004): 220–226. Print.

Owusu-Ansah, Edward K. "Beyond Collaboration: Seeking Greater Scope and Centrality for Library Instruction." *portal: Libraries & the Academy* 7.4 (Oct. 2007): 415–429. Print.

—. "Information Literacy and Higher Education: Placing the Academic Library in the Center of a Comprehensive Solution." *Journal of Academic Librarianship* 30.1 (Jan. 2004): 3–16. Print.

Stanford Study of Writing. "Preliminary Findings." 2006. Web. 11 June 2008.

Swanson, Troy. "A Radical Step: Implementing a Critical Information Literacy Model." *portal: Libraries and the Academy* 4.2 (2004) 259–273. Print.

Wilder, Stanley. "Information Literacy Makes All the Wrong Assumptions." *Chronicle of Higher Education* 51.18 (7 Jan. 2005): B13. Print.

White, Carolyn, and Margaret Pobywajlo. "A Library, Learning Center, & Classroom Collaboration: A Case Study." Elmborg and Hook 175–203. Print.

"WPA Outcomes Statement for First-Year Composition." *Writing Program Administration* 2 3.1/2 (Fall/Winter 1999): 59–63. Print.

Writing Center Research Project. University of Louisville. Web. August 2008. <http://coldfusion.louisville.edu/webs/a-s/wcrp/>.

14 The Peer-Interactive Writing Center at the University of New Mexico

Daniel Sanford

Heading into the second decade of the twenty-first century, the typical university writing center stands firmly on two theoretical feet: expressivism, with its focus on the writer's experience of self-discovery, exploring, and learning in the creation of a text (and the resulting authenticity and writer ownership of the resulting text), and social interactionism, which emphasizes writing as a social act most fruitfully approached as a collaborative exercise. The former advocates the engagement of writing tutors with students' processes for writing and stresses invention and drafting in tutoring sessions (see Peter Elbow, Ken Macrorie, Donald Murray, and Stephen North). The latter, in the context of the writing center, focuses on the role and relationship of the tutor and tutee, advancing a view of the ideal writing tutor as a peer consultant who downplays their own authority in order to act as a co-learner in the enterprise of writing (see Andrea Lunsford). The two approaches work in concert to develop students' sense of authority as writers. Students navigate a process for bringing their own ideas to fruition that includes both individual exploration (as in student-led sessions in which tutors make liberal use of the Socratic method to draw out student writers' intuitions about writing) and participation in a dialogue that brings ideas into beneficial conversation (as in peer-review sessions facilitated by writing tutors).

These ideas are no longer merely formative. While writing program administrators certainly still work every day to combat older notions (grounded in traditional rhetoric and a view of writing programs as remedial) of what a writing center is, what it does, and who it serves, these are the accepted pillars of current best practice, also serving as the axioms upon which newer, writing across the curriculum (WAC)-oriented approaches to writing center theory are based. Every semester, hundreds of new writing tutors across North America and around the globe read Stephen North's "The Idea of a Writing Center" and march off to their first sessions with students under the aegis of "Better Writers, not Better Writing!" Tutors are prepared in meetings and training programs to engage with student texts as outputs of the process whereby they are created rather than as ends in and of themselves, and to provide direction that will help students to grow and develop as writers rather than merely to turn in a more polished piece of writing in the immediate future. While these ideas continue to evolve and to be challenged and expanded upon in the literature, a clear state of the field incorporating the above ideas has arrived, accompanied, within our field, by a broad consensus as to our shared values and goals.

There is one place, however, where outdated views of the writing center persist, and it's not a discipline, academic department, or institution. It's the mainstay of the writing center, the core of our practice, and the main way in which writing tutors engage with student writers: the individual consultation. With respect to both engaging with students' writing processes and to fostering a view of writing as a social act, individual consultations can fall short of (and, in fact, undermine) our goals.

There's an enormous amount of variation in how individual consultations are practiced in different centers, but there are two defining aspects. First, students work one-on-one with writing consultants; sessions may, on occasion, involve more than one student, but this is the exception rather than the rule, and the models for tutoring applied by consultants assume a single student writer. Second, individual consultations have time constraints, either explicitly (being scheduled in set blocks of time) or implicitly (as in a walk-in center where tutors are working with students on a first-come-first-served basis). The first aspect means that students aren't exposed to other students' writing in a typical visit, and that they receive feedback only from a single person. The second aspect encourages students to do their writing outside of the center and to use the limited time of an individual appointment to receive feedback focused on a draft. Wonderful things can and do happen within individual consultations, and there will always be a place for individualized sessions in a center that aptly serves all comers. However, the format, by its nature, encourages draft-centered sessions focused on tutor feedback and discourages engagement earlier in the writing process and across peers.

In an attempt to move away from one-on-one, tutor-tutee sessions that occur within set time limits, we at the University of New Mexico implemented a new format for tutoring, the peer-interactive writing center approach, in order to promote informal peer-review sessions, scaffolded group work, and the direct support of students in the process of writing itself. The methods of the peer-interactive writing center approach, as exemplified in what we've implemented and marketed at the University of New Mexico as the Writing Drop-in Lab, skew things in the other direction: towards peer feedback and engagement with the writing process itself, and away from sessions focused on tutor feedback. Within the writing center at the University of New Mexico, the Writing Drop-In Lab is an experiment with creating a model for peer tutoring that places the goals and methods of the peer-interactive writing center approach at the center of practice. The approach is built on methods already applied in excellent academic writing centers that set aside an important place for student peer reviews[1] as well as for group sessions[2] and is also informed by the methods used in many *non*-academic writing centers that cater to the needs of the writing community at large (foremost among these methods being workshops in which participants share and critique one another's writing).[3] It is kindred in spirit to the movement towards classroom-based writing tutoring (CBT), which advocates for the development of programs that place peer tutors into classrooms where they can engage with a wider cross section of the student population and support and educate faculty on the effective integration of writing into a course (see Margot Iris Soven, as well as Candace Spigelman and Laurie Grobman) in that it challenges long-standing assumptions about the nature of a writing center, and in particular the role of a peer tutor. Whereas CBT, however, focuses on what takes place outside of the walls of the writing center, the peer-interactive writing center approach provides a model primarily for what happens within it. Our name for our version of this approach, the Writing Drop-In Lab, enthusiastically adopts the "lab" designation, adopting a metaphor whereby a writing center is understood not merely as a place to discuss and learn about writing, but to actively engage in it. In what follows, I will describe the Writing Drop-in Lab, report on findings from its assessment, and look ahead to its future.

Overview of the Writing Drop-in Lab

Within typical writing center appointments, students work individually on their writing elsewhere (in computer pods, in their dorm rooms, at their parents' homes, in coffee shops, and many other locations), get feedback and guidance from a tutor in the writing center, and then leave the center again to incorporate the feedback. The focus in the UNM Drop-In Lab, on the other hand, is for students to write, either on their own or in groups, in the lab itself, with tutors close by to offer assistance as it is needed.

As I describe in greater detail below, the Writing Drop-in Lab is located within the UNM Writing Center, which is itself part of the Center for Academic Program Support. Students make no appointments for visits to the Drop-in Lab; students are welcome to come at any point during its hours of operation (these currently totaling 49 hours/week) and stay for as long as they like (sometimes for five minutes to ask a brief question, often for hours at a time to complete a substantial amount of writing). The space is a mix of tables and workstations with a laptop set up at them; a student receptionist logs in students as they enter the space. The Drop-in Lab is staffed at all times by between 1 and 3 (depending on the typically attested utilization of the lab at any given time) peer tutors who circulate among the anywhere from 1 to 20 students in the space, offering input and guidance as it is requested by students. Tutors are trained to frame their input as brief statements of useful principles, based on a review of the student's writing, and then to leave the student to work on integrating the feedback while the tutor attends to other students. Thus, for example, a tutor might respond to a student's raised hand by reviewing the introduction that the student has just written in the lab, briefly talking about general guidelines for introductions and statements of thesis, and then moving on to begin a conversation with another student while the first student writer works to apply the tutor's feedback.

Tutors are also trained to provide scaffolding guidance for students who are writing in groups, and to facilitate peer review among student writers. Most tutors working in the center are undergraduates, in the interest of preserving a peer dynamic. One notable exception to this is the graduate students who act as student managers, greeting students as they come in and orienting them to the space, inviting them to get to work writing, and advising them that tutors are available as needed. The other major role of these student managers is to seat students from the same course or from similar courses together at tables and adjacent workstations. Peer review arises gratifyingly often on an informal basis in this environment, with minimal prompting from student managers and tutors. [4]

Within this environment, then, two activities are explicitly fostered: writing, and talking about writing. Writing forms the greater part of activity in the center; conversations about writing take place both as student-tutor interactions, and as student-student interactions (the latter albeit with frequent scaffolding from tutors).

GOALS OF THE WRITING DROP-IN LAB

The Writing Drop-In Lab was established with the following five goals:

1. To implement in practice a view of writing as a process

Teaching writing as a process rather than as a final product has long been a goal of writing centers, as well as composition and WAC programs. Despite a long-standing policy of advertising the writing center as being available for students at any stage of the writing process, both students and tutors tend to think of a completed draft as prerequisite to a productive tutoring session. The vast majority of individual appointments take place around such a draft, and students often feel that they can't make an appointment until they have a draft completed. The reasons for this are understandable; with a set amount of time during which to conference with a tutor, the choice to complete a draft beforehand so that the session itself can focus on feedback is a shrewd one. The outcome of this, however, can be unfortunate. Sessions focused on completed drafts reinforce an idea that the writing process is important only insofar as it contributes to the final form that the finished product takes. Sessions that take place during the process leading up to a completed draft reinforce another idea: that development as a writer is in large part a matter of developing one's process of writing. Within the Writing Drop-in Lab, students with no constraints on the amount of time that they can spend in the center use the Lab to engage in the actual creation of texts. Rather than being placed into the role of commenting on a completed (for example) statement of thesis or concluding paragraph, diagnosing issues in the drafting process and making suggestions to be implemented in revision, tutors are able to engage students before or while they write, reminding students of key aspects of a statement of thesis or guiding students through rereading the body of a paper as a wind-up to writing a concluding paragraph.

Knowledge of the process of planning, crafting, and revising written communication is an essential aspect of development for college writers. Tutors engaged with a completed draft see only an outcome of this process. In working with student writers who are actively writing and pre-writing, tutors can engage with the writing process itself, diagnosing issues related to how students approach writing. Additionally, students who have completed a draft tend to want sessions focused on surface issues (grammar, spelling, etc.), a preference often shared by tutors who find such issues easier to diagnose than concerns relating to style, organization, and audience. Sessions that take place earlier in the writing process allow tutors to provide input on these deeper issues.

2. To embody a view of writing as a collaborative act

A core guiding principle of the "new" writing center, as informed, in particular, by social interactionism, is that writing is most effective when it is surrounded by conversation and discussion. Facilitating peer interactions among student users creates a dialogue around writing that helps to lead student writers away from a view of writing as a solitary activity and towards a view of writing as a social activity.

In traditional individual appointments, the focus is on one-on-one interactions. Group sessions occur, but they are the exception rather than the rule. In the Drop-In Lab, by contrast, students often casually consult one another on their writing. Tutors facilitate peer review sessions for groups of students working on the same assignment, and students working on group projects meet in the presence of a tutor who can help to facilitate genuine collaboration.

One of the most common laments of writing center directors—the rush of students who come into the writing center in the last few days before an assignment is due—becomes an asset, as students from the same or related courses are seated in groups to discuss the assignment, review one another's work, and create a dialogue around writing.

3. To validate student users as intelligent commenters on ideas and writing

Peer review has the obvious reward of providing feedback for the student writer. The more important benefit is also the more commonly overlooked one: commenting on other students' work validates student writers as authorities on writing, able to formulate meaningful and helpful responses to not only their peers,' but also their own, writing. In addition, the practice fosters students' metacognition, which can facilitate the transfer of skills. The approach addresses the tendency of individual appointments to foster inherently unequal power dynamics (see Michael Joyner, Carol Stanger, Susan Strauss and Xuehua Xiang, and Caroline Walker). Ultimately, by shifting the authority away from tutors and toward the student users, the writing center creates more independent writers.

4. To advance a view of the writing center as a place to write, rather than as a place to get writing fixed

The much-decried view of the writing center as a fix-it shop is, in some ways, a consequence of a model in which students write, then see a tutor, and then revise based on the session. Advertising the writing center as a place for students to work on their writing fosters a healthier image of the center, as a place for the student to learn and engage in good writing practices.

A major approach for tutors in the Drop-In Lab is to work with a student for ten to fifteen minutes, discussing and reading over the assignment and working with the student user to prioritize areas to focus on in the session. The tutor then leaves the student user to work independently for another 15 to 30 minutes (often while the tutor goes to check in with another student) before checking back in to see how feedback has been implemented.

5. To increase student usage by more effectively utilizing tutors' time

By freeing up tutors to work with multiple students simultaneously and allowing students to drop into the writing center without having previously made an appointment and at whatever time is convenient for them, the approach has the potential to greatly increase tutor utilization in the writing center. Writing center staff members make use of workshopping, peer review, and group learning techniques without creating an undue load on themselves—and in doing so, meet each of the above goals more effectively than they can within individual appointments.

Implementing the Writing Drop-In Lab at the University of New Mexico

The University of New Mexico is a large, public, research university serving a highly multicultural and multilingual student population. The writing center, of which the Writing Drop-in Lab is a part, has existed since 1981, as part of a larger learning center. It is a large writing center, currently seeing 2,000+ visits/semester. The Writing Drop-In Lab was instituted with the above-mentioned goals in fall of 2009. The sections that follow provide a brief overview and timeline of the lab's development. Table 1 presents total student visits to the main location of the UNM writing center, from fall 2008 to fall 2010.

Table 1. Visits to the Writing Center, Fall 2008 to Fall 2010

Semester	Individual Appointments	Writing Drop-In Lab
Fall 2008	1643	NA
Spring 2009	1321	NA
Fall 2009	858	536
Spring 2010	525	582
Fall 2010	628	1416

Fall 2008, Spring 2009

In all semesters prior to the fall of 2009, writing tutoring was available in two formats, both standard for writing centers nationally: individual consultations, and walk-in tutoring. In individual consultations, students come to or call the front desk, where they are scheduled for an appointment lasting for 25 or 50 minutes with a writing tutor. During walk-in times, students do not need an appointment to see a tutor. Tutoring is individual, such that at busy times students wait to work with the tutor in the order they arrive. Both formats, as practiced in our center, were and are guided by current best practices: tutors engage the students in a dialogue around writing, working together to negotiate the session.

The writing center at UNM is, as mentioned above, situated within a larger learning center (the Center for Academic Program Support, or CAPS). While at many institutions the struggle to separate writing tutoring from content tutoring has been a positive one, signaling a program coming into its own, the emergence of the peer-interactive writing center approach points to the possibilities for cross-pollination of ideas that can emerge when tutoring programs share a roof. STEM (Science, Technology, Engineering & Mathematics) tutoring at CAPS takes place largely on a drop-in basis, with the student population the program serves being well educated on and well-accustomed to utilization of the tutoring center as a place to come multiple times per week for long stretches of time to do their homework within a scaffolded environment. The exigence for the peer-interactive writing center model was, for the author (at the time, a newly hired professional staff member inheriting a well-regarded program), a combination of the issues noted above with individual sessions, and the propinquitous example of the math and science tutoring labs. I developed a proposal for the Writing Drop-in Lab that operated within the existing budget for the writing center (which remains true through the present, although that overall budget has grown as a response to the highly increased student visits created by the Drop-in Lab), spent the spring and summer of 2009 working with a supportive director to implement it, and assumed administration of it as a part (and, eventually, the core) of the overall writing center in fall 2009.

The writing center has several satellite locations around campus, but its main location (where individual appointments were, and continue to be, held, and the space within which the Writing Drop-in Lab was implemented) is within one of the main university libraries.

Fall 2009

The Writing Drop-In Lab, with its peer-interactive format, was implemented in the fall of 2009. The lab, located within the larger writing center, was open during the hours of 12 pm to 4pm Monday through Saturday: hours of peak utilizations for the writing center (the full hours of which extended from 9 am to 6 pm). During these hours, individual appointments were no longer available for writing, and receptionists at the center were instructed to refer students seeking appointments during these times to the Drop-In Lab.

Six laptops were purchased, placed out during drop-in hours as workstations for students (along with two desktops, providing eight total workstations). In advertising, students were encouraged to bring their own laptops or to work on one of the writing center's computers. In all campus outreach (class visits, e-mails to the faculty listserv, and printed writing center materials), the Writing Drop-In Lab was advertised as a place for students to engage in writing, either alone or in groups, and with or without the assistance of writing center staff. Our primary printed outreach tool, the writing center hours flier, very briefly introduces students to some of these core aspects of the drop-in format.[5]

The major challenge of the first semester was training the writing center staff on the new format of tutoring, which included persuading them of its merits so that their motivation to use the tools it offers was internal rather than enforced from above. Training for fall 2009 began with an introduction to the new format of tutoring. In training the tutors, I focused on the theory behind the approach, and on strategies for the tutors to use in engaging with writing earlier in the writing process (e.g., drawing out critical information from a prompt, reading sources critically, and pre-writing activities such as freewriting and clustering) and diagnosing issues in the writing process (e.g., beginning to write without a plan of action, a linear approach that impedes students who feel that they must write an introduction before they can proceed to body paragraphs, or a crippling focus on surface issues in the early drafting stages). As a rule, new tutors took to the new format readily, while returning tutors (in particular those who had spent more than a few semesters tutoring in the center) were acculturated to intensive one-on-one interactions with students and had a difficult time making the switch to the new format. In many cases, tutors who were enthusiastic about the new approach still had a difficult time not treating the new format as simply first-come, first-served sequential one-on-one sessions.

A veteran graduate student writing tutor was given a position as the student manager for the Writing-Drop in lab, supervised by me. This manager's duties include overseeing the flow of student usage in the Drop-In Lab, assisting with training the tutors on the new format of tutoring, orienting new student users to the new format of tutoring, and providing oversight for the lab.

In terms of usage, as indicated in Table 1, the overall number of visits at the writing center remained steady from previous semesters, as the number of drop-in visits greatly increased and the number of individual visits decreased correspondingly.

Spring 2010

Heading into the second semester of the Writing Drop-In Lab, the writing center staff was composed entirely of tutors who had either spent at least one semester being exposed to the drop-in format, or who were new to the center and therefore saw nothing unusual in the Drop-In Lab format. Time, training, and turnover had effectively dealt with the issue of tutor resistance to moving from the familiarity of one-on-one tutoring to the new system, and the emerging issue became educating student users on the new format of tutoring. The focus of spring 2010 was to counter students' expectations that a visit to the writing center meant one-on-one time focused around a draft and also to address reticence around sharing writing. The tutors, the student manager mentioned above, and I invested time in educating new and returning student users on the Writing Drop-In Lab: what it is, and how to use it. In class visits, tutors explained the concept of the peer-interactive approach to students, explaining that they can use the writing center as their primary place to write, and that tutors can help students to work productively in groups. They also worked to set up the expectation that students will share their writing with other students. Additionally, in-class and out-of-class workshops focused on leading students in peer review, so as to expose students to and model the pedagogy of the center. In the Drop-in Lab itself, the tutors and student manager worked to counter expectations among student users that a draft was prerequisite to visiting the center, and that students needed to leave the center in order to incorporate revisions. In all promotions, a central message is that writing at home alone leaves an author with little recourse when they are uncertain of how to proceed, whereas in the center there are always other students and tutors from whom to seek guidance in moving forward.

In the fall of 2009, all writing tutors were expected to spend some time engaged in drop-in tutoring. In the spring, more attention was paid to tutors' relative aptitudes for drop-in vs. individual tutoring. In some cases, tutors' reticence to circulate among students was a matter to be amended with training; in others it was a better solution to allow tutors for whom one-on-one interactions was a strength to remain focused on individual sessions. Tutors demonstrating a knack for collaborative learning strategies and building a group dynamic had their time prioritized for the Drop-In Lab. However, a major concern voiced by tutors, addressed in training for subsequent semesters, was a need for tools and strategies for engaging students in interactions with one another.

Fall 2010

A central goal for the first year had been to acculturate students to the new format of tutoring, encouraging them to take advantage of it by making it the only format of tutoring available during peak hours. For the fall of 2010, the hours at the Drop-In Lab were expanded considerably, to encompass the full hours of operation of the center: 9am to 7pm Monday through Friday, and 12 pm to 4 pm on Saturday. Whereas in previous semesters the Writing Drop-In Lab and individual appointments were offered on a complementary schedule, for the fall of 2010 both services were offered concurrently throughout the day (such that students could choose between the two services for any given time, and allowing for a preference for one-on-one sessions among individual student users).

Visits increased dramatically. Visits to the Drop-In Lab nearly tripled from fall 2009. Interestingly, this happened without a sharp decrease, over the same period, in visits to individual appointments. Fall 2010 was, by a handy margin, the writing center's busiest semester to date.

Fall 2010 saw three new locations for the Drop-In Lab's peer-interactive format hosted by other campus organizations (a student ethnic center, the university advisement center, and a computer pod in the student union building). These locations were opened with the rationale that writing tutors could most effectively engage with the writing process if they go to where students write, rather than wait for the student to come to them. Tutors in these locations circulate through the space, making themselves available to student writers who, having come to use the computers rather than to engage with a tutor or their peers, may or may not avail themselves of the tutors' services. This program continues to be a success, effectively reaching out to new students and new student populations who may not otherwise have made use of the main location of the writing center. An emerging lesson, however, is that the writing drop-in format (as follows from the interactive model on which it's based) works well only above a certain critical mass, in terms of the volume of student users making use of the space. At slower locations it is less suited.

Fall 2010 was an exciting semester for seeing the fruits of previous semesters' efforts. The major focus in training sessions was developing tools for facilitating groups: peer review, scaffolding group collaboration, and workshopping model texts. Whereas in previous semesters much of the tutoring in the Drop-In Lab remained interactions between one tutor and one student user, it's now increasingly common to see students working in groups—indeed, students now sometimes arrive at the center in groups, something almost unheard of as recently as 2008. Many more students come to the writing center to write, in some cases staying for hours and only occasionally asking for assistance from a tutor.

While signs such as this seemed to indicate that the Writing Drop-in Lab was successfully changing how students used the writing center, a programmatic assessment was clearly called for in order to establish whether the program was meeting the goals that it had been designed to meet. This assessment was created and implemented in the fall of 2010.

Assessment of the Writing Drop-In Lab

The Drop-In Lab's five goals presented above were assessed using a survey distributed to the writing center staff, a survey distributed to student users of the Writing Drop-In Lab, and a survey of usage data.

Writing Center Staff Survey

The tutor survey, administered using http://www.surveymonkey.com, was distributed electronically at a writing center staff meeting on October 8, 2010. In the survey instrument, tutors were first provided with a description of each of the first four objectives. An initial set of questions directly assesses how the tutors rate the effectiveness of the Drop-In Lab vs. the effectiveness of individual appointments at meeting the goal in question, on a scale of 1 to 5, with an increase in numerical quantity corresponding to an increase in perceived effectiveness. A second set of questions asks tutors how often, on a text scale ranging from "all the time" to "never," they utilize particular collaborative strategies on which they have been trained. Tutors were seated at computers and directed to the survey instrument. They were provided brief instructions on the survey, and advised that all responses were confidential. A total of seventeen tutors completed the survey.

Table 2 presents the results of the first set of questions.

Table 2. Mean effectiveness for individual appointments vs. Drop-In Lab, averaging across participants ($n=17$)

On a scale of 1 to 5, with 1 corresponding to "not at all effective" and 5 corresponding to "very effective," how would you rate each format of tutoring at meeting:	Individual Appointments	Writing Drop-In Lab
Objective 1: implementing in practice a view of writing as a process?	3.33	4.47

On a scale of 1 to 5, with 1 corresponding to "not at all effective" and 5 corresponding to "very effective," how would you rate each format of tutoring at meeting:	Individual Appointments	Writing Drop-In Lab
Objective 2: embodying a view of writing as a collaborative act?	2.83	4
Objective 3: validating student users as intelligent commenters on ideas and writing?	3.09	3.64
Objective 4: advancing a view of the writing center as a place to write, rather than as a place to get writing fixed?	2.17	4.27

For each of the objectives 1 through 4, the average rating is higher for the Writing Drop-In Lab than for individual appointments, indicating the writing center staff as a whole perceives the former to be more effective in satisfying the stated objectives.[6] The results of these questions clearly indicate that, from the perspective of the tutoring staff, the Writing Drop-In Lab is meeting its stated goals more effectively than individual appointments.

Objective two, as well as the training needs of the staff, was addressed with an additional set of questions relating to tutors' use of collaborative strategies. Table 3 presents the results of this set of questions:

Table 3. Estimated frequency (percentage) of tutors using key strategies

In the Writing Drop-In Lab, how often do you ...	All the time	Sometime	Not very often	Never
Work in a group with two or more students who have similar issues/papers?	6.3%	37.5%	37.5%	25%

In the Writing Drop-In Lab, how often do you . . .	All the time	Sometime	Not very often	Never
Arrange students into peer-review groups, with students working on the same or similar assignments reviewing and commenting on one another's work?	0%	12.5%	43.8%	43.8%
Run a short, informal workshop on a topic that multiple students in the lab are facing?	0%	12.5%	31.3%	56.3%

Here we see a tutoring staff in transition, with most tutors making at least some use of specific strategies on which they have been trained, and others not. Tutors' comments explaining their answers for this set of questions indicate that, as of the beginning of October 2010 (mid-semester), the lab simply hadn't yet been busy enough to provide an opportunity for these strategies. Nonetheless, these results point to a need for further training, particularly on peer review.

The responses to these questions offer evidence that Objective two is, to some extent, being met, with a number of tutors making use of collaborative strategies (peer review, mini-workshops, collaborative writing) in the Drop-In Lab. They also, however, point to a clear issue for future training: a gap between the pervasiveness of collaborative techniques in the Drop-in Lab as envisioned by its administrator and perceived by tutors (despite, as is indicated in Table 2, the fact that tutors do recognize the collaborative value of the Drop-in Lab), and the Writing Drop-in Lab as implemented in practice.

Student User Survey

The student user survey was administered to all students who visited the writing center (both in individual appointments and the Drop-in Lab) during a two-week long period in the fall of 2010. All 292 students who used writing tutoring during this period received an email, following their first visit during the target period, with a link to the survey instrument. The survey was administered electronically through Student Voice, an online survey design and administration tool that provides an html link, included in the email, for participation. Students were advised that all responses were confidential, and no identifying information was collected in the survey. A total of fifty-six student users completed the survey. Of these, fifty-three were users of the Drop-In Lab, and it is results from these fifty-three students that are analyzed below.

The survey opens with a question on the reason why the student visited the writing center:

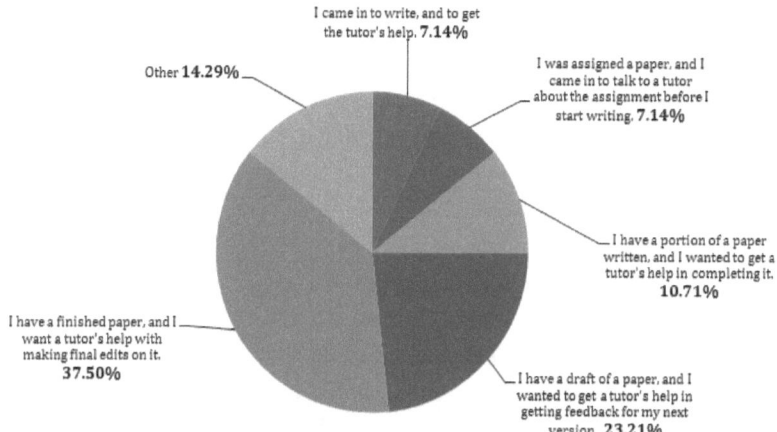

Figure 1. Question 1: Student user visits to the writing center, by reason.

The responses to Figure 1 indicate students' changing perceptions of the writing center. Fully 39.3 percent of respondents arrived at the center prior to having completed an initial draft (this tally includes responses from the "Other" category, none of which involve a completed draft), indicating that the Writing Drop-In Lab is clearly meeting its goal of having tutors engage with students earlier in the writing process (supported as well by the 23.21 percent of respondents who visited the center to seek a tutor's help in revising towards a subsequent draft). In addition, 7.14 percent of student users arrived at the writing center planning to write during their time there, speaking directly to the goal of advancing the writing center as a place for writing.

The remainder of the survey poses attitudinal questions directly related to each of the program goals; these data are presented in Table 4.

Table 4. Percentage of students responding 'true' or 'false' to each attitudinal question

After my session today in the **Writing Center** ...	True	False
2. I think differently about the process of writing a paper.	70.00%	30.00%
3. I'll consider modifying my writing process as I write my next paper.	73.00%	27.00%
4. I could see the benefit of coming in earlier in the writing process the next time I use the Writing Center.	73.00%	27.00%
5. I think of writing as something that needs to be talked about with other people in order to be done well.	86.00%	14.00%
6. I think that feedback from other students is useful in revising my own writing.	93.00%	7.00%
7. I think that looking at other students' writing is useful in making my own writing better.	82.00%	18.00%
8. I feel more confident in my ability to look at, comment on, and evaluate other people's writing.	68.00%	32.00%
9. I feel more confident in my ability to look at, comment on, and evaluate my own writing.	82.00%	18.00%
10. I have developed more of an ability to talk about writing.	70.00%	30.00%
11. I think of the Writing Center as a place to engage in, and not just talk about, writing.	73.00%	27.00%
12. I appreciate the benefit of doing my writing in the Writing Center, as opposed to at home in a computer pod.	68.00%	32.00%
13. I plan to do more actual writing in the Writing Center the next time I'm working on a paper.	62.50%	37.50%
14. I felt that the tutor who I worked with valued me and my writing.	84.00%	16.00%
15. I feel satisfied with my visit to the Writing Center.	82.00%	18.00%
16. I plan to use the Writing Center again.	89.00%	11.00%

Each set of three true or false questions relates to a particular goal. Questions 2 to 4 focus on students' perceptions of writing as a process, 5 to 7 on writing as an act requiring social interaction, 8 to 10 on themselves and their peers as intelligent commenters on student writing, and 11 to 13 on the writing center as a place to engage in the act of writing itself. Responses to all of these questions reflect a clear pattern of students' perceptions of writing being broadly in line with the view of writing advanced by the program, following a visit to the center. The results of the student survey suggest that the Writing Drop-in Lab, at least for those who responded to the survey, is succeeding in its proposed goals.

Questions 14 to 16 are more traditional customer satisfaction questions, included here to assess whether increased tutor utilization (see following section) is resulting in negative perceptions (on the part of student users) of the amount of attention that they are receiving in the writing center. No such relationship is indicated.

Usage Data

Table 5 presents utilization data (student contact hours / tutors' available hours) for fall 2008 through the present, for tutors across all tutoring locations.

Table 5. Writing tutor utilization, 2008 to 2010

Semester	Total Hours available, Individual and Drop-In	% of available hours from Drop-In	Overall Tutor Utilization
Fall 2008	2286	0%	63%
Spring 2009	3573	0%	37%
Fall 2009	4278	61%	57%
Spring 2010	2954.5	66%	59%
Fall 2010	3807	70%	83%

The first column shows the sum of tutoring hours available in a given semester, the second the percentage of these hours that come from Drop-In Lab tutoring. The final column shows overall utilization of the writing center for the semester in question.

Fall 2010, the semester in which the highest percentage of available hours comes from drop-in tutoring, has by a wide margin the highest rate of tutor utilization, indicating that the goal of increasing tutor utilization is being met. Table 6 shows tutor utilization at the main writing center location, contrasting utilization for individual vs. drop-in writing tutoring.

Table 6. Writing tutor utilization by tutoring format, Fall 2010

Tutoring type	Total Hours available	Tutoring contact hours	Tutor Utilization
Individual Appointments	738.4	149.1	20%
Writing Drop-In Lab	1217	1719.2	141%

As the 141 percent utilization suggests, the Writing Drop-In Lab format of tutoring clearly exceeds individual appointments in allowing a tutoring staff to be utilized effectively by maximizing time spent with student users. The results of both surveys demonstrate how, with the appropriate tools and training, this increase in utilization can be accomplished alongside an increase in the overall quality of services being offered.

Conclusion: Reflections from Writing Tutors

No one is better able to address the strengths and pitfalls of the Drop-In Lab's peer-interactive format than the tutors who work in it every day, informing moment-to-moment decisions on the basis of the goals outlined above. I'll close with a few reflections from the writing tutors who staff the center (names in quotation marks have been changed at the consultant's request), before looking ahead to future considerations in the last section. Each of these comments point towards specific ways in which tutors and students have arrived at ways of engaging over writing that are uniquely fostered by a peer-interactive center, emerging from the repositioning of tutors as facilitators of peer-to-peer interactions.

> *Hannah*: I recently had a student come in with the assignment his teacher had just given him for a paper due at the end of the week. He told me in his own words what the assignment was about, then we read through the prompt together. The teacher

had included a rubric, so we then read what an A paper was expected to include. Next, we chose which of the songs and readings from class he wanted to use in his paper. We made up examples, creating a sample outline as we went. Two days later, he came back with a very good paper, which we further revised together. This went well because he came before he had written anything! We were able to get him started on the right track from the beginning.

Michelle: I had a girl come in one evening having to write a letter to a member of congress concerning a local issue she felt passionately about. She had the assignment and its rubric, but she didn't know where to begin. She told me she had read the assignment but didn't understand what the teacher was asking. So, I read the assignment to myself and, in my own words, explained to her what her instructor was looking for in her paper. We went through the five bullet points point by point, and she wrote down her thoughts in her notebook. After that pre-writing process, I let her work on her own for a bit so she could formulate her ideas more clearly and succinctly. Every once in a while, she asked me to check her work, just to make sure she was on track. She was, and I encouraged her to continue doing exactly what she was. The drop-in format worked so well because, had this been an individual appointment, we would have spent the whole time just on pre-writing—which isn't necessarily a bad thing, but because I knew she wanted to get a good start on it before the day was over, she was allowed the freedom to do that. Another advantage of this format was that I was able to let her write by herself for a little bit, giving her the space and time to think and collect her thoughts without my being over her shoulder.

Chris: I was tutoring a student from South Korea; she had a hard time organizing her essay since she felt she could not express herself well in English. I began to talk to her and without focusing on the assignment, after a moment of talking just about her ideas, she was able to elaborate a well- structured essay that covered all points she felt were necessary for the assignment. The interaction was typical for a drop in lab format, since she decided to stay longer and have a conversation instead of just focusing on the assignment.

Sarah: I had one student who had a draft of a paper that she was continuing to work on. It was a longer paper, and when we went through it together, I noticed that it lacked some clarity, and I began to notice that many of her sentences were awkwardly worded. First, we picked one awkward sentence out together and fixed it together. Then we highlighted a couple others together, and I left her to go through the rest of the paper to highlight the rest. When she was done, I looked over them, but then left her to try to fix them on her own. I checked back with her when she was done with that, and she had done a pretty good job fixing them on her own; I might have helped even further clarify some of them, but she had done most of the work by herself. This worked well because she saw her paper as a work in progress. Therefore, I was able to pick out one thing to concentrate on, rather than trying to get everything right. In other words, because she didn't need the paper to be perfect that day, we could start slowly. Also, there was a specific problem that could be fixed in her paper, and that problem had a lot to do with just seeing the issue. Once she started to see what looked awkward, she could easily fix them on her own. This was typical of a good drop in session: being able to step away from the student so that they have to recognize and correct some of their own mistakes makes for the most productive tutoring session. This illustrates the possibilities by showing that students can be active learners in the tutoring process, rather than just passive receivers of corrections.

In each of the above testimonials, tutors speak to how the absence of time constraints allows the session to "breathe," fostering a productive dialogue around writing. Being able to step away from the session to work with other student users gives student writers time to apply feedback from the tutors. In many cases, the approach encourages students in their own ability to apply principles that they're picking up from the tutors, including the collaborative discussion of ideas, as indicated by the following tutor's reflection.

James: I have had many enjoyable tutoring interactions while working at the Writing Center, but one experience that I found to be especially rewarding involved a freshman student who was working on her first college paper. She was excited about her visit to the center and thrilled about her first semester. She was slightly frustrated with her creative writing assignment. She was asked to write about a memorable experience in her life and had run into a mental block. Before I looked at her paper, I asked her to tell me about her experience. Her words were filled with so much passion and excitement that other students began to listen in. When she was done, several students began to ask her questions. As they conversed, I stepped back and watched as they shared stories and their own ideas of which angle she take on her paper. The student was very pleased with this interaction and was able to complete her paper. Aside from a few organizational and grammatical issues at the end of her writing that we had to address, the paper was very captivating. The student user was greatly appreciative and even stayed awhile to talk to me about her semester and family life.

Such "A-ha" moments, enabled in a writing center where everyone present is able and encouraged to speak to everyone else, have been the payoff of the last several years of hard work in changing tutors,' instructors,' and students' perceptions of writing tutoring. But there have been and continue to be struggles, and the tutors have been more than willing to share their critiques of the approach, as well.

Hannah: So far I have enjoyed working with this format. I'm a little nervous about the possibility of having too many students at once who need help and not being sure how to balance helping all of them. What if I can't think of something for one of them to work on while I help the other?

Michelle: Some students come to the center because their teacher requires it; others come just because their teacher simply encourages it. If we could get more teachers more educated about what exactly the writing Drop-In Lab is, then I think we'll see more of these objectives met. Students would be more encouraged to start writing up in the lab rather than just handing their paper over to proofread.

> *Esther*: I think that whenever the concept of a drop in lab is fully understood, a more organic writing process could occur in this drop in labs. I think by letting people know the concept of the drop in lab, people will be aware of it and start using it more as a way to create a well-structured essay instead of a place to revise a finished document.
>
> *Jill*: The drop in lab creates a nice, casual atmosphere, and students don't always need as much intense help. This is nice as a tutor, and I can imagine it makes the writing process less stressful. I do think we need to be reminded that circulating, and letting students do some work on their own, can be really effective, and whenever possible, we should be using that strategy. And I think we need help figuring out how to encourage students to work together, and how to facilitate those interactions.

Responding to the concerns voiced here will be the explicit goal of future semesters.

Looking Ahead

Upcoming semesters in the Writing Drop-In Lab will be marked by further training, with an eye towards providing the tutoring staff with an even broader range of tools for working with groups of writers. They will also be marked by further outreach. Increasingly, students are arriving at the writing center earlier in the writing process and are prepared to share their own work and to read other students' writing. The major goal of future semesters will be to educate faculty, who continue to refer students to the Drop-In Lab almost exclusively to have completed drafts reviewed, on the other possibilities of the center: for students to write, to collaborate with their peers, to participate on both sides of the review process, and to engage in a conversation around writing.

The survey results, as well as the tutor comments provided above, both indicate that the program has been highly successful in meeting the goal of engaging tutors with students' processes for writing. They also, however, indicate a high degree of variation among the tutoring staff, with some tutors taking frequent advantage of opportunities for peer collaboration, and others doing so far less often.

Informal discussions with the tutors point to three main reasons for this issue. First, the use of collaborative strategies requires a critical mass of students who are engaged in the same or similar assignments. When there are five to ten (or more) students in the lab, as is typical during the busy mid-day period, this threshold is readily crossed. During slower times, the prerequisite number of students is reached far less often. One reason, then, for the high number of tutors who do not take advantage of collaborative techniques is that they work at slower times and have less, if any, opportunity to use them. Second, tutors who work together in the Drop-in Lab have often established roles for themselves in the lab: in shifts where two or three tutors are working, one or two of the tutors will circulate among students, while another tutor assumes the role of group facilitator, directing students into and working with groups of students working on related tasks. These roles, while informal, seem to be persistent, such that over time a tutor falls into a more or less permanent role of circulator or facilitator. Third, it appears to be the case that, in some instances, tutors are clear on the goals of the strategies that they are being asked to make use of, but they are simply unclear on or unpracticed in the nuts and bolts of how to apply them, and aren't confident enough in making use of them to direct students into groups.

The first of these, to the extent that it can be addressed, will be responded to in scheduling, attempting as much as possible to ensure that all tutors have at least some hours during peak periods. In the case of the second, while I don't believe that the system that's arisen is a bad one (and may in fact be quite positive in that it draws on the unique strengths of individual tutors), we will be encouraging all tutors to take on the role of group facilitator periodically so that the entire staff is well versed in all relevant approaches. The third, to me, indicates that the trainings that tutors have received on collaborative techniques have erred too far on the side of theory. In future trainings, it will become a greater priority to show tutors models (through observations of the Drop-in Lab, or demonstrations staged by veteran tutors) of collaborative approaches being applied, and to give them practice applying them in mock tutoring scenarios. We will also be working to create more opportunities for in-class peer review workshops, so that more tutors can be given the chance to become more fluent in the approach in a more formal, structured environment.

To writing center directors seeking to implement the peer-interactive model at their own universities, I'd offer one piece of advice above all else: involve the tutors early and often, creating a sense of investment in and ownership of the model during the planning stages. Our center has arrived over time at a very high degree of buy-in to the peer-interactive model on the part of the tutors, who are the most important stakeholders in the format, but I believe that the process of getting to where we are now could have been greatly eased by more consciousness early on of how the tutors might feel discomfited by the shift away from the familiar, cozy environment of the individual appointment, and how this might have been avoided by creating more enthusiasm around the model. In the fall of 2009, I thought that the best path towards full adoption of the model would be to explain the methods associated with it, and to coach the tutors on how to implement them. I've since discovered, in discussing the peer-interactive writing center approach in forums of interdisciplinary faculty and of WPAs at UNM and elsewhere, that I wasn't giving my audience enough credit: nothing needed to be explained. The principles of the peer-interactive center are intuitive. Presented with the nearly universal goals for writing center work that are outlined above, and invited to think through the possibilities for how writing centers can more effectively meet them, people tend to arrive at much the same conclusions that we have about which tools are best to use within a center that fully aligns its practice with its theory.

The intensive feedback that students receive in individual writing appointments is invaluable for student writers, and individual appointments clearly have a place in a writing center that seeks to meet its students' needs in a variety of ways—indeed, in many cases, an individual appointment is an ideal follow-up to a session in the Writing Drop-In Lab. However, individual appointments are often characterized by an inherently unequal dynamic between the tutor and the tutored, such that the tutor is the authority on writing and the student user is a recipient of knowledge. The transfer of such knowledge is one important part of what writing centers do. Just as important, however, is for writing centers to promote their student users as reviewers and commenters of their own and others' work, validating a relationship to writing that will serve them throughout their academic and professional careers well after their direct relationship to the center has ended. The peer-interactive writing center approach, as exemplified by the Writing Drop-in Lab at the University of New Mexico, presents a model for doing so.

Coda: Where We Are Now

Fall 2009 feels, writing this, like a surprisingly long time ago.[7] Since then, the entire staff of the writing center has turned over, replaced by a team that has inherited the peer-interactive writing center as received practice rather than as a new initiative. The student population has for the greater part turned over as well, such that, for most students at UNM, the way that writing tutoring is done here is simply the way that writing tutoring is done. This has had a profound effect on everyday practice, as a population of more than 25,000 students has come to know a university writing center as a place to share their writing, to read and comment on their peers' writing, and above all: to *write*. The average length of students' visits to the center has increased from 1.18 hours (fall 2009) to 1.7 hours (fall 2012), and usage has continued to grow as well (1,649 student visits for fall 2012).

The look and feel of the center has become different as a result of the established pedagogy. In 2009, it was normal to see student users waiting patiently for attention, hands raised. Now, we see student users with their course materials and laptops spread out in front of them, actively engaged with a current project. Often the tutors are now the ones patiently waiting: waiting for students to reach a point when they have a question or would like something reviewed, waiting for a conversation between student writers to reach a lull so the tutor can join the conversation.

Looking ahead, we plan to push the pedagogy in new directions. Another major change of recent years has been the expansion of writing support to a number of non-English languages, with the goal of engaging our student users across languages on issues of writing and literacy (Sanford, "Writing"; Sanford, "Multilingualism"). More recently, we've implemented a number of changes to our online tutoring. As our center seeks a synthesis across all of these initiatives, the five goals outlined in the profile remain central to our decision-making. How can the connectivity of our student population enable a culture of social, collaborative writing, and what role can our center play in that culture? How can we foster students' ethos as intelligent commenters on others' writing in emerging online spaces? As the writing center becomes more distributed, both on the Internet and across campus in locations focusing on different languages, what does it mean for it to be "a place to write?" How do these values intersect with differing, culturally predicated views of writing encountered as we move our services into non-English languages? All of these questions are incredibly exciting; none of them is easy to answer.

As writing program administrators, the answers are only partially ours to give. One of the more potentially unsettling implications of an ecological view is the lack of control that any particular agent has over the system. If we accept writing programs as complex adaptive systems, then we have to imagine them as something akin to termite mounds—complex structures arising from the interplay of many actors, each acting within a limited scope of awareness—rather than as structures bearing the stamp of any one architect or overseer. As writing centers go, the approach to tutoring that I've outlined here makes for a particularly good example of a self-organizing system: students seek their self-interest from the center; tutors act on the basis of a few guiding principles, and from there the specifics of tutoring interactions unfold with no higher-order guidance. But it is no more nor less the case than for any other system that while we can make changes (by tweaking the parameters, by changing the inputs, by altering the rules that govern the system), there are an infinite number of factors that aren't in our control, and we can't possibly do more than make informed guesses at the outcomes of our tinkering. If that's unsettling, it's also a bit thrilling. I'm excited to be on the front lines as the format continues to evolve.

Notes

1. See the University Writing Center at Texas A&M (http://writingcenter.tamu.edu/), which offers a workshop for faculty on leading peer reviews, the Center for Writing at the University of Minnesota (http://writing.umn.edu/), which enables electronic peer reviews for students through their online writing center, and the Writing Center at Pacific Lutheran University (http://www.plu.edu/writingcenter/), which facilitates peer review workshops for specific courses.

2. See the Writing Center at the Boston University Educational Resource Center (http://www.bu.edu/erc/), which encourages group tutoring for students working on group assignments, the Writing Lab at the Citadel University Academic Support Center (http://www.citadel.edu/root/asc) in Charleston, SC, which takes advantage of small group sessions to increase tutor/tutee interaction, and the Chaffey College Writing Center (http://www.chaffey.edu/writingctr/index.shtml), which offers group sessions as a service to faculty who wish to refer a subset of their students to a guided session on a set topic in writing.

3. See, for example, the Writer's Center (http://www.writer.org/), a nonprofit on Bethesda, RI, which caters to the needs of professional writers using

on-line and in-person workshopping sessions in which participants review and critique one another's work, and the Midwest Writing Center (http://www.midwestwritingcenter.org/), a program funded by the Illinois Arts Council that provides a forum for professional and amateur creative writers to share their work.

4. A formal peer review guide is also available for students (to view the peer review guide, see: http://compositionforum.com/issue/25/unm-caps-peer-rev-sheet.pdf).

5. For the current, as of November 2011, flier, see: http://compositionforum.com/issue/25/unm-wc-flier-fall2011.pdf.

6. T-tests for sets of responses for each question were significant, $p < .05$.

7. Many and most sincere thanks to Evan Ashworth, the current Program Specialist for the Writing & Language Center at the Center for Academic Program Support, for his consultations on this section. He and his team of outstanding peer tutors have contributed massively to continuing, at a high level of excellence, the work of the peer-interactive writing center at UNM.

Works Cited

Elbow, Peter and Mary Dean Sorcinelle. "How to Enhance Learning by Using High-Stakes and Low-Stakes Writing." *McKeachie's Teaching Tips: Strategies, Research, and Theory for College and University Teachers.* 12th ed. Eds. Svinicki, Marilla and Wilbert McKeachie. Belmont, CA: Wadsworth, 2005. 213–233. Print.

Elbow, Peter. "Write First: Putting Writing before Reading is an Effective Approach to Teaching and Learning." *Educational Leadership* 62.2 (2004): 8–14. Print.

Joyner, Michael. "The Writing Center Conference and the Textuality of Power." *The Writing Center Journal* 12.1 (1991): 80–89. Print.

Lunsford, Andrea. "Collaboration, Control, and the Idea of a Writing Center." *The Writing Center Journal* 12.1 (1991): 3–10. Print.

Lunsford, Andrea, Lisa Ede, and Corinne Arráez. "Working Together: Collaborative Research and Writing in Higher Education." *Modern Language Association Profession* (2001): 7–15. Print.

Lunsford, Andrea and Lisa Ede. "Collaboration and Concepts of Authorship." *PMLA* 116.2 (2001): 354–370. Print.

Macrorie, Ken. "Words in the Way." *The English Journal* 40.7 (1951): 382–385. Print.

Murray, Donald M. "Teach Writing as Process Not Product." *The Leaflet* (Nov 1972): 1–11. Print.

—. "The Listening Eye: Reflections on the Writing Conference." *College English* 41.1 (1979): 13–18. Print.

North, Stephen. "The Idea of a Writing Center." *College English* 46.5 (1984): 433–46. Print.
Sanford, Daniel. "Writing Tutoring and Language Rights: Navajo and Spanish Writing Tutoring at the University of New Mexico." *National Conference on Peer Tutoring in Writing (NCPTW) 2011: Tutors, Tutoring, and the Teaching of Tutors*. Biscayne Bay, Florida. 2011.
—. "Multilingualism, Writing, and the Academy: Beyond ESL." *Conference on College Composition and Communication (CCCC) 2013: The Public Work of Composition*. Las Vegas, Nevada. 2013.
Soven, Margot Iris. *What the Writing Tutor Needs to Know*. Boston, MA: Thomson/ Wadsworth, 2006. Print.
Spigelman, Candace and Laurie Grobman, Eds. *On Location: Theory and Practice in Classroom-Based Writing Tutoring*. Logan: Utah State UP, 2005. Print.
Stanger, Carol. "The Sexual Politics of the One-to-One Tutorial Approach and Collaborative Learning." *Teaching Writing: Pedagogy, Gender, and Equity*. Eds. Cynthia L. Caywood and Gillian R. Overing. Albany: SUNY UP, 1987. 31–44. Print.
Strauss, Susan and Xuehua Xiang. "The Writing Conference as a Locus of Emergent Agency." *Written Communication* 23 (2006): 355–396. Print.
Walker, Caroline. "Teacher Dominance in the Writing Conference." *Journal of Teaching Writing* 11.1 (1992): 65–88. Print.

15 Writing the Transition to College: A Summer College Writing Experience at Elon University

Jessie L. Moore, Kimberly B. Pyne, and Paula Patch

> "We would like to make a summer section of ENG 110 a part of a larger Transitions Program for our graduating seniors, a key summer experience to help them be best prepared for their life as college freshmen at a variety of schools." (Email from Elon Academy Representative to College Writing Coordinator)

> "The Elon Academy request presents a terrific opportunity to pilot a summer session that extends beyond the traditional summer session time boundaries (5 weeks instead of 3) and provides additional support through a daily two-hour Writing Workshop." (Internal Grant Application led by College Writing Coordinator, in collaboration with Elon Academy Representative and ENG 110 Faculty Member)

What role can first-year writing courses play in supporting underrepresented[1] students' transitions to college? A seemingly simple request from the assistant director of a college access and success program provided an opportunity to explore this question. Numerous studies have described the achievement and opportunity gaps that often exist between dominant and underrepresented groups in college. Students from low-income or first generation homes, in particular, may arrive without access to the same sociocultural capital of their more affluent peers—resources, knowledge, skills, and dispositions that traditionally lead to success in higher education. Instead, they commonly have limited access to rigorous and well-resourced high school preparation; cultural mismatches with traditional curriculums and schooling routines; significant emotional and economic pressures; and inadequate advising about the college pathways, financial aid opportunities, and other resources for success. Once accepted to college, underrepresented students must then compete with peers who may have experienced a greater variety of privileged extracurricular opportunities (e.g., science camps, international travel, extra arts and athletics lessons, etc.), as well as the confidence passed along by family members who possess generations of college experience (Davis; Klugman and Butler; Strayhorn). In addition, colleges tend to reflect dominant cultural modes in their curriculum, climate, and student body culture, and students from underrepresented groups often report feeling marginalized or outside of the college experience for a variety of reasons (Villalpando and Solórzano; Yosso, Smith, Ceja, and Solórzano). Their social worlds and cultural strengths are often unrepresented and unrecognized inside and outside of the classroom. Combined with increased academic and/or financial challenges, it is little wonder that institutes of higher education struggle to retain talented underrepresented students, especially through the first years.

With frequent student-faculty contact and small class sizes, this struggle often becomes visible to faculty in first-year writing classes. Writing faculty often have more opportunities to learn about their students' backgrounds and their college experiences than colleagues in other disciplines who might be teaching first-year students primarily in large, introductory lecture classes. Furthermore, the very writing pedagogies (e.g., one-on-one conferences) that create spaces for students to share their retention-related struggles with faculty (i.e., financial challenges, limited access to resources, feelings of marginalization, etc.) also have potential as retention efforts, since they increase students' contact with faculty (Powell). This program profile examines a collaboration between a college access and success program and a first-year writing program to reimagine a summer section of a required first-year writing course, ENG 110: College Writing, with the goal of preemptively improving students' opportunities for college success—and retention. While we recognize that many aspects of our course are not generalizable to other contexts, the challenges we sought to address (e.g., curriculum design for special populations, adjusting course curricula for summer sessions, providing underrepresented students access to resources essential for success, etc.)

are familiar to both first-year writing and broader higher education contexts. We offer our model and our assessment of the course outcomes as one way universities might reexamine or extend their support programs for underrepresented students. This model gives underrepresented students access to college capital while helping them develop writing process strategies that provide a foundation for continued writing instruction and practice across the curriculum.

CONTEXT FOR THE ELON ACADEMY

Elon University is a mid-size (5,700 students), private university with a primary emphasis on undergraduate education. Recognized as a model for engaged learning, Elon also is committed to the liberal arts and sciences and to service partnerships. Elon students, faculty, and staff routinely are recognized for their volunteer service hours, and this service commitment has led to several deeper community partnerships, including with the surrounding public schools. In 2006, one public high school six miles away from the university's main campus was threatened with closure by the state for continuing poor performance. The school served predominantly working class and minority students, including a burgeoning immigrant population, groups that are of high concern in the research on achievement and opportunity gaps in K-12 education (Delpit; Noguera and Wing).

The university responded with increased support to the local schools on various fronts, especially for the school that had been so critically in danger of closure. As part of this effort, the university president appointed a senior education department faculty member to launch a college access and success program in collaboration with a small team of university faculty and staff, including Kim. The resulting program, Elon Academy, serves academically-promising high school students with significant financial need and/or no family history of college. It builds on models at other schools, including Furman University's Bridges to a Brighter Future (http://www.bridgestoabrighterfuture.org/) and the Princeton University Preparatory Program (PUPP) (http://www.princeton.edu/pupp/), as well as the successes of other community-based and federal college access initiatives (e.g., Talent Search, Upward Bound, and the TRIO programs). Like these other programs, Elon Academy offers mentorship, support, cognitive and social enrichment, and a space to practice college-ready skills (both academic and personal). It develops student and family knowledge of pathways to higher education, advises and supports academic rigor and advancement, and provides opportunities for enrichment, service, and leadership. The program inspires and enables high school students from backgrounds often underrepresented on college and university campuses to pursue higher education, build leadership skills, and develop an active sense of social responsibility. Students are recruited from the county schools during their ninth-grade year and agree to participate in a three-year access program that combines intensive four-week residential experiences on Elon's campus during the summers before the 10th, 11th, and 12th grade years, with college-readiness support during the regular school year. The summer program is a four-week residential experience during which students live in the Elon dorms, take specially designed academic classes taught by Elon faculty and other master teachers, engage in service projects and enrichment opportunities, and begin developing both their knowledge about and individual plans for attaining their dream of attending college. To supplement these summer experiences, the year-round support program provides monthly Saturday sessions on college planning and academic enrichment, regular mentoring by specially trained Elon students and program staff, and tutors who support a more rigorous high school curriculum. Programming is based on a youth development model that supports not only academic skills, but also personal growth—public speaking, self-advocacy, time management, and leadership opportunities, among others. Postgraduation, the Academy continues to follow students through their college years at their chosen college or university, with staff visiting at least once each semester during the first year to offer assistance with the challenges of higher education—from advice about navigating their college resources (academic, social, and financial) to counseling about life decisions that reach beyond the halls of academia. Second-year college students in the program serve as peer mentors to new first-year students, networking with their peers both on their individual campus and across campuses. As the third graduating class heads for college in fall 2012, the program continues to develop in response to the needs of the college participants, with additions such as career- and graduate school-focused programming.

WHY ELON ACADEMY WANTED TO OFFER A SUMMER SECTION OF FIRST-YEAR WRITING

As the first Elon Academy cohort neared graduation during the 2009 to 2010 academic year, the program turned its attention to assisting them more specifically with the difficult transition to the college environment, a well-documented stumbling block for many college students and especially difficult for underrepresented students (Davis). Transition programming included workshops on everything from navigating new campus resources to day-to-day managing of scholarship and personal funds, and from approaching college professors during office hours to setting home-rules for roommates with dramatically different lifestyles. In addition to these college life skills, the program also wanted to attend to their academic development in the summer prior to college. Ample research has shown that the summer months between school years allow significant cognitive regression, and summer learning experiences can go a long way toward bolstering students' abilities and confidence (Alexander, Entwistle, and Olson). Academy graduates had spent the previous three summers being challenged on Elon's campus, and the post-graduation summer looked empty and abandoned in comparison. It would be their first summer since arriving in high school that they would have no academic expectations.

For students historically underrepresented in higher education (low-income, first generation, minority and/or immigrant students), ongoing academic preparation during the transition summer may prove especially beneficial. Underrepresented students often bring additional challenges to the college classroom, including those that stem from weaker academic preparation in the critical skills of reading, writing, and quantitative reasoning, as well as social, cultural, and financial mismatches with college norms and expectations (Tierney, Corwin and Colyar; Conley, *Redefining*). Yet they also bring an array of life experiences to college classrooms when they achieve access, enriching the conversation with perspectives different from their more privileged peers. They bring great excitement and enthusiasm for being in college, have strong desire to succeed, and often believe deeply in the value of an education for themselves and others. By creating a sheltered college-level course that enrolled only Elon Academy graduates and responded both to these students' cognitive, socio-cultural, and financial needs and their strengths, the Academy staff hoped to help Academy graduates transition more thoughtfully and effectively between high school and college and leave them better prepared for the rigors of the college classroom. The modified ENG 110 summer section, while taught and administered separately from the Academy, nevertheless was an extension of the program's high expectations for student academic performance and one more element in the culture of achievement the program seeks to instill in its participants. It also presented students an opportunity to earn college credit towards their degree programs at Elon University or wherever they were matriculating.

Taking any summer term course would have provided a "real" college experience while maintaining Academy graduates' intellectual engagement over the summer months. By offering Elon University's first-year college writing course, however, the Academy hoped to allow students to concentrate exclusively and intensively on developing their writing and research skills, two areas that frequently have been identified as vital to college success (Conley, *Redefining;* Conley, *College Knowledge; Framework for Success*). Since writing is a primary gatekeeper to college achievement, the overarching goal was to give Academy graduates a positive (but rigorous) learning experience with college-level writing, not merely an experience with college-level disciplinary content. With no other classes to make additional demands on student attention, students could concentrate on improving their writing through regular practice and thoughtful, reflective analysis of their work.

In addition to focusing their attention on writing, the course also allowed Academy graduates to experience an ideal student-professor relationship and learn from a faculty member who was genuinely aware of students' strengths and weaknesses, passionate about composition studies and student learning, and able to incorporate best pedagogical practices. The professor for the course, Paula, was selected carefully for her ability to address the developmental skill gaps often seen in underrepresented student writing; for her willingness to address college culture issues as they arose; for her desire to build on the enthusiasm, unique experiences, and strengths brought by the students; and, overall, for working with students to become more successful college students as well as more successful academic writers. In this way, the course made the expectations of college immediate and real while simultaneously providing the necessary scaffolding for students who struggled to rise to those expectations. In this small, sheltered section, students should have been unable to fall between the proverbial cracks and unable to hide their missteps—whether in college success behaviors or in college writing. Rather they could practice navigating college learning without fear of being misunderstood or judged for their lack of resources and experience and would have a chance to focus intently on their individualized writing processes. Beyond the classroom, Academy graduates also could discover and begin to hone strategies for balancing the additional responsibilities many underrepresented students might bring with them to the college classroom. Surrounded by familiar others who shared similar social and cultural struggles, the students might experience less of the stigmas of not being able to afford textbooks, of lacking transportation to and from school, or of lacking computer and Internet access at home. The very design of the class needed to reduce barriers, especially those erected by financial constraints.

Similarly, the design of the class capitalized on the experiences the Elon Academy graduates brought to the class. Because they had engaged in intensive academic enrichment activities for each of the three previous summers, the students were familiar with balancing academic, work, and social activities during the summer. Since those summer activities had taken place on Elon's campus, the students were familiar with campus resources, from the library to dining services. More important, because the students had been carefully selected for and mentored through Elon Academy, they understood that the stakes were high for this course that would set them squarely on the path to college graduation, and they knew that they had much to lose if they were unsuccessful. Paula deliberately developed a course theme and writing assignments that would remind students of their past experiences as high school and Elon Academy students and connect those experiences to ones they would encounter as college students.

WHY THE COLLEGE WRITING PROGRAM COLLABORATED ON THE MODIFIED SUMMER SECTION

At Elon, first-year writing consists of a one-semester, four-credit-hour course, typically taught over a 15-week semester. All sections of College Writing (ENG 110) aim to develop the following:

- A more sophisticated writing process—including invention, peer responding, revising and editing—that results in a clear, effective, well edited public piece;

- A more sophisticated understanding of the relationship of purpose, audience, and voice, and an awareness that writing expectations and conventions vary within the academy and in professional and public discourse; and

- An appreciation for the capacity of writing to change oneself and the world.

Additionally, a shared experiences document for the course indicates that ENG 110 should emphasize persuasive writing, while giving students opportunities to work with sources and to practice writing for a variety of academic and non-academic audiences.[2]

In 2008, the department of English stopped offering summer sections of ENG 110 because program assessment data suggested that students in the accelerated summer sections were not as successful at meeting the program's intended learning outcomes as students who took the course during a regular fall or spring term. In addition, faculty who taught the course expressed frustration with the tenor of the summer sections. The summer sections primarily attracted students who had not met the C- or above graduation requirement the first time they took ENG 110, so students often entered the course with a guarded attitude. Further, learning rhetorically grounded strategies for writing in a variety of contexts requires opportunities for practice and reflection over an extended period of time, and the traditional summer section's condensed schedule simply had not provided enough time for that development.

At the time, the associate provost asked the department to consider resuming the summer section but acknowledged that the parameters needed to change for the learning experience to be more successful. To that end, Jessie, as the college writing coordinator, and the associate department chair brainstormed possible modifications to the summer section and proposed piloting a section of ENG 110 that met longer than the three to four weeks allotted for the traditional summer sections. The idea was put on hold, though, until the specific parameters could be planned in more detail.

Elon Academy approached the department of English in late October 2009 about offering a sheltered summer section of ENG 110 for Elon Academy graduates who would be matriculating at universities (including Elon) in fall 2010. Because the Elon Academy's summer program traditionally involved having students on campus for a longer period than a traditional summer session and all of the students involved were living in the local community during the summer, the request presented a terrific opportunity to pilot a summer session that would extend beyond the traditional summer session time boundaries (five weeks instead of three) and would provide additional support through a daily two-hour writing workshop.

Drawing partially on a college access grant from North Carolina Independent Colleges and Universities (NCICU), Elon Academy funded the faculty salary for the course, as well as paid for books/materials for the students and a stipend for one teaching assistant (a student with experience consulting in the writing center and with coursework in writing studies and secondary education/pedagogy). An internal grant funded a stipend for a second teaching assistant with similar qualifications and a small stipend for Jessie to facilitate assessment of the course, since the course ran outside the timeframe of her annual contract.

This design enabled the College Writing Program, in collaboration with Elon Academy, to examine students' progress towards meeting the ENG 110 shared objectives in the modified summer section, using the program's existing indirect and direct assessment measures. Additionally, the project used other qualitative measures, including teaching journals, writing consultant logs, and participant interviews, to examine the impact of the five-week timeline and the additional writing workshop hours. A primary goal of these assessments and the data collection was to assess how well the modified course structure supported students' achievement of the learning outcomes and the viability of this course structure for broader implementation in future summers. Although the Academy graduates are not representative of the first-year Elon University student population (which is much less economically and ethnically diverse), developing a summer ENG 110 that successfully adapted to the logistical needs and academic scaffolding needs of this specific group could inform future non-Academy summer sections of ENG 110. In previous summer sections of ENG 110, the non-Academy students enrolled in the course often have struggled to balance other time commitments (e.g., work, athletics, etc.) and have entered the course with lingering apprehensions about previous writing experiences. Additionally, Elon has recently committed to further diversifying its campus, including new programs and scholarship funding for first generation and limited-income students. While the Academy students could not be said to be truly representative of Elon as it was in 2010, they do reflect an increasing number of students matriculating at Elon. Therefore, we hypothesized that a successful Academy summer section of 110 could inform future non-Academy summer sections. We also wanted to identify other outcomes that might be a byproduct of this experience for Academy graduates.

In the following sections, we extend our discussion of the students we enrolled; share our course design, research methods, and primary results; and offer reflections on and recommendations for designing transitional writing courses for underrepresented students, based on our experiences.

Understanding Our Students

The first cohort of Elon Academy graduates were offered the opportunity to take this special extended section of ENG 110 in summer 2010 for transfer credit to their chosen universities, including Elon. Eleven of twenty-two graduates chose to participate. The class included five males and six females, a diverse mixture of racial backgrounds (six African American, three Hispanic, and two Euro-American), a wide range of prior academic successes and struggles, and a typical trepidation about college-level writing. All but one would be the first in their family to attend college; one would follow in his older sister's footsteps. Three planned to matriculate at Elon in the fall; seven planned to matriculate at other four-year colleges in North Carolina, including East Carolina University, North Carolina A&T University, UNC-Chapel Hill, and Saint Augustine's College; and one planned to attend Alamance Community College for nursing.

The students in our ENG 110 course were very representative of low-income, first-generation students today, a group which is the largest growing segment of the K-12 student population. Almost half of all school children (44 percent) come from low-income families, and these numbers continue to rise. More than 4.5 million low-income, first-generation students are currently enrolled in postsecondary education today, approximately 24 percent of the total undergraduate population (Engle and Tinto). But the statistics for college completion for this group are disturbing, even when controlling for academic achievement levels, with only 29 percent of talented low-income students obtaining their bachelor's degree, compared to 74 percent of talented high-income students (Fox, Connolly, and Snyder). Access programs can increase the numbers of underrepresented students entering college, but access alone is insufficient.

Once admitted, underrepresented students like the Elon Academy graduates face additional challenges academically, socially, emotionally, and financially during their college career, further increasing the likelihood that these students will be denied a college degree. According to the Pell Institute's report on college success for low-income, first generation students, nearly half leave college without earning their degrees, with 60 percent of those leaving before their second year. During the first year of college, low-income, first-generation students often report that the hidden costs of college can become overwhelming (including costs for books, photocopying, phone calls, and even the means to travel home for holidays and emergencies). Most must learn to balance additional responsibilities alongside their classes, including needing to live off-campus and maintain a paying job. Many find themselves less prepared academically for college-level work and unaware of opportunities for assistance and enrichment offered by colleges, such as writing and tutoring centers and study abroad experiences. Even though high school academic preparation is considered the most significant predictor of college achievement, low-income students, first-generation students, and students of color are disproportionately tracked into less rigorous courses in K-12 schools, leaving them with unaddressed skills deficits despite their high grades (Davis; Strayhorn). The research on persistence for these students is ample and clear. They face barriers their more advantaged peers do not, including financial instability, competing family pressures, cultural mismatch with mainstream student expectations, weaker academic preparation, lack of opportunities to integrate and engage in the campus community, and inefficient or absent advising, among others. The academic barriers are especially significant when it comes to literacy skills, including managing the reading and writing loads, comprehending complex texts across a variety of disciplinary genres, and developing more sophisticated abilities in research and writing (Conley, *Redefining*; Burke). Leaving aside the ethics of equity—including higher education's ability to address historic social stratification and economic inequalities–these numbers alone suggest that underserved populations deserve special attention by higher education faculty and officials.

The Elon Academy students enrolled in the summer ENG 110 section brought the full range of these challenges to the classroom, balancing work, care for siblings, and other family responsibilities alongside their class assignments. Notably, they also brought the more positive hallmarks of underrepresented students, including excitement about college, desire to succeed, and a rich diversity of life experiences (Alvez). They chose to participate in the course for a variety of reasons and entered with a wide range of attitudes and expectations. Some felt acceptably strong in their skills and looked forward to delving deeper into this strength; some loathed writing and just wanted to "get the class [out] of the way" before their freshman year; others believed they might have a higher chance of success in a college class supported by the familiar college access program and "because of the resources available" that might not exist at their future schools. All of them recognized the opportunity to earn free college credit as a tangible financial benefit, especially given the cost of a regular Elon summer session course (waived for these summer section students by Elon as in-kind support for the Academy).

Their participation in the comprehensive college access program certainly filled some of the college knowledge gaps, encouraged improvements in their academic preparation, and helped students mediate obvious financial barriers for this particular experience, but the Elon Academy graduates were still strangers to the higher expectations of a college classroom and inexperienced in some of the behaviors that would foster success in this new environment—from the appropriate ways to interact with faculty and teaching assistants to the elevated standards for organization, study habits, written work, and research.

Designing an Extended Summer Section of First-Year Writing for Underrepresented Students

As a summer transitional course initially proposed by the college access and success program for implementation within another program (the first-year writing program), the Academy summer section reflected some dual-enrollment challenges, even though it was not a dual-enrollment course. Because the summer option was geared toward pre-college students completing the Academy but tapping into the writing program's existing first-year course, the programs had to negotiate mutually agreeable goals, accountability, and assessment plans (Anson; Farris); tension between earning credit and learning rhetorically-grounded writing strategies (Hansen); and questions about the economic incentive overshadowing whether students are adequately prepared for future college writing (Schwalm). Attentive to these challenges, we negotiated the parameters under which the course would run, meet the expectations of the first-year writing program, and thus carry the first-year writing course's designator (i.e., ENG 110). As the writing program administrator, Jessie agreed to support the Academy's summer section staffing request (Paula), since Paula was an established ENG 110 instructor with a record of success supporting the course objectives. In turn, Kim, as the Academy representative, agreed to participate in a research project that would extend the standard ENG 110 program assessment (described more below) and collect additional data to inform decisions about offering the Academy summer section beyond the initial pilot.

At the same time, we focused our course design on helping our specific student population achieve the course goals. Research on the experiences of low-income students in higher education emphasizes the need for additional academic support, personal advising, and ongoing financial assistance (Alvez). In designing the ENG 110 summer course, we also incorporated strategies for promoting academic success as described in Engle and Tinto's research analysis:

- attentive monitoring of student progress;
- well-aligned, highly visible support for developmental learning needs;
- use of proven pedagogical practices in the classroom that foster active student engagement and learning (to better capitalize on the limited time students spend on campus); and
- clear commitment to student success at the institutional level.

Perhaps unsurprisingly, pedagogical and institutional practices that best serve underrepresented students also serve other students equally well, no matter their socio-economic, racial/ethnic backgrounds, or prior academic strengths and weaknesses.

With our understanding of who the Academy graduates were, and drawing on scholarship from both writing studies and college success for underrepresented students, we focused on two major modifications to our previous summer section design: extending the length of the course and adding writing workshop time to the daily schedule. Both adjustments extended the contact hours for the course, but facilitated different goals and activities. Extending the length of the course gave Paula, the instructor, flexibility to plan longer assignment arcs, enabling more focus on and practice of writing process strategies. Adding writing workshops ensured that all students had access to a computer outside of class time and facilitated small-group and one-on-one interactions with the course teaching assistants. With these adjustments, the summer section met for three hours each morning, and students attended the writing workshop for two hours each afternoon after lunch (also funded for students via a small stipend).

An Overview of the Class

While supporting the shared course objectives (listed above), Paula attempted to engage students in mapping their own paths through first-year writing and connecting their experiences in the course to their previous academic and Elon Academy experiences, as well as to their future academic experiences, adopting a "Maps, Legends, and Signs" theme for the section. The course design, from the theme to the scaffolding of assignments to the daily agenda, was mindful of the transitional nature of the summer section for the Elon Academy graduates (see the section syllabus in Appendix).

An *individual research project and presentation* required students to investigate the first-year writing requirement at the college or university they would be attending and to present their findings to the class; as part of that assignment, students drafted, revised, and sent a *formal e-mail* to someone affiliated with a writing program at their future school in order to learn more about the program. Next, students wrote a *personal essay* analyzing significant moments in their path to college; this assignment was designed not only to tap into the students' rich lived experience, but to give students extensive practice in process strategies, including drafting, revising, and editing. A *critical analysis essay* moved students' subject matter from self to text, and required students to summarize and analyze scholarly texts related to the course theme. Finally a *research essay* prompted students to extend their research and develop an argument related to the theme. There was no time for students to create a new research project from start to finish, so this assignment was purposely designed to grow out of the previous course material, as well as to teach students how to handle the pressure associated with learning under a time constraint. In fact, in end-of-course reflections, two students even mentioned how finding out they *could* write a pretty good research paper in a short period of time might help them in future classes.

Students also created personal *blogs* and contributed to a class blog, a requirement that allowed students to practice reflecting and communicating with their peers, professor, and teaching assistants in an electronic medium. Students used their personal blogs both to create scholarly identities and to showcase their work from the semester, including a final reflective essay in which they explained and used evidence from their work to illustrate how their writing expertise developed over the course of the term. The class blog functioned throughout the course as a shared space for discussion; at the end of the course, students practiced digital literacy skills by revising their research essay (originally formatted as a traditional written text) into a multimedia Web 2.0 text posted to the class blog.

Daily agendas varied across the arc of the course, but students often had reading and a written response due at the start of class. The daily homework assignments were purposely rigorous—reading assignments averaged fifteen pages, and writing assignments averaged two–three pages—to prepare students for the pace and work required at the college level and, more importantly, to foster persistence, or "the ability to sustain interest in and attention to short- and long-term projects" (*Framework for Success in Postsecondary Writing*). Research indicates that students learn more when faculty have high, clearly articulated expectations, and that students in classes that require a significant amount of both reading and writing invest more time into their classes and develop better critical thinking and writing skills as a result of "high impact practices" (*Framework;* NSSE). Instead of being a challenge, the length of the class period and the daily class meetings allowed Paula to create and deploy a variety of active learning activities, as well as provide frequent oral and written feedback that she often does not make or have time for in a regular semester. The first part of class was devoted to discussion of the reading. After a short break, students had time to complete guided work in support of their individual assignments or complete group activities that extended the earlier discussion. Daily activities introduced key terms—rhetorical situation, audience, purpose, conventions, etc.—and prompted students to apply their understanding of these concepts. As the term progressed, discussions often were interspersed with more small group and writing process activities.

An Overview of the Writing Workshop

The writing workshop was described to students as "designed to enhance your writing performance. The course teaching assistants will be available for writing conferences as you complete the course assignments. You also will be able to use this time to access computers and campus resources, as well as to complete your homework." It was intended to substitute for the writing center, which is not open during the summer; to provide students with additional structured practice with process activities under the guidance of English teacher licensure majors who had completed writing center training; and to offer time, space, and computer access for students to complete their assignments.

Jessie and Kim anticipated that the daily two-hour writing workshops also would facilitate the implementation of strategies outlined by Engle and Tinto, particularly monitoring student progress and following disciplinary best practices, by integrating additional opportunities for feedback through regular conferences with Paula and the teaching assistants during the writing workshops. We hoped that writing workshops would provide a venue for scaffolding the development of students' strategies for eliciting feedback, tailoring writing instruction to each individual student, and motivating students (Harris). If Academy graduates struggled with organization, the writing workshops would be a space to help students focus their drafts, learn to compose reader-based texts, and experiment with transitions—all with the support of a more experienced student writer (Trupe). We expected that the writing workshops might, as Stephen North envisions in "Revisiting 'The Idea of a Writing Center,'" present a situation in which Paula, the teaching assistants (as writing consultants), and student writers could really get to know each other and talk repeatedly about writing and about college.

Paula and the teaching assistants implemented minor changes as the term progressed to keep the students on track and to move them along in their practice or learning of course concepts during the required two hours of workshop time, resulting in the following workshop activities:

1. *Individual conferences with the course instructor.* Each week, Paula met with each student for at least twenty minutes during workshop time. These conferences, which took place in her faculty office rather than in the classroom, allowed her to spend one-on-one time with each student outside of the regular classroom, show the students how to interact with a professor during office hours, get and provide feedback on how the class was going, work with students on their current assignment, and very simply, get them out of the classroom for a while.
2. *Individual consultations with the teaching assistants.* Each week, the students also met with each teaching assistant at least once to discuss a writing project. These consultations were scheduled so that students knew at what point in the workshop they would need to be prepared to discuss their writing and so that they could manage their time before and after the consultations. In

addition, students completed "writing consultation forms," similar to those used in the university writing center, in which they indicated the feedback they sought on their work or the questions they had about the assignment.
3. *Small tasks for which there was no time in class or that would jumpstart homework or the next day's class.* The students completed these tasks, such as commenting on the class discussion board or posting on the class blog, at the beginning of the workshop, which seemed to help them transition back into a work mode after taking a break for lunch.
4. *Collaborative work.* The workshop proved to be an ideal environment for students to collaborate and provide feedback on one another's work.

The course focused on Elon's goals for first-year writing classes, but it also served Elon Academy's goals for students in transition to college—maintaining academic skills across the summer months, developing a more sophisticated toolkit of academic literacies, inspiring critical thinking and metacognition about their journey toward and into college, and raising student awareness (and, for many, confidence) in their ability to belong and be successful on a college campus. It served not only as a writing development opportunity, but as a key transition experience to help bridge students into the world of real college classrooms, expectations, and responsibilities.

Assessing the Summer Section

Our study used a mixed-methods approach to assess the outcomes of the modified summer section. Our ongoing program assessments include both an indirect assessment, an online survey of students' participation in activities that support the shared objectives,[3] and a direct assessment of students' writing samples, paired with their reflections on their writing processes and rhetorical choices. The direct assessment is scored using an internally developed rubric keyed to the course objectives and to the university's related general education goals.[4]

Paula and the teaching assistants kept journals, and we collected all student work for the course. Jessie and Kim conducted focus groups with the students and the teaching assistants, as well as interviews with Paula.

Learning Outcomes

Looking at our program assessment data described above, the impact on student learning was mixed. Students scored higher than the previous academic-year 110 averages for articulating their understanding of their own writing processes, but lower on the two other outcomes currently evaluated by our direct assessment (see Table 1). This trend holds when we average the first summer (2010) assessment results with those from a subsequent pilot.[5] The assessments are on a five-point scale, with a score of five reflecting excellent work.

Table 1. Outcomes Assessment

Goal	Standard Semester Overall Average for 2008–2011	Summer Elon Academy Average for 2010	Summer Elon Academy Average for 2011
Articulates an understanding of his/her own writing process	3.77	4	3.82
Displays a sophisticated understanding of the relationships between purpose, audience, and voice	3.73	3.45	3.63
Supports own ideas by selecting, using, and properly documenting relevant and credible resources	3.63	3.18	3.04

Given the design of the course, we are not surprised that students excelled at the writing process outcome. The course structure emphasized self-assessment and peer- and instructor-feedback; it also integrated time for revision and intensive activities for practicing writing process strategies. Further, the class environment enabled at least one student to feel comfortable throwing out a "completed" draft and starting over to better meet the goals of an assignment. Overall, class activities and frequent opportunities for students to reflect on the strategies they were trying led to an emphasis on writing process, so end-of-term assessment results showing that students excelled in this area are not surprising.

For this group of students, that emphasis also was probably a strength of the course. None of the students were identified as needing the program's developmental writing course, based on our placement rubric, but most had a different starting point than their eventual peers. Based on our familiarity with the students' high school writing experiences and on their early reflections about their writing processes, we knew they entered the first-year writing course with a narrower range of experiences with writing process strategies than most students who matriculate into the first-year writing program. As a result, for this group of students, and particularly for those students moving on to universities with two-course, first-year writing sequences, the modified summer section likely adequately prepared them for their future writing instruction. The course helped them hone writing process strategies that worked for them, while introducing them to rhetorical concepts and research strategies—introductions that might be extended in second semester writing courses. Many students also reported writing process strategies as their more significant "take away" from the course, the insight most likely to be carried into later courses.

For students like those enrolling at Elon who would not have a second semester writing course, we have more reservations. The accelerated summer section, even in its modified form for this collaboration, did not meet our goals for student learning outcomes related to understanding the relationships between purpose, audience, and voice or to selecting, using, and documenting sources. We suspect that students simply need more time to successfully achieve these outcomes, and we can only hope that writing across the curriculum initiatives will inspire pockets of opportunity for students to further develop their rhetorical awareness and rhetorical and information literacy strategies.

Curricular Feedback: Student, Faculty, and Teaching Assistant Experiences

The research and teaching team's conflicting understandings of the writing workshop presented several challenges, which Paula, the teaching assistants, and the Academy graduates repeatedly discussed in focus groups and interviews. Students initially perceived the workshop time as study hall and resented not being recognized (they perceived) as responsible enough to complete the assignments on their own time. In a mid-term focus group, students told Jessie, "When I'm trying to write, I don't like people hovering over me, reading what I'm saying," and "It feels like I'm in pre-K or something." In addition, the teaching assistants' multiple roles in the summer program[6] and their close proximity to the students' ages led to students challenging the teaching assistants' expertise, potentially undermining the teaching assistants' efforts to offer mini-lessons and activities on writing technologies, writing process strategies, and related topics. Focus group sessions held during the workshop time also highlighted other topics for the faculty member, teaching assistants, and students to negotiate: tolerance for regional diversity,[7] expectations for rigor in a university class, expectations for in-class interactions, and uses/misuses of technology.

As the summer term progressed, frank discussions among all participants about competing expectations led to a happier balance of scheduled writing center-style consultations and time to work on assignments, with more flexibility regarding where students worked and how they used the time. Despite the initial struggles regarding this component of the course, several students recognized its necessity, with one commenting in our final focus group, "For me, I *need* workshop time because I really don't have Internet at home. Some of our stuff required Internet for research." For this reason, several students indicated they missed the workshop when Paula cancelled it during the last week of class; not having access to the Internet and other technology resources created anxiety for students who depended on the workshop time for that access. Another student expressed appreciation for the conferences with Paula, noting, "It helped me be more focused with what I actually need help with . . . instead of just saying, 'Oh, check this.'" Students also had impromptu lessons on representative qualifying/training programs for writing center consultants, rich discussions about types of diversity, and eye-opening experiences with the default lack of privacy in many social media platforms. All of these experiences inform the reflections and recommendations we share at the end of this article.

Despite our concerns about the course's writing-related learning outcomes, we did note another noteworthy outcome: access to more foundational college capital. Students learned about the benefits of attending class, the challenges of being responsible for deadlines, and how to self-monitor. They also gained experience interacting with college classmates, with their professor, and awareness of how communication both in class and via electronic means could alter their college experiences. Most significantly, students repeatedly commented on navigating their new-found independence. In a mid-term focus group, students told Jessie, "We're not children. We know when to work," and "It's our decision whether we want to do this work or not." While the program strove to monitor student progress (Engle and Tinto), students wanted both a recognition of their independence and the freedom to make mistakes—including college experience mistakes.

For the Elon Academy's larger goals, this alone made the ENG 110 experience worthwhile. As one student said in our final focus group, the length expectations for papers surprised her, and "the reading was pretty intense, sometimes. . . . This one class was kicking my butt for a while. It's not like it wasn't doable. It was just I didn't want to do it sometimes, like some of the homework assignments. Since we had to get them in, you had to stay focused." Students grew in their academic literacy skills, but also began developing the skills and dispositions of successful college students—learning to better navigate the classroom and other college resources (including TAs, writing centers, office hours, and available technology), balancing their time and external responsibilities, and growing their individual levels of confidence and sense of belonging in the college environment.

REFLECTIONS AND RECOMMENDATIONS

As program administrators, we had several take-aways that informed a revised pilot summer section of first-year writing for Elon Academy students and that will influence Elon Academy's decisions about future transitional experiences.

Faculty must be able to relate to and engage with underrepresented students. We recognized early on that faculty selection was key to successful outcomes, and Paula went above and beyond to connect with students. She displayed genuine interest in their success, excelled at accentuating positive outcomes, and ensured that students understood what they needed to do differently in the future when the outcomes were not successful. Jessie and Kim believe that her commitment to the first-year writing course objectives and her willingness to learn more about the unique needs of underrepresented students were key to the course's success in fostering students' development of personalized writing processes and their access to college capital. The confidence to approach a college professor for help with writing assignments, no doubt facilitated by Paula's accessibility, is itself a powerful outcome for the experience.

Even well prepared teaching assistants benefit from additional training. We erringly assumed that our undergraduate teaching assistants would transfer their writing center experiences and writing studies and education knowledge to their writing consultations as teaching assistants. Initially, our teaching assistants did not make the connections between their writing center coursework and consulting experiences and their roles assisting in the ENG 110 classroom and facilitating the writing workshop, in part because we were exposing them to a student group with whom they had minimal previous experience and in part because they, too, were still learning how to support writing in a classroom and would not have full-fledged teaching experiences to draw from until later in their senior year. The teaching assistants were most successful in planning and leading instruction, the knowledge for which they had developed in their prior coursework (one TA, in fact, chose to create a unit for this summer course to fulfill an assignment in a course she took the preceding spring). For example, one teaching assistant developed and taught a unit on writing for blogs, while the other teaching assistant developed and taught a unit on organizing research materials. However, the teaching assistants had a more difficult time managing the less-structured workshop time. Therefore, any summer section that integrates teaching assistants should actively facilitate their transfer of prior knowledge from writing studies courses and writing center consulting experience, while also extending their knowledge base through additional professional development, such as working with the course instructor to plan specific activities and develop classroom management strategies.

All participants need to engage in discussions about class and workshop structures and their alignment with pedagogical scholarship. Everyone involved—program administrators, the faculty member, and the teaching assistants—needed to be clearer about what the writing workshop was supposed to do. As administrators, Jessie and Kim thought all participants were philosophically aligned with writing center scholarship for what we anticipated would be a prominent writing consultation component in the writing workshops. Yet, initially, the workshop fostered more disengaged learning, with students identifying it as a "study hall" that they resented. Eventually, the workshop time aligned more closely with the intended disciplinary pedagogy, utilizing more supportive and engaging writing consultations, but it took us a while to get there because we had not recognized we were starting with different, unspoken conceptions of how the workshop time would be used.

The programs' assessment and research process built in opportunities for reflection that should be integral to transitional experiences. By participating in focus groups, students had opportunities to reflect on the course as a college experience and to share their study strategies. As students voiced their own strategies for and reflections on the class, their peers often made note of practices to try in the future. For instance, students often shared time management strategies that they found effective, while others identified study locations that they would keep in mind as alternatives to their dorm rooms. We recommend intentionally integrating opportunities for these types of shared college capital reflections to best support underrepresented (and all) students, perhaps as part of end-of-assignment reflection activities.

Longer days might not be tenable for summer transition experiences, regardless of the student population. If we were to repeat this program, or a summer section of first-year writing for any student population, we would opt for an even longer term, rather than longer days, and we might alternate class days and writing workshop days. The long days (five hours, plus a lunch break, which most students took with their class peers) left everyone exhausted, and the hyper-structured days did not allow for the independent college experience that students were (appropriately) envisioning.

Transportation needs present significant challenges. We underestimated the challenges presented by our students' transportation needs. While this may seem disconnected from the practices of the first year writing classroom, such struggles are central to many underrepresented students' college experiences. The teaching assistants understandably found their dual-role of teaching assistant and transportation provider challenging. We had recruited the teaching assistants as respondents and co-teachers with an expectation of also serving as driver, but driving students to and from campus devalued their disciplinary knowledge and writing center experience in the eyes of the students—even as it improved the teaching assistants' understanding of the students' lives, interests, and issues (a natural outgrowth of the conversations that happened while traveling). While budget limitations during the first summer necessitated this dual-role, we recommend separating these responsibilities.

For students, the transportation route also meant that they spent an exorbitant amount of time in a van (sometimes as long as two hours), when they would have preferred to work on homework or needed to tend to siblings or meet job responsibilities. These types of details, while not often part of writing program administrators' responsibilities, can impact significantly the outcomes of transitional courses for underrepresented students.

Overall, we are pleased with the experience we provided for the students in the summer pilot sections of the first-year writing course, even though we have not resolved all the challenges of offering accelerated sections or addressed all of the myriad needs common to underrepresented students. We recognize a need to rethink which learning outcomes faculty emphasize in an accelerated summer section, but for this student group, the heavy emphasis on writing process served many of them well. Furthermore, the opportunity to participate in a more deliberately scaffolded first college course helped them gain important college capital as they prepared to matriculate into their university programs. Rather than beginning college with a more limited understanding about expectations and lifestyle than their more affluent, more experienced peers, these students begin college with a successful classroom experience already completed and clearer sense of their personal strengths and areas of challenge in a higher education environment as well as in college-level writing. They reported being far more likely to communicate with faculty, to use TAs as resources, and to seek assistance in general—behaviors often linked to success in higher education for underrepresented (and for all) students. As apparent in one student's final reflection, students looked forward to their first fall semester in college with increased confidence, based on real experiences and self-reflection, and able to envision themselves as college students in more concrete and meaningful ways—an important foundational achievement for their future success: "Weeks before I had even set foot in the class, I caught myself doubting my writing abilities. I thought the worst of the class before it even began, but that was because I feared the level of work expected. I wasn't sure I could deliver college-level work. . . . I am happy to say now that my strength is developing new and bright ideas that can be discussed from several different angles. . . . After taking this class, the lessons I will take with me are trust in your abilities, take responsibility for your work, and . . . never be afraid to ask for help."

Coda: Where We Are Now

The College Writing/Elon Academy summer partnership at Elon University supported an underrepresented group of students in their transition to college and helped them develop more nuanced writing processes. At the time of our pilot program, the summer partnership also responded to the writing program's dynamic and interrelational potential within our local ecology to provide a through-line for students, building on Elon Academy graduates' prior writing experiences and connecting them to the students' future college-level writing in the disciplines and across the curriculum. Nevertheless, the abbreviated summer session did not enable students to successfully meet the College Writing program's other assessed learning outcomes: displaying a sophisticated understanding of rhetorical situations and supporting ideas with relevant and credible resources. Five weeks simply is not enough time to develop these strategies to the same extent that students in a regular semester session can. As a result, we discontinued the summer partnership and embraced the inherent fluidity in our programs' partnership in order to explore alternate support structures for Elon Academy students and other local students anticipating the transition to college-level writing.

Although we no longer offer the summer section of first-year writing, College Writing faculty continue to teach in Elon Academy's summer program for high school students, and Elon Academy has explored offering other dedicated sections of summer courses—in content areas that perhaps are less dependent on extended time to develop and practice strategies. The university also recently expanded its concurrent-enrollment course offerings, including College Writing, so that high school seniors in the local community can earn college credit while still in high school. Elon Academy scholars who participate in this program receive free tuition and textbooks and are mentored through their first college classes by a college success mentor, a trained undergraduate who coaches them on college readiness habits and helps them access content area tutoring, writing support, and other resources. Graduated seniors are assisted in the transition to college through both a summer-before-college retreat and mentoring support: an Elon Academy staff mentor visits scholars on their individual campuses to assist with needs—academic, social, and otherwise—as they arise. For scholars who choose to attend Elon University, the program makes sure they participate in developmental writing support opportunities, as needed. In this way, both college transition strategies and college-level writing skills receive more timely, intensive, and extensive focus.

In addition, the College Writing program is participating in other initiatives in the local ecology of programs designed to support students' transitions to college, including offering a semester-long writing seminar for local high school juniors. The seminar helps students develop process strategies, practice composing texts for authentic audiences and purposes, gain rhetorical knowledge, and learn how to conduct college-level inquiry. It focuses on two genres: an admissions or scholarship application essay and a research argument.

Ultimately, these fluctuations in the College Writing program's offerings reflect a complex balancing act: an attempt to retain connections to the disciplinary ecology and its frameworks and outcomes for first-year writing, as illustrated by course-level learning outcomes, and an effort to be responsive to emerging needs in the local ecology. The College Writing program is intricately connected to, and must adapt responsively to, Elon's other writing programs and general education curriculum. Likewise, Elon Academy is embedded in a broader collection of evolving partnerships between the university, the local school systems, and the many colleges and universities where scholars matriculate.

Since completing our study of the College Writing/Elon Academy partnership, our own roles in the local ecology also have evolved. Jessie has rotated out of the College Writing coordinator role and assumed responsibilities as the associate director of the Center for Engaged Learning. Paula was selected as the College Writing program's next coordinator and has had an active role in a university-wide writing excellence initiative. Kim continues to commit her time and energy to Elon Academy and its students. The scholars who participated in the research are all rising juniors and seniors at universities across the state; many return to serve as college role models for their younger peers during the summer program, mentoring and speaking about their journeys to and through college.

Appendix

Syllabus (Excerpts) for Elon Academy Summer Section of ENG 110

ENG 110: College Writing
Summer 2010
M-F 8:30–11:30am Carlton 321

Writing Workshop hours: M-F 1:00–3:00 p.m.

The Writing Workshop is designed to enhance your writing performance. The course teaching assistants will be available for writing conferences as you complete the course assignments. You will also be able to use this time to access computers and campus resources, as well as to complete your homework. Attendance at the workshops is the same as the course attendance policy (described below).

Required Texts

- Bullock & Goggin, *The Norton Field Guide to Writing with Readings and Handbook,* 2nd ed., 2010. This text has an online site available at wwnorton.com/write/fieldguide.
- Orienting Reading (this will provide the metaphorical framework for things we do in the course): Essays on Signs and Maps from slate.com. We will read Parts 1, 2, 4, and 6, as well as this essay on hand-drawn maps.

Maps, Legends, and Signs

In architecture, urban planning, and seafaring, the concept of "wayfinding" is understood to mean the strategies that people use to navigate or orient themselves in new and unfamiliar surroundings. For centuries, people have used maps, signs, and legends to both "find their way" and "point the way" to and from places, a process that can be applied metaphorically to what will happen in our writing class this summer. This intensive Summer Term section of English 110: College Writing will focus on how in writing, we use rhetoric and process to map and find our way or lead others in the right direction. As Elon Academy alumni, you will be asked to think about where you've been and where you're going—providing an "I Am Here" placemark every so often to let your peers and former instructors know what you're experiencing in those new places. And we want to create a map for those who follow you: other Elon Academy classes, your siblings, your friends, your classmates, your neighbors. How can they get to where you are or to where you are going? In what directions would you point them? From what obstacles or perils might you warn them to steer clear?

We'll be able to discuss and write about all of these things in relation to the following course objectives:

- A more sophisticated writing process including invention, peer responding, revising and editing that result in a clear, effective well edited public piece.
- A more sophisticated understanding of the relationship of purpose, audience, and voice, and an awareness that writing expectations and conventions vary within the academy and in professional and public discourse.
- An appreciation for the capacity of writing to change oneself and the world.

To achieve these objectives, you will have the following kinds of learning experiences:

1. Write to persuade by analyzing, interpreting, researching, synthesizing, and evaluating a wide variety of resources.
2. Understand how to approach a variety of writing assignments.
3. Read and understand writing assignment documents.
4. Make decisions about how to approach particular writing situations based on the context and your preferences as a writer.
5. Understand and adapt to the differences in style, purpose, audience, and context when writing in different academic disciplines, in public and professional writing, and in informal and personal writing.
6. Manage large writing and research assignments.
7. Manage timed or "on-the-spot" writing assignments.
8. Be an engaged and reflective writer and reader.
9. Reflect on who you are as a writer, including your writing process, your strengths and weaknesses, and your ability to set and meet personal writing goals.
10. Respond to college-level reading assignments in a sophisticated manner.

11. Offer and receive effective, constructive feedback about writing.
12. Revise and edit your writing, and the writing of others, applying grammar, style, and citation concepts appropriate to the writing situation.
13. Conduct library and online research, and use source material to support an argument.
14. Apply MLA/APA style requirements and documentation to your writing.
15. Select, evaluate, synthesize, and integrate outside sources into your writing.
16. Present your ideas orally.

Assignments

Because of the intensive nature of the course, you should be prepared to work at a rapid pace. You will be writing every day and will usually have something due each class session. The course calendar and detailed information about each assignment are available on Blackboard.

Formal projects (55–60% of total course grade)
These formal projects are designed to teach you and allow you to practice and, especially, refine skills associated with the course goals and experiences. Work on these assignments will take place both in and out of class and over the course of several days. The products of these projects must be carefully edited and proofread—like Mary Poppins, practically perfect in every way.

1. Individual Research Project and Presentation: College Writing Where You're Going (*5% of course grade*). This assignment will require you to research the writing requirements, courses, and assignments at your chosen college, and present your findings to your classmates via an 8–10 minute presentation.
2. Formal E-mail (*5%*). As part of the College Writing Where You're Going assignment, you will create and refine a formal e-mail to send to someone affiliated with some aspect of writing at the college.
3. Personal Essay (*15%*). This assignment will ask you to personalize the course theme, analyzing significant moments on the road to college and beyond. The final product will be posted to your blog.
4. Critical Analysis Essay (*10%*). This assignment will require you

to summarize, analyze, and develop an argument in response to a scholarly text related to our course theme to your research project.
5. Research Essay *(20%)*. The product of this project will be a traditional academic research paper. The topic will be related to our course theme and/or something you wrote about in your personal essay.

In-class activities (15%)
In-class activities will function as informal writing projects (more concerned with practice than with perfection) and will usually be completed over one or two class periods, either individually or in groups. These activities will be graded according to the requirements of each activity (e.g., for application of concepts taught in class related to the activity, for ability to collaborate in a group) and will include, among others, the following:

- A group project in which you will work in small groups to evaluate the reliability of a Wikipedia article. You will present this evaluation as an argument, written collaboratively by all members of your group.
- A comparison of your writing to the writing of a student who has completed his or her first year of college, a college senior, a professional writer, and/or a scholarly writer.
- Illustrating an argument. You will examine how illustrations or visuals can be argumentative texts and how they can enhance a written text; you will also create a short illustrated text.

Class blog posts (10%).
For each class meeting (after the first week), you will create a blog post, usually in response to a prompt provided by Prof. Patch or one of the teaching assistants and usually related to that day's reading assignment or to the project we're working on at the time.

Blog/Electronic Portfolio (10%)
Your blog will be the electronic home for the products of your formal writing projects, daily posts, and informal activities. Each of these products will receive a separate grade, as explained in above. Along with posting your projects and activities, you will create

- A blog design

- An About Me page that explains who you are
- And, at the end of the course, a self-assessment of the blog contents—basically, an assessment of your work over the course of the semester

All of the above blog contents will be graded for thoughtfulness, thoroughness, and correctness.

Final Exam (5% of grade).
On the last day of class, you will write an in-class essay on a topic to be provided later in the course. This exam will assess your ability to draft, revise, and edit an argumentative under a time constraint.

Active course participation and preparation (-5%).
Much of our coursework will be completed in class; homework assignments will prepare you for this in-class work *or* give you chance to complete work begun in class. I expect you to come to class prepared to actively engage and participate in all class activities, both individually and in groups. Your participation/preparation grade will include the following:

- completion of homework (other than blog posts) and drafts.
- participation in peer response/review. Each major project and some of the in-class activities will go through the peer review process.
- completion of activities related to formal projects.
- completion of note-taking duties. Each day, two students will be responsible for taking notes on class lecture, discussion, and activities. See details about this on Blackboard.
- contribution to class discussion.

Grading
A grade of "C-" or higher is required to pass ENG 110. This is a graduation requirement. Successful completion of the course is also a prerequisite for all other English courses.

Special Information about Research Activities
This class is part of a research project on the effectiveness of time-intensive courses for teaching writing. At the beginning of the semester, you (or your parent or guardian, if you are not yet 18) will be asked to sign

a permission form, indicating whether or not you will participate in the study. If you give your permission, some of the work you do in this class will be used to assess the effectiveness of learning writing in a shortened course. I won't know if you have or have not given permission for your work to be included in the study until after the final grades have been posted, so there is no way that your participation or lack of participation in the study can help or hinder you performance in this class. Regardless of your decision to participate, you will still do the same work as the rest of the class. Please contact me if you have any questions or concerns about this research.

Notes

1. We use "underrepresented" to encompass the many labels that could be applied to our program's students. They are all first-generation college students, and most are racial minority students (African-American or Hispanic). Some are members of the local immigrant communities and the primary speaker of English in their homes, and all come from low-income, working-class families.

2. To view the shared experiences document, see: <http://compositionforum.com/issue/27/elon.php#appendix1>.

3. To view online survey, see: http://compositionforum.com/issue/27/elon.php#appendix3.

4. To view the rubric, see: http://compositionforum.com/issue/27/elon.php#appendix4.

5. In 2011, the College Writing program and Elon Academy partnered on a second summer pilot that reduced the contact hours slightly in comparison to the first pilot and refocused the afternoon writing workshop time as regularly scheduled conferences with the faculty instructor and an undergraduate teaching assistant. We offer the assessment results for this additional pilot to highlight the continuing trend: summer section students outperformed their standard semester peers on articulating their own writing processes, but they continued to perform lower on the other two measured learning outcomes.

6. Both teaching assistants were required to drive the university vans that picked up students and took them home. Due to a lack of public transportation, most students, who lived at home during the summer program, needed rides to campus. The route might take only 20 minutes or, when more students needed transport, could run nearly 2 hours. This logistical issue was an ongoing challenge.

7. Many students initially were put off by one of the teaching assistant's New York personality traits, leading to a discussion of regional differences in word choices, pacing, humor, etc.

Works Cited

Alexander, Karl L., Doris R. Entwisle, and Linda Steffel Olson. "Summer Learning and Its Implications: Insights from the Beginning School Study." *New Directions for Youth Development* 114 (2007): 11–32. Print.

Alves, Julio. "Elite Colleges Must Give Low-Income Students the Tools to Succeed." *The Chronicle of Higher Education.* 28 September 2007. Web. 15 August 2011.

Anson, Chris M. "Absentee Landlords or Owner-Tenants? Formulating Standards for Dual-Credit Composition Programs." *College Credit for Writing in High School: The "Taking Care of" Business.* Eds. Kristine Hansen and Christine R. Farris. Urbana, IL: National Council of Teachers of English, 2010. 245–271. Print.

Burke, Jim. *ACCESSing School: Teaching Struggling Readers to Achieve Academic and Personal Success.* Portsmouth, NH: Heinemann, 2005. Print.

Conley, David T. *College Knowledge: What It Really Takes for Students to Succeed and What We Can Do to Get Them Ready.* San Francisco: Jossey-Bass, 2005. Print.

—. *Redefining College Readiness.* Eugene, OR: Educational Policy Improvement Center, 2007. Web. 8 October 2012.

Council of Writing Program Administrators, National Council of Teachers of English, and National Writing Project. *Framework for Success in Postsecondary Writing.* CWPA, NCTE, and NWP, 2011. Print.

Davis, Jeff. *The First-Generation Student Experience: Implications for Campus Practice, and Strategies for Improving Persistence and Success.* Sterling, Virginia: Stylus, 2010. Print.

Delpit, Lisa. *Other People's Children: Cultural Conflict in the Classroom.* New York: New Press, 1995. Print.

Engle, Jennifer, and Vincent Tinto. *Moving Beyond Access: College Success for Low-Income, First-Generation Students.* The Pell Institute for the Study of Opportunity in Higher Education, Washington D.C., 2008. Print.

Ferris, Christine R. "Minding the Gap and Learning the Game: Differences that Matter Between High School and College Writing." *College Credit for Writing in High School: The "Taking Care of" Business.* Ed. Kristine Hansen and Christine R. Farris. Urbana, IL: National Council of Teachers of English, 2010. 272–282. Print.

Fox, Mary Ann, Brooke A. Connolly, and Thomas D. Snyder. *Youth Indicators 2005: Trends in the Well-Being of American Youth.* Washington, D.C.: U.S. Department of Education—National Center for Education Statistics, 2005. Web. 8 October 2012.

Hansen, Kristine. "The Composition Marketplace: Shopping for Credit Versus Learning to Write." *College Credit for Writing in High School: The "Taking Care of" Business.* Eds. Kristine Hansen and Christine R. Farris. Urbana, IL: National Council of Teachers of English, 2010. 1–39. Print.

Harris, Muriel. *Teaching One-to-One: The Writing Conference.* Urbana, IL: National Association of Teachers of English, 1986. Print.

"High-Impact Practices and Experiences from the Wabash National Study." *Wabash National Study of Liberal Arts Education.* Center of Inquiry in the Liberal Arts of Wabash College. n.d. Web. 8 October 2012.

Kelly, Patrick. *As America Becomes More Diverse: The Impact of State Higher Education Inequality.* Boulder, CO: National Center for Higher Education Management Systems (NCHEMS), 2005. Web. 8 October 2012.

Klugman, Jason R., and Donnell Butler. *Opening Doors and Paving the Way: Increasing College Access and Success for Talented Low-Income Students.* Princeton, NJ: Princeton University Preparatory Program, 2009. Web. 8 October 2012.

National Survey of Student Engagement (NSSE). "Selected Results: Writing Matters." NSSE Annual Results 2008. 21–22. Print.

Noguera, Pedro A., and Jean Yonemura Wing, eds. *Unfinished Business: Closing the Racial Achievement Gap in Our Schools.* San Francisco: Jossey-Bass, 2006. Print.

North, Stephen M. "Revisiting 'The Idea of a Writing Center.'" *The Writing Center Journal* 15.1 (1994): 7–19. Print.

Powell, Pegeen Reichert. "Retention and Writing Instruction: Implications for Access and Pedagogy." *College Composition and Communication* 60.4 (2009): 664–82. Print.

Roksa, Josipa, and Richard Arum. "The State of Undergraduate Learning." *Change: The Magazine of Higher Learning* 43.2 (2011): 35–38. Print.

Schwalm, David E. "High School/College Dual Enrollment." *WPA: Writing Program Administration* 15.1–2 (1991): 51–54. Print.

Strayhorn, Terrell L. "When Race and Gender Collide: Social and Cultural Capital's Influence on the Academic Achievement of African American and Latino Males." *Review of Higher Education* 33.3 (2010): 307–332. Print.

Tierney, William G., Zoe B. Corwin, and Julia E. Colyar, eds. *Preparing for College: Nine Elements of Effective Outreach.* Albany: SUNY P, 2004. Print.

Trupe, Alice L. "Organizing Ideas: Focus is the Key." *A Tutor's Guide: Helping Writers One to One.* 2nd ed. Ed. Ben Rafoth. Portsmouth, NH: Heinemann, 2005. 98–106. Print.

Villalpando, O. & Solórzano, D.G. "The Role of Culture in College Preparation Programs: A Review of the Research Literature." *Preparing for College: Nine Elements of Effective Outreach.* Ed. William G. Tierney, Zoe B. Corwin, and Julia E. Colyar. Albany: SUNY P, 2005. Print.

Yosso, Tara J., William A. Smith, Miguel Ceja, and Daniel G. Solorzano. "Critical Race Theory, Racial Microaggressions, and Campus Racial Climate for Latina/o Undergraduates." *Harvard Educational Review* 79.4 (2009): 659–90. Print.

Index

academic literacy, 70, 256–259, 383
accreditation, 6, 233, 291, 293–294; reaccreditation, 296
Accuplacer, 25, 30, 34–35
active learning, 167–169, 174, 176, 181, 186, 377
activity system, 164, 240
adjunct, 19, 27, 59, 93, 96–98, 101, 106–109, 112–113, 157, 187, 235, 245, 274, 286, 288, 294
Adler-Kassner, Linda, 12
advocates, 37, 96, 110–111, 334, 336
agency, 5, 7, 21, 60, 251, 255, 261–262, 266, 273, 277, 327
agents, 3, 7, 10, 150, 258, 261, 263, 274
Amidon, Stevens, 13, 19, 88
Andersen, Wallace May, 214
Anderson, Benedict, 242
Aronowitz, Stanley, 257
assessment, 15, 20–21, 37, 43–44, 48, 57, 103, 105, 118, 177, 192, 204–206, 221, 226, 243, 249–250, 252–254, 258, 261–266, 274–278, 282–284, 289–292, 294–298, 306, 310–311, 319, 346, 365, 371, 374, 379–381, 385, 393
assignment construction, 46
assignment sequence, 27, 49, 60, 70, 75–76

Ballif, Michelle, 10, 14, 122, 136
Balzhiser, Deb, 212
Bamberg, Betty, 54
Barlow, John Perry, 311
Bartholomae, David, 31, 237
basic writing, 13, 19, 22–24, 26–35, 37–38, 284, 293, 297
Bawarshi, Anis, 10, 148, 170
Bazerman, Charles, 228–230, 238–239
Beaufort, Anne, 69–70, 73, 82, 147, 170
Belanoff, Pat, 261, 278
Bizzell, Patricia, 133, 241
Blakesley, David, 30
Bloom, Lynn Z., 22
Blythe, Stuart, 30, 38
Boff, Collen and Toth, Barbara, 311
Borrowman, Shane, 12, 54
Bräuer, Gerd, 12
Britton, James, 166
Broad, Bob, 295
Brooke, Collin, 5
Brown, Stuart C. and Enos, Theresa, 12
Bruffee, Kenneth, 167

Carlino, Paula, 12
Carroll, Lee Ann, 125–126, 170, 181, 203
CCCC Writing Program Certificate of Excellence, 87, 223
Clouse, Barbara Fine, 31

397

college capital, 16, 307, 365, 382, 384–386
Common Core, 37, 249, 297
common syllabus, 41, 45, 50–51, 53–54, 56–57, 59–60, 146, 150, 152–153, 155–156
Commoner, Barry, 6
competency measures, 262, 264, 268–271
complexity, 4–6, 8–10, 15, 21, 53, 121, 135, 171, 176, 206, 249, 251, 275, 282, 284, 287–288, 297–298; complex systems, 6, 8, 9
Composition Forum, 9–10, 16, 58, 157, 164
composition studies, 7, 13–14, 23, 27, 96, 197, 232, 234, 237–239, 241, 291, 368
Conference on College Composition and Communication, 156, 159
Conley, David T., 367–368, 373
Cooper, Marilyn, 3, 6
Costello, Kristi Murray, 14, 42, 60, 66, 122, 146–147, 158–159
Council of Writing Program Administrators, 12, 19, 83, 89, 277, 282, 303
creative writing, 44, 71–72, 101, 156, 200–201, 208, 214, 236, 355
critical inquiry, 82, 282
critical literacy, 145
Crowley, Sharon, 143, 160
Csikszentmihalyi, Mihaly, 47

Delpit, Lisa, 256–257, 365
Devitt, Amy, 23, 148
Dew, Debra, 97
direct assessment, 283, 371, 379, 380
directed self-placement, 250, 252–254, 257, 259–260, 264–266, 270–271, 273–275, 277, 279; guided self-placement, 19, 30, 34
disciplinary knowledge, 21, 87, 94–97, 111, 126, 135, 385; disciplinary writing conventions, 134
discursive system, 4–5, 10
diversity, 23, 25, 27, 36, 61, 129, 144, 148, 151, 177, 186, 206, 241, 374, 382
Dobrin, Sidney, 5, 8
Donahue, Christiane, 14, 123, 171, 191
Downs, Doug, 95, 102, 117, 147
drop, withdraw, fail, 24, 35–36
Dynamic Criteria Mapping, 295

ecological, 3–11, 13–15, 19–21, 87, 112, 114, 121–124, 135, 189, 195–197, 206, 223, 244, 249–250, 296–298, 305, 307, 326, 360; ecological balance, 123, 297
ecology, 3–6, 8–9, 11, 16, 36, 59–60, 87, 123–124, 157, 195–197, 242–246, 275, 297, 307, 387–388; institutional, 14, 20–21, 249, 305, 307; material ecologies, 4
Edbauer, Jenny, 5
Ede, Lisa, 167
emergence, 5, 9–10, 12–15, 20, 88, 122–124, 135, 189, 195–198, 296–297, 305–307, 326, 341; emergent, 9–10, 20–21, 88, 121, 123, 135, 223, 250, 306
Enos, Theresa, 12
ESL, 73, 146, 151–152
ethnography, 102, 203, 225
first-generation students, 147, 159, 254, 257, 265, 272, 363, 367, 371–373, 394
first-year composition, 13–15, 19, 21, 24, 26, 28–31, 33, 35, 37,

60, 69, 72, 86, 88, 143, 147, 149, 164–165, 174, 182–183, 198, 203, 218–219, 233, 235, 239, 250, 282, 284–285, 287–289, 296, 318, 326, 363, 369, 379, 385–386; conditions, 101; curriculum, 70, 75, 87, 106, 282, 284; curriculum revision, 70, 73, 147, 286–287, 290, 292–293; experience 14, 37, 122, 125; outcomes, 29–30, 92, 303, 383, 388; problems, 93, 95; program, 19, 21, 69, 71, 74, 88, 108, 136, 151, 180, 198, 212, 223, 232, 252, 305, 307, 364, 374; requirement, 19, 69, 121–123, 219, 250, 282–283, 286, 376; sequence, 66, 73, 87, 122–123, 172, 178, 289, 381; students, 19, 24–25, 33, 35, 87, 250, 282, 293, 296, 307, 319–320, 381; teachers, 69, 72, 74, 125, 154, 185, 282, 285, 289, 293, 295, 297, 364; texts 75, 80, 88
Fleming, David, 198
Flower, Linda, 46
fluctuation, 5, 9–10, 13–15, 19–20, 135, 157, 189, 195–197, 249–250, 277, 306, 326, 388
Freedman, Aviva, 240
Freire, Paulo, 168, 252–253
Frost, Samantha, 5

Gaines, Robert, 240–241
Ganobcsik-Williams, Lisa, 12
Gee, James Paul, 256
general education, 20–21, 23, 26, 30, 36–37, 66, 69, 72–73, 111, 113, 142–143, 145–146, 159, 170, 214, 286–287, 308, 318, 379, 388
general writing skills instruction, 287

genre, 9, 23, 29, 32–33, 35, 41–44, 46, 48, 52, 54, 56, 70–71, 73, 78, 80–81, 85, 95, 99, 102–103, 108, 114–117, 123, 142, 145, 147–151, 153, 170, 229, 231, 240–241, 243, 299–300, 302, 388; disciplinary genres, 373; genre knowledge, 23, 76, 85
Gilles, Roger, 30, 260
Giroux, Henry, 257
Glau, Gregory, 29, 273
grade-norming, 43, 140, 186, 279
grading contract, 266, 276–279
Graduate Teaching Assistants, 13, 19, 41–45, 48–55, 58–60, 66, 84, 109, 130, 289,
Graff, Gerald, 80, 85
Graff, Gerald and Birkenstein, Cathy, 256, 258, 275, 278
Gramsci, Antonio, 277
Greene, Nicole Pepinster, 38, 144
Grego, Rhonda and Thompson, Nancy, 22
Griffith, Kevin, 47
Gunner, Jeanne, 11–12, 17

Handa, Carolyn, 12
Hawking, Stephen, 8
holistic, 5, 63, 283, 295; holistic scoring, 283–284, 291–292, 295–296, 299
Holliday, Wendy, 311
Howard, Rebecca Moore, 142, 146, 151, 167, 192, 212
Huot, Brian, 30, 277, 283

information literacy, 306, 308, 309, 310–323–328, 381
institutional change, 13, 19, 72, 121
institutional ecologies, 14–15, 20–21, 249, 305, 307

institutional structures, 100, 169, 207
interconnectedness, 5–6, 9–10, 13–15, 19, 36, 121–122, 135–136, 157, 189, 195–197, 223, 250, 305, 307
internship, 217–218, 222, 226

kairos, 27, 97, 99–100, 102, 109; kairotic, 13, 21, 27, 97, 101, 111
Kinney, Kelly, 14, 42, 122, 142, 145–147, 158–160
Kitzhaber, Albert, 165–166, 191, 234

labor conditions, 15, 92, 94, 282
Langer, Judith, 126, 138
Latour, Bruno, 239, 240; Latour, Bruno and Woolgar, Steve, 240
Lave, Jean and Wenger, Etienne, 171, 173
libraries, 13, 15, 62, 81, 89, 101, 116, 228, 241, 305, 308–312, 314–317, 320–324, 326, 329, 342, 369, 391
literacy, 52, 70–71, 115, 146, 204, 208–209, 212, 236, 242, 252–254, 256–262, 308, 359, 373; academic 256–260; assessment 250; critical 145; digital, 216, 230; embodied, 89; instruction 122,142; learning 70, 150; literacy studies, 203, 207, 229; literacy tasks, 126, 143; narrative, 48, 52, 62, 65, 219; skills, 373, 377, 383
Lunsford, Andrea, 167, 170–171, 241, 334

Macrorie, Ken, 334
Maid, Barry M., 310
Martin, Wanda, 42, 45

material conditions, 10, 16, 19, 93–94, 96–97, 101, 104, 108–109, 208
McAlexander, Patricia J., 38
McLeod, Susan, 12, 212
meta-awareness, 83
metacognition, 96, 168, 170, 173, 175, 340, 379
Mill, John Stuart, 9
Miller, Carolyn, 100, 109
Miller, Richard, 149
Monroe, Jonathan, 129, 133

National Council of Teachers of English, 91, 197, 219, 277, 282, 285, 288, 303,
Nelms, Gerald and Dively, Ronda Leathers, 43, 66
New London Group, 168
new media, 207, 215–217, 219, 221–222, 225–226, 230–231, 239–240
Newell, George, 169
non-credit, 22, 26–28, 34, 38
Norgaard, Rolf, 312
North, Steve, 237, 334–335, 378
Nyquist, Jody D. and Wulff, Donald H., 42

Ong, Walter, 235, 242, 246
Owusu-Ansah, Edward K., 312

Paine, Charles, 42, 45
passing rates, 265, 270–272
peer review, 33, 63–64, 79, 127, 151, 175, 319, 325, 336–340, 344–345, 348, 357, 360–361, 393; peer-interactive, 16, 306, 328, 336, 339, 341–342, 344, 345, 352, 357–359, 361
Petraglia, Joseph, 287
Phelps, Louise Wetherbee, 234–235, 237
Pigg, Stacey, 78, 89

placement, 13, 19–20, 26, 30–31, 34–38, 250, 252–253, 254–255, 257, 259–260, 264, 266, 277–278, 294, 381
portfolios, 48, 103, 105, 154, 189, 243, 250, 261–263, 265–269, 271, 275–279, 319; portfolio assessment, 48, 105–107, 118, 262–263; portfolio evaluation, 151
Prior, Paul, 172
professional development, 8, 54, 93, 97–98, 103–104, 154, 156, 175, 177, 185, 187, 219–220, 251, 294, 296, 310, 384
professional writing, 11, 61, 200–202, 215–217, 220, 225, 231, 233, 235, 239, 243, 390, 400
program assessment, 11, 12, 15, 221, 224, 250, 253–254, 263–266, 274–275, 278, 370, 375, 379–380

reflection, 10, 16, 32, 45, 53, 83,151, 179, 190, 220, 240, 243, 250–251, 258–259, 261–262, 265, 267, 305, 307, 321, 325, 328, 354, 370, 376, 379, 381–382, 385–386
Regaignon, Dara Rossman, 12
Reiff, Mary Jo, 10, 13, 20, 73, 89, 147, 170
remedial, 13, 20, 22, 24, 26–28, 36–37, 152, 213–214, 233, 273, 278–279, 334; remediation, 22, 37, 214, 271, 279, 318
research logs, 315–317, 323, 327, 332
research methods, 70, 79–83, 186, 222, 224, 264, 372
revision, 32, 50, 53, 61, 63, 65, 95, 98, 114, 117, 127, 137, 139, 150–151, 184, 338, 344, 380; curriculum/program, 12–13, 15, 19–20, 27, 59–60, 70, 74–75, 81, 84–85, 107–108, 196, 202, 216, 219, 221, 223–224, 249–250, 282–283, 285–290, 293–294, 301–302, 317, 321, 323
Reynolds, Erica, 270, 278
rhetorical analysis, 62, 77–79, 83, 87, 102, 105, 117, 150, 203, 314–315
rhetorical awareness, 73, 76, 84, 381
rhetorical knowledge, 20–21, 32, 69–70, 73, 85–88, 314, 388
rhetorical tasks, 25, 32–33
Rodriguez, Richard, 40
Rose, Mike, 22, 152
Royer, Daniel, 30, 259–260
rubrics, 26, 51, 65, 75, 99, 103, 105, 140, 182, 204–206, 209, 224, 291, 295, 317, 332, 353, 379, 381, 394

Sawyer, Paul, 146
scaffolding, 60, 123, 145, 169–170, 174, 185, 190, 337–338, 345, 368, 371, 376, 378
Schilb, John, 236
Scribner, Sylvia, 261
self-assessment, 204, 209, 249–250, 253–254, 257–259, 261–263, 266, 274–275, 277, 380, 393; self-assess, 253, 263
self-placement, 13, 19, 30–31, 34–35, 250, 252, 259–260
Sinha, Aparna, 12
Smit, David, 68–69, 85, 143, 183
Smoke, Trudy, 22
social practice, 69, 256
Soderlund, Kelly, 24
Strain, Margaret, 233–234
stretch program, 29, 261, 269, 273, 277–278
student achievement, 92

student evaluations, 34, 127
Stygall, Gail, 22, 278
Swanson, Troy, 311
Syverson, Margaret, 5–6, 9, 124

Thaiss, Chris, 12
transfer, 14, 21, 23, 43–44, 56–57, 66, 68–69, 71, 73, 74–77, 80, 82–83, 85–88, 123–124, 136, 163–191, 305, 340, 358, 384; backward-reaching knowledge transfer, 56; far transfer, 170; high road transfer, 68–69, 75–76, 79, 85–86; low road transfer, 68, 75, 79; near transfer, 170; transferable, 72, 82, 86, 93, 168, 243; transferability, 70–71, 74, 83–85, 89
Trimbur, John, 167
Troyka, Lynn Quitman, 22
tutoring, 16, 56, 151, 215, 217, 306, 308, 318, 320, 326, 328, 334–336, 338, 341–348, 351–357, 359–360, 373, 387; peer tutoring, 215–217, 225, 336
two-year colleges, 22

underrepresented students, 307, 364, 367, 369, 372–373, 375, 383, 385–386

Vygotsky, Lev, 169

Wardle, Elizabeth, 13, 21, 37, 83, 87, 91, 118, 147, 161, 181, 184
ways of knowing, 14, 122, 126, 129
Webb-Sunderhaus, Sara, 13, 19, 22, 88
Weisser, Christian, 10, 11, 18, 39
White, James Boyd, 200
Wikipedia, 315–316, 321, 327, 392
Wilder, Stanley, 309

workshops, 33–34, 43, 51, 82–83, 103, 107–109, 133, 145, 151–152, 156, 172, 175, 185–186, 219–220, 283, 294–295, 301–302, 320, 336, 344, 348, 357, 360, 363, 367, 371, 375–379, 381–382, 384–385, 389; faculty development, 82, 172, 175, 283, 301 ; professional development, 220, 295; workshopping, 340, 345, 361
WPA Outcomes, 73, 90, 93, 288, 314, 333; WPA Outcomes Statement, 73, 90, 93, 288, 314, 333
writing about writing, 13, 21, 95, 97, 99, 101–103, 105–107
writing across the curriculum, 4, 11, 69, 74, 106, 121, 123, 128, 142, 198, 284, 335, 381; WAC, 14, 69, 80, 108, 112, 121–122, 124, 128–129, 142–147, 153, 196, 276, 284–285, 287–288, 293–294, 297, 305, 335, 338
writing centers, 11, 13, 15, 69, 134, 305, 308, 310, 318, 336, 338, 341, 358, 360, 383
writing initiative, 14, 106, 118, 121–125, 135
writing majors, 195–196, 212, 243; undergraduate writing majors, 4, 14, 195–197, 211
writing outcomes, 20, 23, 26, 28–30, 32–33, 38, 60, 70, 79, 92–93, 95–99, 101–103, 105, 108, 110, 114, 155, 172, 175, 177–178, 180–182, 185, 190–191, 204, 253–254, 258–263, 266–267, 270, 275, 278–279, 282, 285, 295–296, 301–303, 360, 365, 370–371, 379–383, 386–388, 394; course goals, 139, 375, 391; course objectives, 44,

58, 282, 295–296, 375–376, 379, 383, 390
writing program administration, 4, 9, 11–13, 45, 207; WPA, 5, 12–13, 16–17, 20, 39, 45–46, 60, 73, 87– 88, 90–91, 93, 97, 102, 118, 159, 250, 277, 282, 288, 293, 303–304, 314, 333, 396
writing studies, 3–5, 15, 20, 42, 45, 52–53, 55–58, 91, 94, 96, 107–108, 142, 146–147, 150, 153, 156–158, 160, 169, 197, 207, 212–213, 228–232, 235–238, 241–245, 371, 375, 384
writing tutoring, 16, 306, 341, 348, 352, 355, 359; classroom-based, 336

writing tutors, 16, 284, 306, 313, 334–335, 344–345, 352
writing workshop, 175, 371, 375, 377–378, 381, 384–385, 394
writing-intensive, 14, 41, 43, 73, 122, 127, 135–136, 138, 202, 214, 286, 293, 318, 320
Writing in the Disciplines, 14, 69, 80, 121, 124, 128–129, 287–288, 297, 305
Wysocki, Anne Frances, 71

Yancey, Kathleen Blake, 66, 71, 73, 168, 212–213, 277–278

About the Editors

Mary Jo Reiff is Professor of English at the University of Kansas, where she teaches courses in rhetoric and composition theory, public rhetoric, writing research, and composition pedagogy. Her books include *Approaches to Audience: An Overview of the Major Perspectives* (2004); *Genre: An Introduction to History, Theory, Research, and Pedagogy* (with Anis Bawarshi, 2010); and the textbooks *Scenes of Writing: Strategies of Composing with Genres* (with Amy Devitt and Anis Bawarshi, 2004) and *Rhetoric of Inquiry* (with Kirsten Benson, 2009). Articles related to her research on writing programs, writing knowledge transfer, audience theory, critical ethnography, and public genres have appeared in *Written Communication*, *Composition Studies*, *Composition Forum*, *College English*, *JAC*, and *WAC Journal*.

Anis Bawarshi is Professor of English and former Director of the Expository Writing Program at the University of Washington, where he teaches courses in composition theory and pedagogy, rhetorical genre theory, discourse analysis, rhetoric, and knowledge transfer. He is currently series co-editor for Reference Guides to Rhetoric and Composition with Parlor Press and Program Profiles Co-Editor for the journal *Composition Forum*, and serves on the editorial board for the journal *College Composition and Communication*. His publications include *Genre: A Historical, Theoretical, and Pedagogical Introduction* (2010; with Mary Jo Reiff); *Genre and the Invention of the Writer* (2003); *Scenes of Writing: Strategies for Composing with Genres* (2004; with Amy J. Devitt and Mary Jo Reiff), *A Closer Look: A Writer's Reader* (2003; with Sidney I. Dobrin); and articles and book chapters on genre, uptake, invention, and knowledge transfer in composition. He is currently co-editing a book that examines genre and the performance of publics.

Michelle Ballif is Associate Professor of English at the University of Georgia, where she teaches courses in rhetoric, composition, and contemporary theory, and where she directs the campus-wide writing-in-the-disciplines program that she founded. She is the former managing editor of the journal *Composition Forum* and the current Associate Editor of *Rhetoric Society Quarterly* as Editor for Special Issues. Her research focuses on the intersections between classical rhetoric and poststructuralist theory. She is the author of *Seduction, Sophistry, and the Woman with the Rhetorical Figure*; co-author of *Women's Ways of Making It in Rhetoric and Composition*; co-editor of *Twentieth Century Rhetoric and Rhetoricians* and *Classical Rhetorics and Rhetoricians*; and editor of *Theorizing Histories of Rhetoric*. She is currently finishing a book-length manuscript entitled *Paranormal Investigations into the History of Rhetoric*.

Christian Weisser is Associate Professor of English at Penn State Berks, where he coordinates both the Professional Writing Program and the Writing Across the Curriculum Program. Christian teaches courses in technical writing, environmental rhetoric, and the discourse of sustainability. He has served as Editor of *Composition Forum* since 2005. He is the author or editor of numerous publications including *Moving Beyond Academic Discourse: Composition Studies and the Public Sphere (2001)*, *Ecocomposition: Theoretical and Pedagogical Perspectives* (2001, with Sid Dobrin), *Natural Discourse: Toward Ecocomposition* (2002, with Sid Dobrin), *The Locations of Composition* (2007, with Christopher Keller), and *Sustainability: A Bedford Spotlight Reader* (2013). Christian's current work is focused on the discourse and rhetoric of sustainability in U.S. higher education.

www.ingramcontent.com/pod-product-compliance
Lightning Source LLC
Chambersburg PA
CBHW031411230426
43668CB00007B/274